TM 9-751
WAR DEPARTMENT TECHNICAL MANUAL

155-MM GUN MOTOR CARRIAGE M12

AND

CARGO CARRIER M30

TECHNICAL MANUAL

RESTRICTED DISSEMINATION OF RESTRICTED MATTER. The information contained in restricted documents and the essential characteristics of restricted material may be given to any person known to be in the service of the United States and to persons of undoubted loyalty and discretion who are cooperating in Government work, but will not be communicated to the public or to the press except by authorized military public relations agencies. (See also par. 18b, AR 380–5, 28 Sep. 1942.)

by WAR DEPARTMENT • 28 JANUARY 1944

©2013 Periscope Film LLC
All Rights Reserved
ISBN#978-1-937684-39-6
www.PeriscopeFilm.com

DISCLAIMER:

This manual is sold for historic research purposes only, as an entertainment. It contains obsolete information and is not intended to be used as part of an actual operation or maintenance training program. No book can substitute for proper training by an authorized instructor.

©2013 Periscope Film LLC
All Rights Reserved
ISBN#978-1-937684-39-6
www.PeriscopeFilm.com

WAR DEPARTMENT TECHNICAL MANUAL
TM 9-751

This manual supersedes TM 9-751, 15 September 1942; TB 751-1, 6 July 1943; and so much of TC 47, 13 April 1943; TC 59, 6 May 1943; TC 61, 5 May 1943; TB 700-21, 13 October 1942; TB 700-32, 18 January 1943; TB 700-34, 25 February 1943; TB 700-38, 12 March 1943; TB 700-40, 20 March 1943; TB 700-53, 15 May 1943; TB 700-67, 15 June 1943; TB 700-70, 21 June 1943; TB 700-72, 24 June 1943; TB 700-73, 1 July 1943; TB 700-81, 21 July 1943; TB 700-86, 26 July 1943; TB 700-87, 29 July 1943, and TB 700-91, 11 August 1943, as pertains to the 155-mm Gun Motor Carriage M12 and Cargo Carrier M30.

155-mm GUN MOTOR CARRIAGE M12 AND CARGO CARRIER M30

WAR DEPARTMENT • 28 JANUARY 1944

RESTRICTED DISSEMINATION OF RESTRICTED MATTER. The information contained in restricted documents and the essential characteristics of restricted material may be given to any person known to be in the service of the United States and to persons of undoubted loyalty and discretion who are cooperating in Government work, but will not be communicated to the public or to the press except by authorized military public relations agencies. (See also par. 18b, AR 380-5, 28 Sep 1942.)

UNITED STATES GOVERNMENT PRINTING OFFICE
WASHINGTON : 1944

WAR DEPARTMENT,
Washington 25, D. C., 28 January 1944.

TM 9-751, War Department Technical Manual, 155-mm Gun Motor Carriage M12 and Cargo Carrier M30, is published for the information and guidance of all concerned.

[A. G. 300.7 (22 Apr 43).]

By order of the Secretary of War:

G. C. MARSHALL,
Chief of Staff.

Official:
J. A. ULIO,
Major General,
The Adjutant General.

Distribution:
X.

(For explanation of symbol see FM 21-6.)

CONTENTS

CHAPTER 1. OPERATING INSTRUCTIONS   *Paragraphs*   *Page*

	Paragraphs	Page
Section I. General	1-4	1
II. Operation and controls	5-12	8
III. Inspection and preventive maintenance	13-17	20
IV. Lubrication	18-20	28
V. Tools and equipment stowage on 155-mm gun motor carriage	21-22	34
VI. Tools and equipment storage on cargo carrier	23-24	43
VII. Operation under unusual conditions	25-31	52
VIII. Matériel affected by chemicals	32	57

CHAPTER 2. VEHICLE MAINTENANCE INSTRUCTIONS

	Paragraphs	Page
Section I. General	33	58
II. Organization preventive maintenance	34	64
III. Organization tools and equipment	35	73
IV. Engine and accessories	36-70	80
V. Fuel system	71-90	158
VI. Lubrication system	91-100	177
VII. Cooling system	101-102	189
VIII. Clutch	103-110	190
IX. Propeller shaft	111-115	205
X. Power train	116-127	209
XI. Suspension and tracks	128-147	226
XII. Electrical system	148-162	245
XIII. Fire extinguishers	163-167	264

CHAPTER 3. ARMAMENT

	Paragraphs	Page
Section I. General	168-170	271
II. Description and function	171-173	275
III. Operations	174-187	279
IV. Sighting equipment	188	289
V. Ammunition	189-191	292

APPENDIX. References   299

INDEX   305

ILLUSTRATIONS

		RA PD No.	Page
1.	155-mm gun motor carriage—right front	44932	3
2.	155-mm gun motor carriage—left rear	44672	4
3.	155-mm gun motor carriage—top side	45439	4
4.	155-mm gun motor carriage—top rear	45438	5
5.	Cargo carrier—left front	45440	6
6.	Cargo carrier—right front	45437	6
7.	Cargo carrier—top	45550	7
8.	Cargo carrier—right rear	45436	7
9.	Cargo carrier—tail gate lowered	13317	8
10.	Driving controls	44944	9
11.	Shift lever positions	12401	10
12.	Installed instrument panel	44935	11
13.	Instrument panel—removed	44606	12
14.	Position of battery switch knob and right tank fuel shut-off valve handle	44937	13
15.	Plate on instrument panel giving oil dilution instructions	12523	17
16.	Lubrication guide, Gun Motor Carriage M12	301906	29
17.	Lubrication guide, Gun Motor Carriage M12	301907	30
18.	Lubrication guide, Cargo Carrier M30	301908	31
19.	Lubrication guide, Cargo Carrier M30	301909	32
20.	Left rear of engine with support beam	44950	81
21.	Right rear of engine with support beam	44688	82
22.	Front of engine with clutch and flywheel installed	44626	83
23.	Front of engine with clutch and flywheel removed	44638	83
24.	Engine compartment guards, top plate and rear plates	44617	90
25.	Engine compartment top plate removal	44705	90
26.	Engine compartment rear plates removal	44618	91
27.	Engine through engine rear plates opening	44608	92
28.	Engine installed—from upper rear of engine compartment	44648	93
29.	Engine installed—from upper front of engine compartment	44669	94
30.	Rear terminal box and disconnected wires	44603	96
31.	Rear terminal box and terminals	44624	98
32.	Engine compartment—looking toward rear of vehicle	44939	99
33.	Engine compartment—looking toward front of vehicle	44938	100
34.	Oil filter installed	44609	101
35.	Engine support beam	44621	103
36.	Engine lifting sling	44674	104
37.	Engine partially lifted from engine compartment	44620	105
38.	Engine lifted from engine compartment	44667	106
39.	Tail pipe removal	44603	108

		RA PD No.	Page
40.	Engine oil pump and oil pump finger strainer (governor removed)	44684	111
41.	Magneto with left half of distributor cover removed	56505	113
42.	Breaker point gap adjustment on Scintilla magneto	2738	113
43.	Magneto breaker mechanism installation (Scintilla)	12503	115
44.	Checking piston location with top dead center indicator	56506	116
45.	Use of timing disk	56507	117
46.	Scribe marks	12448	117
47.	Holding alignment of scribe marks while installing magneto	56508	119
48.	Bosch magneto	12538	120
49.	Breaker point gap adjustment on Bosch magneto (before installation)	56509	120
50.	Breaker mechanism of Bosch magneto	45955	122
51.	Parts of the Bosch magneto	45957	123
52.	Ignition harness (exhaust manifold removed)	12427	127
53.	Removing booster coil retaining screws	44678	130
54.	Booster coil and leads	44679	130
55.	Booster coil, cover removed, showing low-tension lead to rear terminal box	44680	131
56.	Starter	44676	132
57.	Left magneto removed to show retaining nuts for starter and generator	44706	132
58.	Generator	2733	134
59.	Fuel pump removal	44698	136
60.	Left side of carburetor	44685	138
61.	Right side of carburetor	44708	138
62.	Carburetor screen removal	44686	139
63.	Air cleaner installation	44615	142
64.	Component parts of air cleaner	44640	143
65.	Air cleaner parts requiring cleaning	44645	143
66.	Governor installation	44707	145
67.	Governor removal	44703	145
68.	Use of special wrench to tighten intake pipe packing nut	12482	146
69.	Interior of rocker box	45952	148
70.	Adjusting valve clearance	5485	149
71.	Rocker arm and push rod removal	45953	150
72.	Compressing valve springs to remove split lock	12475	152
73.	Fuel lines diagram	44675	159
74.	Spade winch installation (155-mm gun motor carriage)	44613	161
75.	Disconnecting fuel shut-off valve	44605	162
76.	Right fuel tank top plate removal	44612	162
77.	Lifting right fuel tank top plate	44641	163
78.	Fuel tank installed	44647	164
79.	Lifting out right fuel tank	44662	165
80.	Position of wooden spacers in fuel tank compartment	44677	166
81.	Primer pump	14375	169
82.	Priming distributor and lines	12502	169
83.	Auxiliary fuel pump removal	44607	171
84.	Fuel filter installed	44940	174
85.	Lubrication system	44713	178
86.	Checking oil supply line with bayonet gauge	44629	180
87.	Parts of disk-type filter	56515	181
88.	Installation of oil tank and oil cooler	44614	183

		RA PD No.	Page
89.	Oil cooler guard removal	43604	186
90.	Oil cooler removal	44942	187
91.	Engine steady-rest tube support removal	44941	188
92.	Front view of engine with clutch and flywheel installed	44954	189
93.	Flywheel and clutch assembly	44714	193
94.	Clutch release bearings	53597	194
95.	Clutch linkage	44635	195
96.	Removing clutch companion flange	44637	196
97.	Clutch spring housing nuts removal	44663	197
98.	Clutch spring housing removal	44664	198
99.	Adjustment of plate separators	300042	198
100.	Clutch spindle removal	44665	199
101.	Ventilated clutch and fan assembled	53588	200
102.	Flywheel hub nut removal	44668	201
103.	Flywheel removal	44671	201
104.	Arrow marks on drive plate	44660	203
105.	Propeller shaft with slip joint removed	44627	206
106.	Propeller shaft	44610	206
107.	Propeller shaft guard	44623	207
108.	Power train	44700	210
109.	Brake band	14366	211
110.	Steering brake adjustment	14365	211
111.	Disconnecting brake band from link	14364	212
112.	Pulling brake band out from below drum	12468	213
113.	Steering lever linkage and installation of parking brake	44952	216
114.	Adjusting left steering brake	45960	217
115.	Adjusting right steering brake	45959	217
116.	Final drive removed from power train	45958	219
117.	Power train on jacks prior to removal	14360	221
118.	Power train separated from vehicle	45951	224
119.	Bogie wheel lever and lever rubbing plate	56521	226
120.	Bogie wheel lubrication fittings	44719	227
121.	Pulling wheel gudgeon to remove bogie wheel	44934	227
122.	Pulling arm gudgeon preparatory to removing volute springs	44720	231
123.	Wheels and arms ready for removal preparatory to removing volute springs	44704	231
124.	Volute springs, lever, and bottom seat removed and partially disassembled	44715	232
125.	Example of loose track	45545	233
126.	Example of proper track tension	45455	234
127.	Clamping and spreader bolts of rear idler	44622	234
128.	Track adjusting wrench in position	44636	235
129.	Wedge nut removal	44931	237
130.	Disconnecting track shoe connector	44658	237
131.	Connecting track	44631	239
132.	Detail of track connecting fixture	44659	240
133.	Dead track link assembly dropping out of line on top of track	45956	240
134.	Idler	14356	243
135.	Installing grouser	14355	244
136.	Wiring diagram	44953	246
137.	Hydrometer correction chart	5851	247

		RA PD No	Page
138.	Battery compartment top plate removal	44943	248
139.	Position of storage batteries	44661	249
140.	Installed battery switch and solenoid starter switch	44628	250
141.	Generator regulator	44702	253
142.	Fuse box, showing circuits controlled by fuses	44639	254
143.	Headlight removal	44601	258
144.	Disconnecting blackout stop and service light	44634	259
145.	Blackout stop and service light removal	44633	259
146.	Siren removal	44602	261
147.	Installation of fixed fire extinguishers	44936	265
148.	Fixed fire extinguishers	44642	267
149.	Fixed fire extinguisher control head	44625	268
150.	155-mm gun motor carriage—top rear	317006	272
151.	Cargo carrier—top	317005	273
152.	Phantom view of breechblock carrier operating lever and operating lever handle	37252	275
153.	Firing mechanism block latch and percussion mechanism	37585	276
154.	Sectional view of counterrecoil and recuperator cylinders	37594	277
155.	Measuring oil index to determine amount of counterrecoil reserve oil	37595	278
156.	Air relief valve in countetrecoil cylinder front head	37596	278
157.	155-mm gun motor carriage in firing position—gun at maximum elevation	44681	280
158.	Installation of gun mount and spade assembly—rear view of vehicle	44718	281
159.	Arrangement of direct and indirect sighting equipment—left rear	55578	290
160.	Arrangement of direct and indirect sighting equipment—right rear	55579	291
161.	Projectile, A. P., 100-pound., M112, w/fuze, B. D., M60, 155-mm guns, M1917-17A1-18M1, M1, and M1A1	53904	296
162.	Shell, HE, M101, unfuzed, 155-mm guns, M1917-17A1-18M1, M1, and M1A1 (adapted for fuze, P. D., M51, w/booster, M21, or M51A1, w/booster, M21A1, or fuze, time, mechanical, M67, w/booster, M21A1)	RA FSD 1290	296
163.	Charge, propelling, NH Powder, 155-mm guns, M1917-17A1-18M1	RA FSD 1295	297

(This manual supersedes TM 9-751, 15 Sept. 1942; TB 751-1, 6 July 1943; and so much of TC 47, 13 April 1943; TC 50, 4 May 1943; TC 61, 5 May 1943; TB 700-21, 13 October 1942; TB 700-32, 18 January 1943; TB 700-34, 25 February 1943; TB 700-38, 12 March 1943; TB 700-40, 20 March 1943; TB 700-53, 15 May 1943; TB 700-67, 15 June 1943; TB 700-70, 21 June 1943; TB 700-72, 24 June 1943; TB 700-73, 1 July 1943; TB 700-81, 21 July 1943; TB 700-86, 26 July 1943; TB 700-87, 29 July 1943, and TB 700-91, 11 August 1943, as pertains to the 155-mm Gun Motor Carriage M12 and Cargo Carrier M30.)

CHAPTER 1
OPERATING INSTRUCTIONS

SECTION I
GENERAL

1. SCOPE.

a. This manual is published for the information and guidance of the using arms and services.

b. In addition to a description of the 155-mm Gun Motor Carriage M12 and Cargo Carrier M30, this manual contains technical information required for the identification, use, and care of the matériel.

c. Disassembly, assembly, and such repairs as may be handled by using arm personnel will be undertaken only under the supervision of an officer or the chief mechanic.

d. In all cases where the nature of the repair, modification, or adjustment is beyond the scope or facilities of the using arm personnel, the responsible ordnance service should be informed in order that trained personnel with suitable tools and equipment may be provided, or proper instructions issued.

2. CONTENT AND ARRANGEMENT. Chapter 1 contains information chiefly for the guidance of operating personnel. Chapter 2 contains information intended chiefly for the guidance of personnel of the using arms doing maintenance work. Chapter 3 contains all information relative to armament.

3. CHARACTERISTICS.

a. Both the 155-mm Gun Motor Carriage M12 and the Cargo Carrier M30 are armored, track-laying vehicles, having the general characteristics of the Medium Tank M4.

(1) A 155-mm gun is mounted on the rear of the Gun Motor Carriage M12.

(2) The Cargo Carrier M30 is of similar design, except that no large gun is mounted. Space is provided for carrying cargo.

b. The vehicles are powered by a 380-horsepower, nine-cylinder, radial aircraft type gasoline engine, which is located in a compartment directly behind the driver's compartment.

4. TABULATED DATA.

a. General.

	155-mm Gun Motor Carriage M12	Cargo Carrier T14
Weight, without armament fuel and crew__pounds__	47,276	39,500
Weight, fully equipped_____do	58,770	45,580
Ground pressure, fully equipped_____do	12.6	9.8
Height, over-all_____inches	96	96
Over-all width_____do	103½	103½
Ground clearance_____do	17	19¼
Tread (center to center of tracks)_____do	83	83

b. Engine.

	155-mm Gun Motor Carriage M12	Cargo Carrier T14
Make and model_____	Continental R-975C-1	Continental R-975C-1
Rated horsepower_____	400	400
Number of cylinders_____	9	9
Weight of engine w/accessories __pounds__	1,370	1,370

c. Fuel and oil.

	155-mm Gun Motor Carriage M12	Cargo Carrier T14
Fuel capacity_____gallons	200	200
Number of miles without refueling_____	190	190
Octane rating of fuel_____	(¹)	(¹)
Oil consumption____quarts per hour (approximate)	2	2
Engine oil capacity_____quarts	36	36

¹ 80 or higher.

d. Performance.

	155-mm Gun Motor Carriage M12	Cargo Carrier M30
Maximum sustained speed on hard road__mph	20	20
Cross-country speeds for various terrains__do	4 to 20	4 to 20

	155-mm Gun Motor Carriage M12	Cargo Carrier M30
Maximum allowable engine speed rpm	2,400	2,400
Maximum grade ascending ability	30°	30°
Maximum grade descending ability	30°	30°
Maximum width of ditch vehicle will cross feet	7 7/12	7 7/12
Maximum vertical obstacle vehicle will climb over (without grousers) inches	18	18
Maximum turning speed mph	10	10

   e. Crew. Six men.

   f. Tracks.

Track shoe width inches	16	16
Track pitch do	6	6
Ground contact square inches	3,432	3,432
Number of blocks per track	79	79

   g. Ammunition.

   (1) 155-mm Gun Motor Carriage M12.

      10 projectiles, 155-mm gun.
      10 propelling charges, 155-mm gun.
      25 fuzes, artillery.
      50 primers, artillery.
      360 rounds, caliber .45.
      300 rounds, caliber .30 carbine.
      60 rounds, caliber .30 rifle.
      10 grenades, AT, M9A1.

Figure 1. 155-mm gun motor carriage—right front.

(2) **Cargo Carrier M30.**
  40 projectiles, 155-mm gun.
  40 propelling charges, 155-mm gun.
  50 fuzes, artillery.
  50 primers, artillery.
  1,000 rounds, caliber .50.
  360 rounds, caliber .45 (submachine gun).
  21 rounds, caliber .45 (automatic pistol).
  360 rounds, caliber .30 carbine.
  60 rounds, caliber .30 rifle.
  10 grenades, AT, M9A1.
  12 grenades, hand.

Figure 2. 155-mm gun motor carriage—left rear.

Figure 3. 155-mm gun motor carriage—top side.

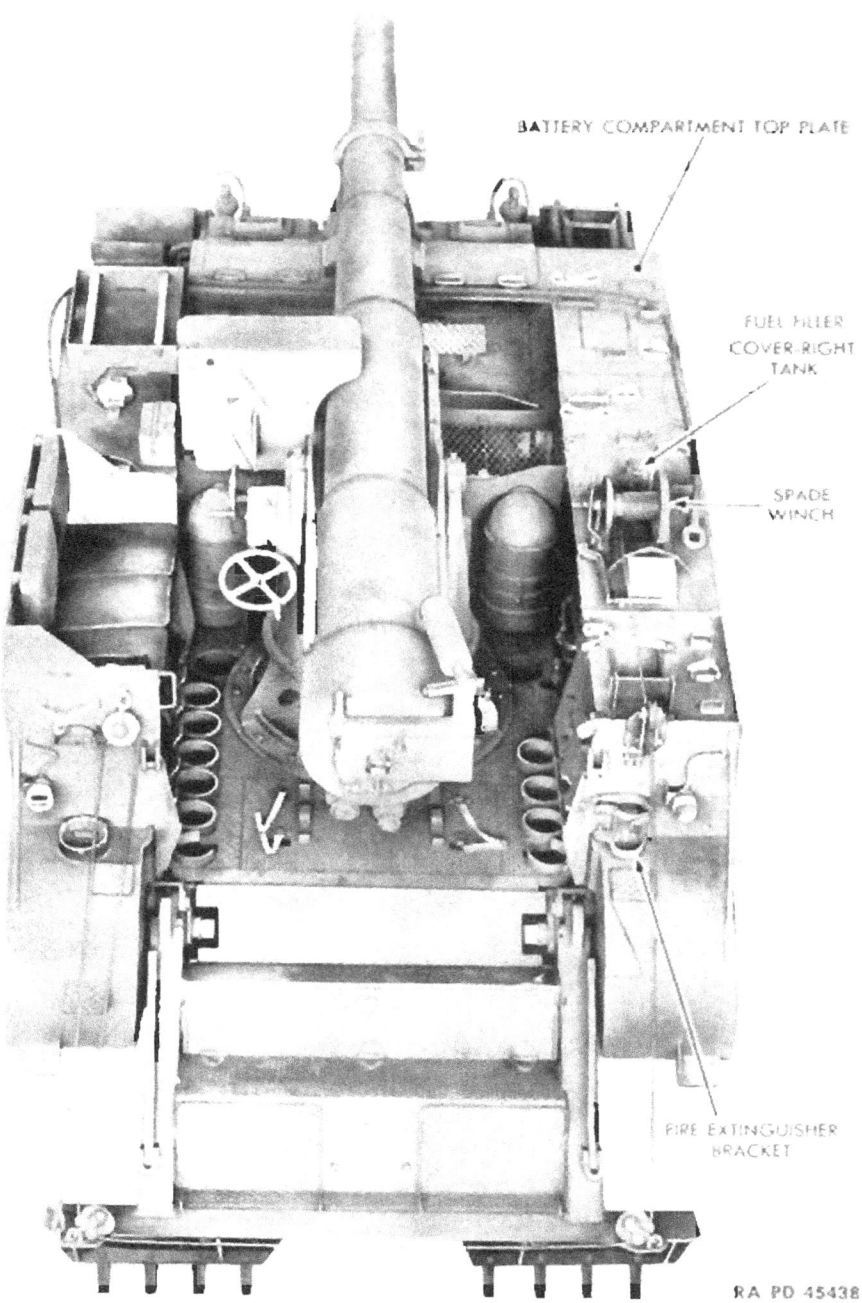

Figure 4. 155-mm gun motor carriage—top rear.

Figure 5. Cargo carrier—left front.

Figure 6. Cargo carrier—right front.

Figure 7. Cargo carrier—top.

Figure 8. Cargo carrier—right rear.

Figure 9. Cargo carrier—tail gate lowered.

SECTION II
OPERATION AND CONTROLS

5. GENERAL. The driver's controls in the 155-mm Gun Motor Carriage M12 and the Cargo Carrier M30 are similar to those used in the Medium Tank M4 except that positioning of the controls has been rearranged. The chassis is also generally similar to that of the Medium Tank M4, except that the engine has been moved forward to a position just behind the driver.

6. CONTROLS.

   a. Position of Driver. The driver sits on the left side of the front compartment, directly under the left front hatch.

   b. Steering Levers. Two steering levers provided with rubber grips are mounted directly ahead of the driver's seat (fig. 10). To

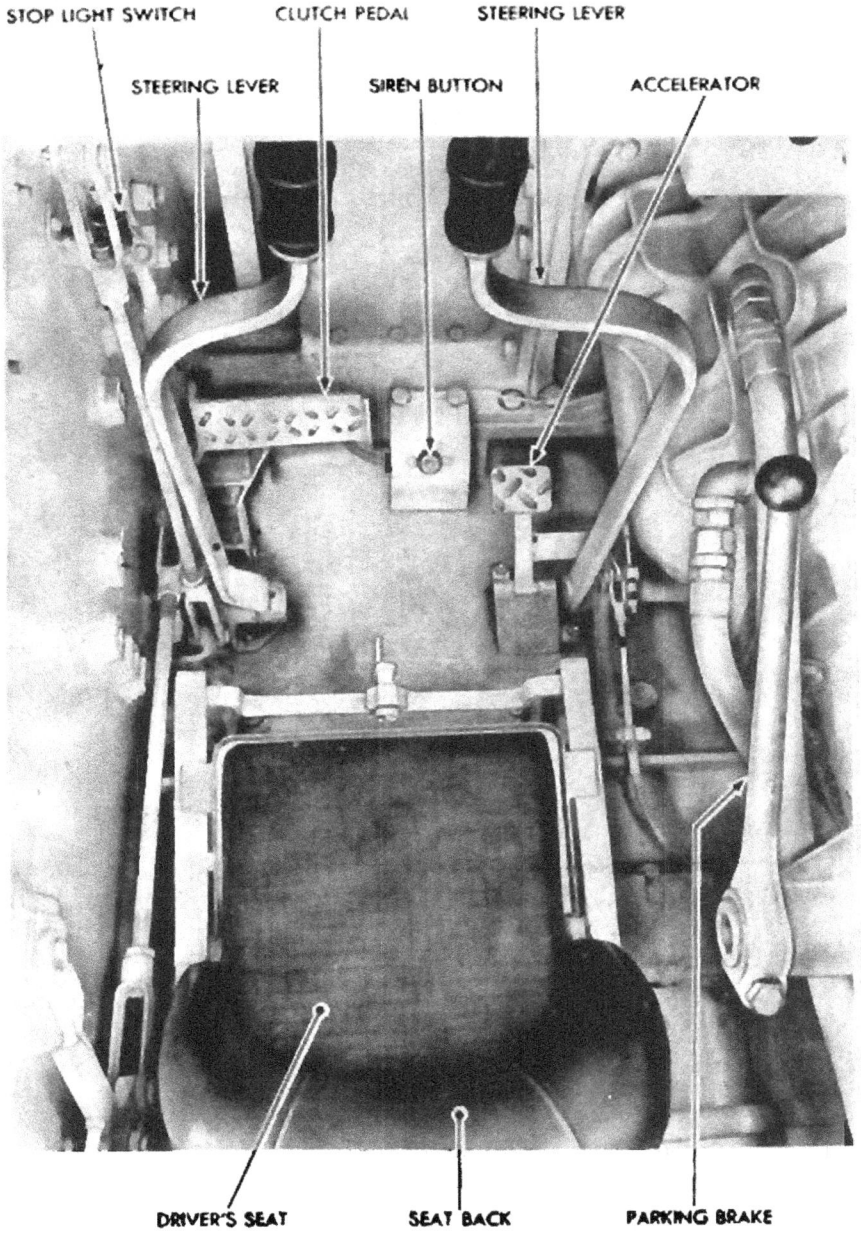

Figure 10. Driving controls.

steer the vehicle, pull the steering lever on the side toward which it is desired to turn. Pulling back either one of the levers slows down the track on that side, while the speed of the other track is increased. Thus the vehicle turns with power on both tracks at all times.

c. Accelerator and Hand Throttle. An accelerator is located to the right side of the steering levers (fig. 10). A hand-operated throttle is also provided (fig. 12).

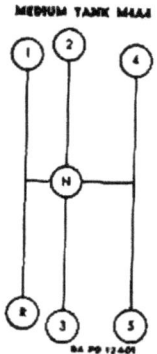

Figure 11. Shift lever positions.

d. Brakes.

(1) **Service Brakes.** Pulling back simultaneously on both steering levers slows down or stops the driving sprockets, depending on the effort applied, and thus slows or stops the vehicle.

(2) **Parking Brake.** The parking brake lever is located on the right side of the driver, at the back of the transmission (fig. 10). It is a transmission type brake, and should never be used for any purpose other than parking.

e. Clutch. The clutch pedal is located to the left of the steering levers, convenient to the driver's left foot (fig. 10). To permit shifting of gears, the clutch is disengaged by depressing the clutch pedal. When the pedal is depressed, engine and power train are disconnected.

f. Gear Shifting. Shifting of gears in the transmission for speed changes is accomplished by the gear shift hand lever, located on the transmission, to right of the driver. The gear shift lever is equipped with a latch which prevents accidental shifting into first speed or reverse. The latch must be released by pressing down on the button at the top of the lever before shifting into first speed or reverse.

g. Instrument Panel. The instrument panel is located in the center of the driver's compartment (fig. 12). It carries the fuel gauge, speedometer, tachometer, engine hour meter, engine oil tem-

Figure 12. Installed instrument panel.

perature gauge, engine oil pressure gauge, ammeter, voltmeter, ignition and starter switches, light switches, fuel cut-off switch, clock, and two auxiliary sockets (fig. 13).

h. Siren Control. The foot-operated siren button is located on a plate in front of the steering levers, within easy reach of the left foot of the driver (fig. 10).

i. Stop Light Switches. Stop lights are operated by two switches connected to the steering brake arms, and are turned on only when both steering levers are pulled back (fig. 10).

7. ENGINE STARTING AND WARM-UP.

a. General. Easy starting, operating efficiency, and the effective life of the engine are greatly influenced by the care used by the driver in starting and warming up the engine. For that reason it is essential that the procedure in **b** below be followed every time the engine is started, even though the vehicle is to be moved only a short distance.

Note: Some of the engines used in the 155-mm gun motor carriages and cargo carriers are equipped with Bosch magnetos, which have no provision for a booster coil. The starting instructions below include references to the booster switch,

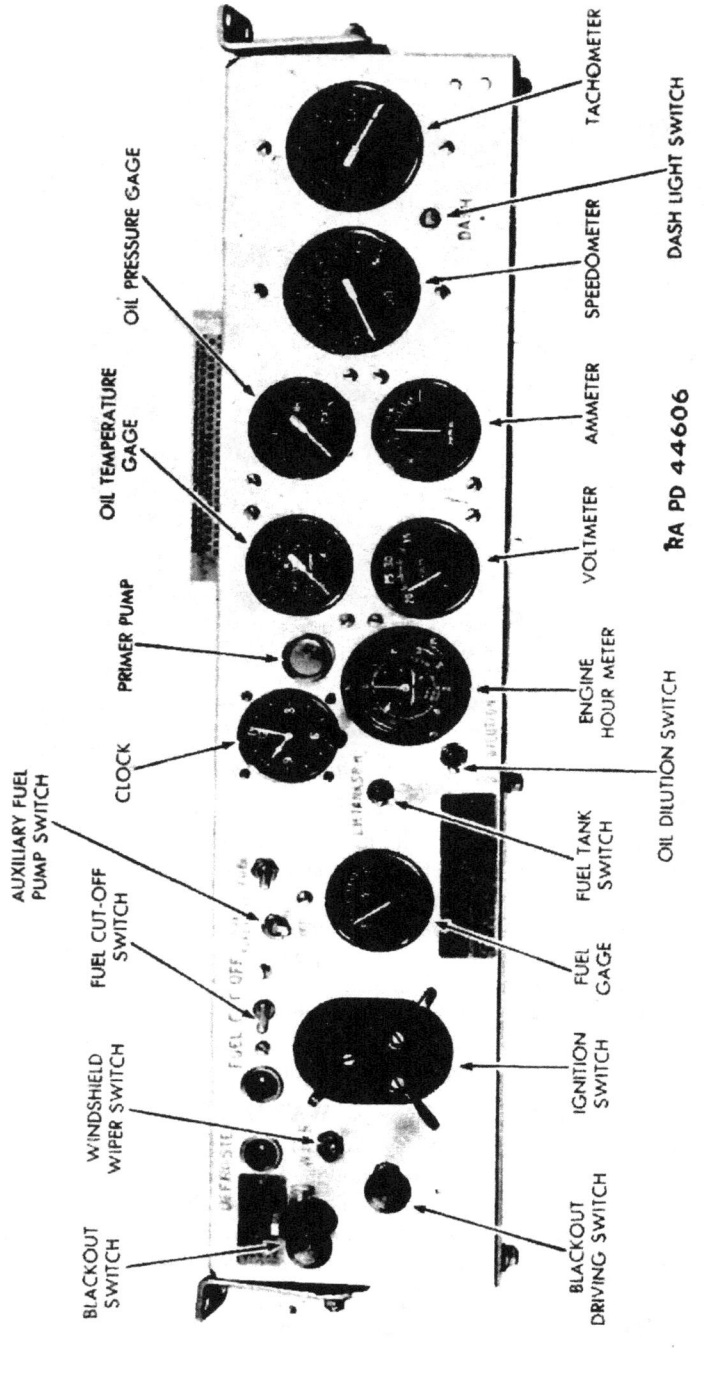

Figure 13. Instrument panel—removed.

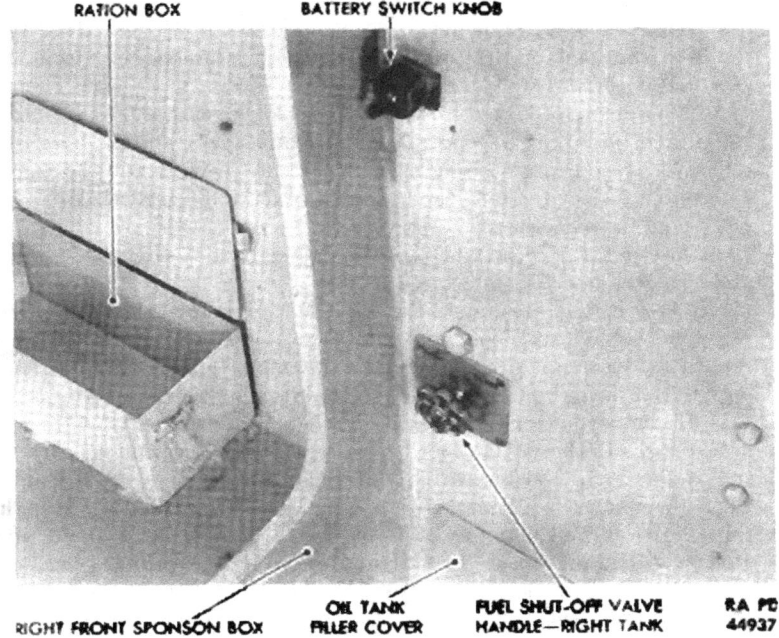

Figure 14. Position of battery switch knob and right tank fuel shut-off valve handle.

used to actuate the booster coil operating in conjunction with the Scintilla magnetos used on most engines. When starting engines that are not equipped with a booster coil, disregard any reference to the booster switch.

b. Procedure.

(1) See that the gear shift lever is in neutral (fig. 11).

(2) Turn engine crankshaft by hand four complete engine revolutions (about 50 turns of the hand crank) to clear combustion chamber of oil or water (from condensation). If excessive resistance is felt, making the engine crankshaft difficult to turn, remove one spark plug from each of the two lower cylinders to allow any trapped fuel, oil, or water to drain out. This is to eliminate any possibility of hydrostatic lock. After clearing the engine of trapped fluid, install the spark plugs.

(3) Open both fuel shut-off valves (figs. 14 and 73).

(4) Turn on the booster switch and listen for the buzzing which indicates operation (fig. 13).

(5) Unless the engine is already warmed up, in cold temperature of 50° F. or lower, prime the engine by a few strokes of the priming pump (pull plunger out slowly, push in briskly) (fig. 13). Avoid overpriming, which tends to wash the oil off cylinder walls. *Caution:* Never use accelerator for priming.

(6) Open the hand throttle slightly (fig. 12).

(7) Depress the clutch pedal to disengage the clutch (fig. 10). This removes the load of the propeller shaft and transmission from the starting motor.

(8) If engine is not already warmed up, prime it with a few strokes of the primer pump, by pulling the plunger out slowly and pushing it in quickly. Avoid overpriming as it will result in wear and damage if sufficient to wash the lubricant from cylinder walls and pistons. Do not prime through the spark plug holes.

(9) Close the starter switch, turn the magneto to the BOTH position, and close the booster coil switch in that order.

(10) When engine fires, open the throttle slowly until the tachometer registers 800 revolutions per minute. Opening the throttle too rapidly, after engine fires, may make the engine stop. An additional stroke from primer may be desirable to keep engine running.

(11) If engine does not start, prime it again and repeat the starting procedure. If the engine has been flooded by overpriming, turn off the magneto and booster switches and hold the accelerator fully open. Close the starter switch and turn engine a few revolutions. Close the throttle to normal starting position (hand throttle open about one-quarter inch) and repeat the starting procedure with the exception of the priming operation. *Caution:* Never hold the starter switch closed for more than 30 seconds at a time. Always allow the starter to cool before turning it on again. If failure to start continues see paragraph 39.

(12) Watch the oil pressure gauge and if oil pressure does not register within one-half minute, the engine should be stopped and an investigation made. In cold weather, if the oil has not been preheated before starting, the engine should be throttled to 800 revolutions per minute until the oil temperature rises at least 10° F. as indicated by the oil temperature gauge.

(13) Warm the engine at 800 revolutions per minute for 5 minutes (fig. 13). The oil pressure should be between 50 and 80 pounds, which is normal for this speed.

(14) After a 5-minute period, increase to 1,000 revolutions per minute. If the oil pressure drops off, go back to 800 revolutions per minute for more complete warm-up.

(15) Check operations of instruments (fig. 13) while warming up, and note results on commander's report. The ammeter reading immediately after starting may be as high as 50 amperes, depending upon the condition of the battery, and will gradually drop as the battery becomes charged. The voltmeter, after starting, may be as high as 28½ volts.

(16) Increase engine speed to 1,800 revolutions per minute and try magneto switch in "L" and "R" positions. A drop in engine speed of

100 revolutions per minute is permissible at this speed, when operating on only one magneto. *Caution:* Do not run engine on one magneto for more than a 30-second interval, as this will cause the inoperative plugs to become carbonized.

c. Precautions in Engine Operation.

(1) Never idle below 800 revolutions per minute at any time, including military ceremonies. Continuous idling will appreciably shorten the useful life of the engine.

(2) Note engine oil temperature prior to starting. Do not operate the vehicle until a 10° temperature rise is noted in the engine oil temperature.

(3) Normal oil pressure is 50 to 80 pounds at operating speeds.

(4) Never lug engine below 1,600 revolutions per minute.

(5) Maximum oil temperature is 190° F.

(6) Operate engine from 1,800 to 2,100 revolutions per minute. Minimum operating speeds for brief periods only, 1,500 to 1,600 revolutions per minute.

(7) Care must be taken not to allow dirt to collect on cylinders, or to permit any object to block the flow of air to or from the cylinders, as this may cause overheating and preignition.

(8) When descending a steep grade, shift to a gear low enough to control the speed of the vehicle. Use the steering brakes to keep the engine speed below 2,400 revolutions per minute.

8. INSPECTIONS DURING ENGINE WARM-UP.

a. The inspections in **b** below are to be made before and during the time required for engine warm-up.

b. Procedure.

(1) **Inspect for Oil Leaks Under the Hull.** This is done from both the front and rear, and is one of the first inspections made or ordered by the commander. Trace possible cause of any leaks, and correct condition.

(2) **Inspect Track for Tension.** If the track shows noticeable sag, it must be tightened (par. 139).

(3) **Inspect Condition of Track.** Inspect end connectors for wear or for bent or broken guide lugs. Inspect all wedges and nuts for presence and tightness. Check presence of self-locking nuts. Inspect for bottomed wedges and replace. Clearance between wedge and connector should be not less than $\frac{1}{32}$ inch. If grousers are being used, inspect their condition and tightness. Inspect all slack portions of the track for the presence of dead track blocks and replace any such blocks (par. 142). The presence of a dead track block is indicated by a block which has dropped definitely out of line.

Note: To be detected, a dead track block must be on top of the track. This necessitates moving the vehicle to make a complete inspection. At this time, the inside wedges and connectors can also be inspected.

(4) **Inspect Sprockets.** Inspect for sprung or worn teeth. Inspect cap screws and hub nuts for tightness. Inspect inside cap screws when vehicle is being moved.

(5) **Inspect Bogie Wheels.** Inspect condition of tires on all wheels. Look for evidence of outer spacer turning, and tighten gudgeon nut if spacer is turning. Inspect gudgeon nuts for presence, and see that cotter pins are properly installed. Inspect condition of grease fittings and release valves, and replace any that are damaged or missing (fig. 120).

(6) **Inspect Bogie Assemblies.** Inspect for broken or weak volute springs. Inspect wheel arm and lever wear plates for wear, and replace if worn. Inspect bogie bracket for presence of bolts, nuts, and lock wire.

(7) **Inspect Support Rollers.** Look for evidence of rollers not turning, and free up all rollers by cleaning out mud, rocks, etc. Check grease fittings, and replace missing fittings. It is essential that all rollers turn freely since inoperative rollers will develop flat spots.

(8) **Inspect Idlers.** Check for security of idler cap and grease fittings.

Caution: It is highly important that the entire track and suspension system be kept as clean as possible and free from dirt, rocks, and sticks. This will protect the life and efficiency of the system.

9. OPERATING VEHICLE.

a. With the driver in driver's seat, the engine at idling speed, and all instruments showing normal readings, the driver is then ready to drive the vehicle.

b. Release parking brake (located on rear end of transmission) (fig. 10).

c. Disengage clutch by pressing clutch pedal down to floor and holding it down (fig. 10).

d. Move gear shift lever into second gear (fig. 11).

e. After speeding up engine, gradually release clutch pedal, at the same time depressing accelerator (fig. 10). Except when under fire, it is very dangerous to attempt to move vehicle in or out of close quarters without aid of personnel outside of vehicle serving as a guide.

f. Correct gear for running is that which enables the vehicle to proceed at the desired speed without causing the engine to labor. Do not ride clutch. The driver's left foot must be completely removed from clutch pedal while driving, to avoid unnecessary wear on clutch.

g. To place vehicle in reverse gear, a complete stop must be made, and throttle closed to idling speed. Shift briefly into second gear to stop the propeller shaft. Then, depressing the latch, move gear shift lever to reverse position (fig. 11). Do not back up vehicle unless an observer is stationed in front to guide the driver.

h. To steer, pull back the right-hand steering lever to make a right turn, and the left-hand lever to make a left turn. This action keeps one of the tracks from turning as fast as the other track and more power is needed. As the driver anticipates making a turn, he must be ready to apply the accelerator to a greater extent, depending on the sharpness of the turn.

i. Tachometer, oil temperature gauge, and oil pressure gauge give the most satisfactory indications of the engine's performance (fig. 13). If there is any indication that these instruments are irregular, stop the engine and investigate the cause.

10. STOPPING VEHICLE.

a. To stop the vehicle, release throttle and pull back on both steering levers at the same time, depressing clutch when vehicle has slowed down to approximately 2 to 5 miles per hour, depending upon which gear is being employed before stopping. It is desirable to shift down as low as possible, preferably to second gear, and use engine drag to slow the vehicle, to facilitate stopping.

b. The parking brake is located on the back of the transmission to the right of the driver's seat (fig. 10). Use this brake only for parking; never use it while the vehicle is in motion. Always make sure that this brake is released before putting the vehicle in motion.

11. STOPPING ENGINE.

a. To stop the engine, close the throttle until the engine is idling at 800 revolutions per minute and run the engine at this speed for 4 minutes to permit a gradual and uniform cooling of the engine parts. Advance engine speed to 1,000 revolutions per minute and throw the fuel cut-off switch on the instrument panel to the OFF position (fig. 13), shut off both the main fuel supply valves, turn main battery switch to OFF position (figs. 14 and 73). To prevent blowing of fuses and damage to regulator contact points, always turn main battery switch to OFF position before firing the gun.

OIL DILUTION

EXPECTED TEMP., DEG. F.	+18	+8	ZERO	-8	-15	-20
HOLD SWITCH ON, MINUTES	1	1½	2	2½	3	3½

1. SHUT OFF ENGINE IMMEDIATELY AFTER DILUTION.
2. STARTED ENGINE HAVING DILUTED OIL MUST BE RUN AT LEAST 30 MINUTES.

RA PD 12523

Figure 15. Plate on instrument panel giving oil dilution instructions.

b. Always stop the engine with the fuel cut-off switch, rather than by switching off the magnetos (fig. 13). This eliminates the possibility of the engine continuing to run due to preignition. After engine has stopped, cut off magnetos.

c. If the vehicle is to remain out of operation for periods of 5 hours or more, in freezing temperature, engine oil must be diluted. Dilute before shutting down engine. With engine idling 1,000 revolutions per minute and oil temperature 140° to 160° F., hold oil dilution valve toggle switch in closed position in accordance with oil dilution instructions on instrument panel (figs. 13 and 15).

NOTE: On later type vehicles, no provision has been made for oil dilution procedure.

d. After stopping the engine, the inspections of engine and vehicle should always be made (par. 17).

12. OPERATING PRECAUTIONS.

a. General Instructions. Do not allow an untrained driver to operate the vehicle, except with personal help and supervision from a competent instructor. (See pars. 13 to 17, incl.) Operation requires definite technique, which can be learned correctly only by continued selection of gears for various terrains, and knowledge of when and how to shift down into lower gears in order to secure maximum performance from both vehicle and engine (fig. 11)

b. Precautions in Driving.

(1) Know the vehicle, its capabilities, and limitations. Learn to judge engine speed by sound. Listen for unusual noises in engine and transmission, as well as the rest of the power train.

(2) Keep engine speeds up to 1,800 to 2,100 revolutions per minute. Never run engine at wide-open throttle below 1,600 revolutions per minute. Limit operating range of engine from 1,600 to 2,200 revolutions per minute. Do not operate the vehicle at high loads below 1,500 revolutions per minute for continuous duty. The vehicle may be operated in fourth and fifth gears over hilly or rolling terrain, but when the engine speed drops to 1,500 revolutions per minute in climbing hills, shift to a lower gear. Do not remain in high gear until the engine threatens to stall during hill climbing, and similar low speed and high load operation, as the engine speed will drop as low as 800 revolutions per minute.

(3) Know approximate speeds in each gear and corresponding revolutions per minute of engine. This will aid in shifting and permit driver to keep vehicle under control at all times.

(4) Always keep power on tracks when turning. Shift down if necessary to keep engine speed up.

(5) Always get into correct gear before attempting hills, muddy areas, or long pulls. Once vehicle has entered difficult terrain, it is difficult to shift down into a lower gear and maintain the desired speed of the vehicle during the gear shifting operation.

(6) Do not turn when climbing a hill or just after shifting to a higher gear. Let engine have every chance to regain speed before using steering lever. Even a slight turn will drag down the speed.

(7) Use engine as a brake when going down hills. Never disengage clutch and coast down. This may result in vehicle reaching a speed higher than the engine is capable of, causing a ruined clutch or sheared propeller shaft when clutch is engaged. Never allow the engine speed to exceed 2,200 revolutions per minute.

(8) Use engine to slow down vehicle by shifting down to lower gears and let engine act as a brake. This saves wear on brakes.

(9) "Double-clutching", when shifting down, will make shifting quicker and lessen wear on cones in transmission.

(10) Remove hands from steering levers when not actually turning vehicle. "Riding" the levers wears the brakes.

(11) Keep left foot on floor and off clutch pedal, except when actually changing gears. Whenever vehicle is moving, clutch must be fully engaged.

(12) In moving around buildings, shops, or confined spaces, use first or reverse gear. It is better to let engine idle down and move slowly than to have the engine speed too high and slip the clutch. In confined spaces, have trained guide outside vehicle to direct movement with hand signals.

(13) Learn how to cross obstacles correctly, keeping clutch engaged. In crossing ditches, let vehicle settle gently, then give it full power as soon as bottom is reached.

(14) In breaking over an obstacle, let vehicle rise and settle down over obstacle instead of applying full power in surmounting the obstacle.

(15) Except when necessary, never drive a vehicle that needs adjusting.

(16) Constantly watch oil pressure gauge (at least 60 pounds at operating speeds), tachometer, and oil temperature (maximum 190° F.).

(17) Never hold engine at wide-open throttle (2,400 revolutions per minute) for more than a few seconds at a time. This is an emergency speed, not a driving speed.

(18) Always set parking brake after stopping vehicle. It is bad practice to leave it in gear.

c. Inspection. After stopping engine, always inspect engine and vehicle (par. 17).

SECTION III

INSPECTION AND PREVENTIVE MAINTENANCE

13. PURPOSE.

a. Regularly scheduled preventive maintenance provides an opportunity for the driver and the using arms maintenance personnel to observe the condition of the vehicle and prevent mechanical difficulty. The services set forth are divided into three phases: first, those performed by the driver, Before Operation, During Operation, At Halt and After Operation Service and Weekly Operation; second, those performed by the company or similar unit mechanics at 50-hour intervals; third, those performed by battalion, regimental, or similar unit mechanics at 100-hour intervals.

b. Driver preventive maintenance services are outlined on the back of Driver's Trip Ticket, P. M. Service Record (W. D. Form No. 48) in a general form to accommodate vehicles of all classes. Because the 155-mm Gun Motor Carriage M12 is a special type of vehicle, it has become necessary to include items of inspection not mentioned on the trip ticket. These additional items must be performed as faithfully as those noted on the trip ticket and all organizations must thoroughly school each driver in performing the maintenance set forth herein.

c. The services listed below are arranged to facilitate inspection and conserve the time of the driver and are not necessarily in the same numerical order as shown on Driver's Trip Ticket, P. M. Service Record (W. D. Form No. 48). The item numbers, however, are identical with those shown on W. D. Form No. 48.

d. The general inspection of each item applies also to any supporting member or connection and usually includes a check to see whether the item is in good condition, correctly assembled, secure, or excessively worn.

e. The inspection for "good condition" is an external visual inspection to determine whether the unit is damaged beyond safe or serviceable limits. The term "good condition" is explained further by the following terms: not bent or twisted, not chafed or burned, not broken or cracked, not bare or frayed, not dented or collapsed, not torn or cut.

f. The inspection of a unit to see that it is correctly assembled is usually an external visual inspection to determine whether it is in its normal assembled position on the vehicle.

g. The check of a unit to determine if it is secure is usually an external visual inspection, a hand-feel, or pry-bar check for looseness in the unit. Such an inspection should include any brackets, lock washers, lock nuts, locking wires, or cotter pins used in the assembly.

h. "Excessively worn", which is a frequently used term, will be understood to mean worn close to or beyond serviceable limits and likely to result in failure if not replaced before the next scheduled inspection.

14. BEFORE OPERATION SERVICE.

a. General. This inspection schedule is designed primarily as a check to see that the vehicle has not been tampered with, or sabotaged since the After Operation Service was performed. Various combat conditions may have rendered the vehicle unsafe for operation and it is the duty of the driver to determine whether or not the vehicle is in condition to carry out any mission to which it may be assigned. This service will not be entirely omitted, even in extreme tactical situations. When thoroughly trained, the driver will be able to quickly determine the condition of this vehicle.

b. Procedure. The following are the procedures to be followed in performing the Before Operation Service:

(1) **Item 1, Tampering and Damage.** Inspect for damage to exterior of vehicle, armament, and equipment by falling debris or collision. Examine tracks, sprockets, driver's instruments and levers, gun firing, elevating and traversing mechanisms for tampering or sabotage. Examine exhaust pipe to see that it is clear.

(2) **Item 2, Fire Extinguishers.** Examine for full charge. See that release handle is in locked position.

(3) **Item 3, Fuel and Oil.** Inspect in engine and driver's compartment for leaks. Add fuel or oil as required. See that fuel tank valves operate freely. Investigate any noticeable loss since last operation.

(4) **Item 4, Accessories.** Inspect generator, starter, magnetos, regulator and carburetor for damage, loose mountings and loose connections.

(5) **Item 6, Leaks, General.** Inspect engine and driver's compartment and ground under vehicle and in for indications of fuel or lubricant leaks.

(6) **Item 7, Engine Warm-up.** Start engine noting if starter has adequate cranking speed. If oil pressure gauge does not register within 10 seconds, engine should be stopped and trouble corrected, or reported. Set throttle at fast idle speed and during warm-up period proceed with the following Before Operation Services. *Caution:* Do not idle below 800 revolutions per minute on warm-up.

NOTE: During warm-up, listen for unusual noises and note engine performance.

(7) **Item 8, Primer.** While starting engine, observe if primer operates properly to deliver fuel.

(8) **Item 9, Instruments.**

(a) **Oil Pressure Gauge.** Normal reading of gauge is 50 to 80 pounds at fast idle speed.

NOTE: Stop engine if zero or excessively low pressure is indicated. Investigate cause and report.

(b) **Ammeter.** The ammeter should show high charge (45 to 55 amps) for a short time; then drop to zero or slight positive charge.

(c) **Oil Temperature Gauge.** Normal reading of gauge is 165° F. to 196° F. *Caution:* Do not operate vehicle until a 10° temperature rise is indicated.

(d) **Tachometer.** The tachometer should indicate approximate engine revolutions per minute.

(e) **Fuel Gauge.** The fuel gauge should indicate approximate amount of fuel in both tanks.

(f) **Voltmeter.** The voltmeter may indicate high reading (28½ volts) after starting; then drop to approximately 24 volts.

(g) **Hour Meter.** The hour meter should register accumulating engine running hours.

(h) **Compass.** See that compass is serviceable.

(9) **Item 10, Siren.** If tactical situation permits, test siren for normal signal.

(10) **Item 12, Lights.** If tactical situation permits, see that all lights operate properly. Inspect for loose mountings, loose connections, and broken or dirty lenses.

(11) **Item 13, Sprockets and wheels.** See that all ring and hub bolts are present and secure. See that bogie wheels are not damaged or loose.

(12) **Item 14, Tracks.** Inspect tracks for broken, bent, or cracked guides and damaged or unserviceable links.

(13) **Item 15, Springs.** Examine springs for weak or broken volute springs.

(14) **Item 17, Fenders.** Examine fenders for looseness or damage.

(15) **Item 18, Towing Connections.** Examine towing eyes for loose mounting and damage.

(16) **Item 20, Decontaminator.** The decontaminator must be secure and fully charged.

(17) **Item 21, Tools, Equipment, and Armament.** Examine for presence and serviceability of tools and equipment. See that they are properly mounted or stowed. Check stowage of ammunition and be sure gun is locked in carrying position.

(18) **Item 22, Engine Operation.** The engine should idle smoothly. Accelerate and decelerate and listen for any unusual noises that might indicate compression or exhaust leaks, worn, damaged, loose, or inadequately lubricated engine parts or accessories. Note excessive smoke from exhaust. With the engine running at 1,800 revolutions

per minute, using both magnetos, the tachometer should not drop more than 100 revolutions per minute if a switch is made to the right magneto. Test left magneto similarly.

(19) **Item 23, Driver's Permit and Form No. 26.** Examine forms for presence and legibility.

(20) **Item 25, During Operation Service.** Start the During Operation Service immediately after the vehicle is put in motion.

15. DURING OPERATION SERVICE.

a. General. While vehicle is in motion, listen for any sounds such as rattles, knocks, squeals, or hums that may indicate trouble. Look for steam from radiator and smoke from any part of vehicle. Know and watch for odor of an overheated generator, brakes or clutch, boiling coolant, fuel vapor from a leak in fuel system, exhaust gas, or other such signs of trouble. Any time the brakes are used, gears shifted, or vehicle turned, consider this a test and notice any unsatisfactory or unusual performance. Watch the instruments constantly. Notice promptly any abnormal instrument indication that may signify possible trouble in system to which the instrument applies.

b. Procedure. During Operation Services consist of observing items listed below according to the procedures, following each item and stopping vehicle when serious trouble develops. Notice minor deficiencies to be corrected or reported at earliest opportunity, usually next scheduled halt.

(1) **Item 26, Steering Brakes.** Apply brakes separately and see that vehicle responds satisfactorily. With brakes released, note tendency of vehicle to creep to either side. Pull both levers and see that vehicle stops effectively.

(2) **Item 27, Parking Brake.** Stop vehicle. The parking brake, when applied and locked, should hold vehicle securely.

(3) **Item 28, Clutch.** The clutch should have 1-inch free pedal travel and should not chatter, squeal, or slip.

(4) **Item 29, Transmission and Final Driver.** Gears should shift smoothly, operate quietly, and not slip out of mesh. Note excessive foaming from transmission breather.

(5) **Item 31, Engine and Controls.** Observe for power on acceleration, and tendency to misfire, stall, knock, or overheat. Note excessive exhaust smoke.

(6) **Item 32, Instruments.** Observe instruments regularly during operation.

   (a) **Oil Pressure Gauge.** Normal gauge reading is 50 to 80 pounds pressure.

   (b) **Ammeter.** Zero or slight positive charge.

(c) **Oil Temperature Gauge.** Normal gauge reading is 165° F. to 190° F.

NOTE: Only under extreme heat or operating conditions should temperature rise to 200° F.

(d) **Tachometer.** The tachometer should register engine revolutions per minute with no excessive fluctuation or noise.

(e) **Fuel Gauge.** The gauge should register approximate amount of fuel in tanks.

(f) **Voltmeter.** Normal reading of voltmeter is approximately 24 volts.

(g) **Hour Meter.** Meter should register accumulating engine running hours.

(h) **Compass.** Compass should react to change of direction.

(i) **Speedometer.** The speedometer should indicate speed without excessive fluctuation or noise.

(7) **Item 34, Running Gear.** Listen for unusual noises from tracks, bogie wheels, and support rollers. Observe any indication of looseness in propeller shaft.

(8) **Item 36, Guns.** Inspect gun for loose mountings and sighting equipment.

16. AT HALT SERVICE.

a. General. At Halt Services may be regarded as minimum battle maintenance and should be performed under all tactical conditions even though more extensive maintenance services must be slighted or omitted altogether.

b. Procedure. At Halt Services consist of investigating any deficiencies noted during operation, inspecting items listed below according to the procedures following the items, and correcting any deficiencies found. Deficiencies not corrected should be reported promptly to chief of section or other designated individual.

(1) **Item 38, Fuel and Oil.** Fill fuel tanks if necessary. Replenish engine oil supply if required.

(2) **Item 39, Temperatures.** Examine bogie wheel and support roller hubs, transmission, and final drive for abnormal heat.

(3) **Item 40, Breathers.** See that transmission and final drive breathers are not clogged. Note excessive foaming.

(4) **Item 42, Springs and Suspensions.** Examine for weak or broken volute springs. Examine suspension arms, levers, brackets and gudgeons for breakage, and loose parts. Remove all stones or debris.

(5) **Item 43, Steering Brake Linkage.** In driver's compartment, examine for loose or excessively worn linkage connections, and loose mountings.

(6) **Item 44, Sprockets and Wheels.** Examine for loose, missing, or damaged hub and ring nuts. See that bogie tires are not damaged or separated from the wheel.

(7) **Item 45, Tracks.** Examine for bent, broken, or cracked guides, and damaged or unserviceable links. Normal sag is 1½ to 2 inches.

(8) **Item 46, Leaks; General.** Look for fuel and lubricant leaks in engine and driver's compartment and under vehicle.

(9) **Item 47, Accessories.** Inspect all accessible engine accessories for damage and loose mountings.

(10) **Item 48, Air Cleaner.** Examine cleaner for loose mounting. When operating under extreme dust and sand conditions, examine filter element for clogged condition. Remove, clean, and refill oil reservoir as necessary.

(11) **Item 49, Fenders.** Examine fenders for looseness or damage.

(12) **Item 50, Towing Connections.** See that all connections are present and securely mounted.

(13) **Item 52, Appearance and glass.** Clean port windows and light lenses.

17. AFTER OPERATION AND WEEKLY SERVICE.

a. General. After Operation Service is particularly important because at this time the driver inspects his vehicle to detect any deficiencies that may have developed and corrects those he is permitted to handle. He should report promptly, to his chief of section or other designated individual, the results of his inspection. If this schedule is performed thoroughly, the vehicle should be ready to roll again on a moment's notice. The Before Operation Service, with a few exceptions, is then necessary only to ascertain whether the vehicle is in the same condition in which it was left upon completion of the After Operation Service. The After Operation Service should never be entirely omitted even in extreme tactical situations, but may be reduced to the bare fundamental services outlined for the At Halt Service if necessary.

b. Procedure. When performing the After Operation Service, the driver must remember and consider any irregularities noticed during the day in the Before Operation, During Operation, and At Halt Services. The After Operation Service consists of inspecting or testing the following units and correcting or reporting any deficiencies. Those items of the After Operation Service that are marked by an asterisk (*) require additional weekly services, the procedures for which are indicated in (b) of each applicable item.

(1) **Item 54, Fuel and Oil.** Fill fuel tank. Check engine oil supply and replenish if necessary.

(2) **Item 55, Engine Operation.** Investigate and report unsatisfactory engine performance noted during operation.

(3) **Item 56, Instruments.** Report any unusual performance observed during operation. Observe fuel gauge for FULL reading after refueling.

(4) **Item 57, Siren.** If tactical situation permits, test siren for normal signal.

(5) **Item 59, Lights.** Inspect lights for damage, loose mountings, and broken lenses. Clean if necessary. See that all lights operate properly.

(6) **Item 60, Fire Extinguisher.** Examine extinguisher for loose mountings and nozzles. See that engine compartment extinguisher handle is in locked position. If extinguishers have been used, report for refill or replacement.

(7) **Item 61, Decontaminator.** Examine decontaminator for full charge, loose mounting, and damage.

(8) **Item 62, *Battery.**

(a) Check electrolyte level. See that vent caps are clean and secure. Inspect for loose mountings and connections, and for leaks.

(b) **Weekly.** Clean dirt from top of battery. Add water to ½ inch above plates. If posts or terminals are corroded, clean and apply fresh coat of grease. Tighten terminal and mounting bolts. Clean battery carrier if corroded.

(9) **Item 63, *Accessories.**

(a) Inspect generator, starter, magnetos, and regulator for loose mountings and connections.

(b) **Weekly.** Tighten all accessory mounting bolts, and line connections if loose.

(10) **Item 64, *Electrical Wiring.**

(a) See that ignition wiring is securely connected, clean, and not damaged.

(b) **Weekly.** Inspect all accessible wiring in driver's and engine compartment. Examine insulation and shielding for damage and excessive wear. Report unserviceable wiring.

(11) **Item 65, *Air Cleaners.**

(a) See that oil in air cleaner reservoir is at correct level. If oil is excessively dirty, drain, clean, and refill reservoir. See that air duct connections are secure.

(b) **Weekly.** Disassemble and clean air cleaner body and elements. Drain, clean, and refill reservoir. Inspect gaskets for damage. Examine air ducts for damage and looseness.

(12) **Item 68, *Tracks and Tires.**

(a) Examine tracks for broken, bent, or cracked guides, loose wedge nuts, and excessive track sag. See that tires are not separating from bogie wheels.

(b) **Weekly.** Tighten all wedge nuts. Check track connectors and wedges for excessive wear.

(13) **Item 82, *Sprocket.**

(a) Inspect sprocket for damaged or missing hub and ring nuts.

(b) **Weekly.** Tighten bogie crab, connector arm, and roller bolt nuts. Report any defects in rollers or idler.

(14) **Item 70, Steering Brake Linkage.** Examine brake for loose or disconnected linkage, excessively worn parts, and damage.

(15) **Item 72, *Vents.**

(a) Inspect for damaged or clogged breathers.

(b) **Weekly.** Remove and clean breathers thoroughly.

(16) **Item 73, Leaks, General.** Look for fuel and lubricant leaks in engine and driver's compartment and under vehicle.

(17) **Item 74, Gear Oil; Oil Levels.** After unit has cooled, check differential, transmission, and final drive lubricant levels and inspect leaks. Proper level of final drives is 1 inch below filler plug hole.

(18) **Item 76, Fenders.** Inspect fenders for damage and looseness.

(19) **Item 77, Towing Connections.** Inspect connections for damage and looseness.

(20) **Item 79, Armor.** Inspect armor for damage, broken welds, and loose mounting bolts. Examine windshield and compartment armor shields. Check their hinges and latches for free operation.

(21) **Item 80, Protectoscopes.** Inspect protectoscopes for damage to cover and lens, loose mounting, and free operation of focusing lever. Clean if necessary.

(22) **Item 81, Gun.** Examine gun elevating, traversing, and firing mechanism for free operation. Report any deficiencies. See that gun is locked in carrying position.

(23) **Item 83, *Lubricate as Needed.**

(a) Lubricate clutch release bearing and bogie wheels, compartment hatch cover hinges, and windshield armor shield hinges.

(b) **Weekly.** Perform regularly scheduled lubrication if service is due.

(24) **Item 84, *Clean Engine and Vehicle.**

(a) Clean dirt and trash from engine and driver's compartment. Wipe off excessive dirt and grease from engine. Drain any accumulation of water from hull.

(b) **Weekly.** Wash exterior of vehicle as necessary making sure identification markings are visible.

(25) **Item 85, Tools and Equipment.** See that all tools and equipment are present and secure. Inspect ground spade for damage and loose mounting. See that winch ratchet and brake operate satisfactorily. Inspect cable for frayed or broken strands and spade pulleys for damage and loose mounting.

SECTION IV
LUBRICATION

18. General. Lubrication is an essential part of preventive maintenance, determining to a great extent the serviceability of parts and assemblies.

19. LUBRICATION GUIDE.

a. General. Lubrication instructions for 155-mm Gun Motor Carriage M12 and Cargo Carrier M30, are consolidated in Lubrication Guides (figs. 16, 17, 18, and 19). These specify the points to be lubricated, the periods of lubrication, and the lubricant to be used. In addition to the items on the guide, other small moving parts, such as hinges and latches, must be lubricated at frequent intervals.

b. Supplies. In the field it may not be possible to supply a complete assortment of lubricants called for by the lubrication guides to meet the recommendations. It will be necessary to make the best use of those available, subject to inspection by the officer concerned, in consultation with responsible ordnance personnel.

c. Lubrication notes. The following notes apply to the lubrication guides (figs. 16, 17, 18, and 19). All note references in the guides correspond to the subparagraph having the corresponding number.

1. **Fittings and Oilers.** Clean before applying lubricant. Lubricate bogie wheels, idler and track support rollers until lubricant overflows relief valve. Lubricate other fittings until new lubricant is forced from the bearing, unless otherwise specified. *Caution:* Lubricate suspension points and gun and carriage oiler points after washing gun and motor carriage. Do not use high pressure washing system for cleaning artillery materiel.

2. **Intervals.** Intervals indicated are for normal service. For extreme conditions of speed, heat, water, sand, mud, snow, dust, etc., reduce interval by one-third or one-half, or more, if conditions warrant.

3. **Air Cleaners.** When operating on dirt roads or cross country, or every 250 miles when operating on paved roads, or during wet weather, remove air cleaners daily, clean all parts and refill with used crankcase oil or oil, engine, seasonal grade. *Caution:* Keep all air pipe connections clean and tight. Every 100 hours or more often under dusty conditions, remove filter section of engine breather air cleaner, wash entire section in dry-cleaning solvent. *Caution:* Do not dip filter section in oil. Pour three or four teaspoonfuls of oil, engine, SAE 10, over filter section, remove surplus oil and replace. Proper maintenance of air cleaners is essential to prolonged engine life.

4. **Engine Oil Tank.** Check oil tank level daily, and add oil if necessary. Drain and refill tank every 250 miles or 25 hours. Drain only when engine is hot. Every 1,000 miles, clean tank and oil fill tube strainer. Refill oil tank to FULL mark on bayonet gauge, located under fill cap. *Caution:* Do not remove strainer when filling tank. Do not run engine while cleaning tank.

5. **Oil and Fuel Filters.** Turn handle daily on top of filter one full turn. Every 250 miles, drain.

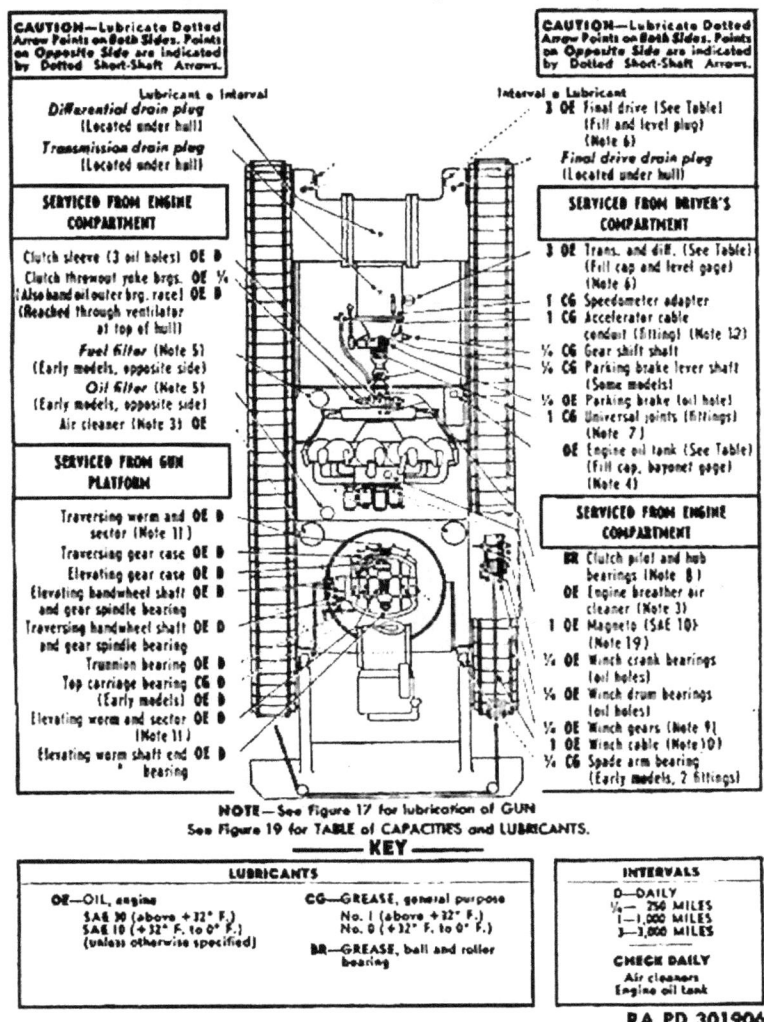

Figure 16. Lubrication guide, Gun Motor Carriage M12.

6. Gear Cases. Check case level weekly, with motor carriage on level ground. Drain, flush, and refill as indicated at points on guide. Fill through transmission filler to mark on bayonet gauge with fill cap resting on top of fill pipe, and through each final drive filler to within ½ inch of fill plug level. Drain through transmission, differential, and final drive drain plug holes. When draining, drain immediately after operation. Every 3,000 miles, clean transmission and differential filler strainer. *Caution:* Do not remove strainer when filling. To flush,

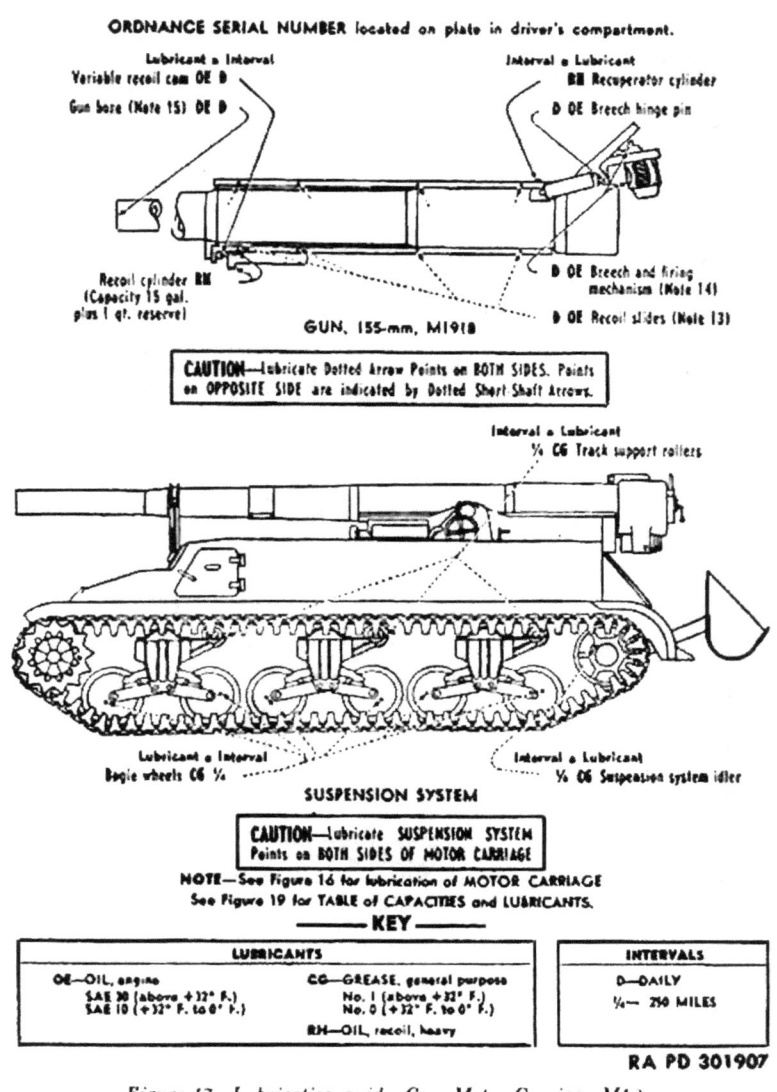

Figure 17. Lubrication guide, Gun Motor Carriage M12.

fill cases to about one-half capacity with oil, engine, SAE 10. Operate mechanism within cases slowly for several minutes and redrain. Replace drain plugs, and refill to correct level with oil, engine, seasonal grade.

7. Universal Joints. Remove tunnel shield sections and apply grease, general purpose, seasonal grade, until it overflows at relief valve.

8. Clutch Pilot and Hub Bearings. At time of disassembly of clutch for

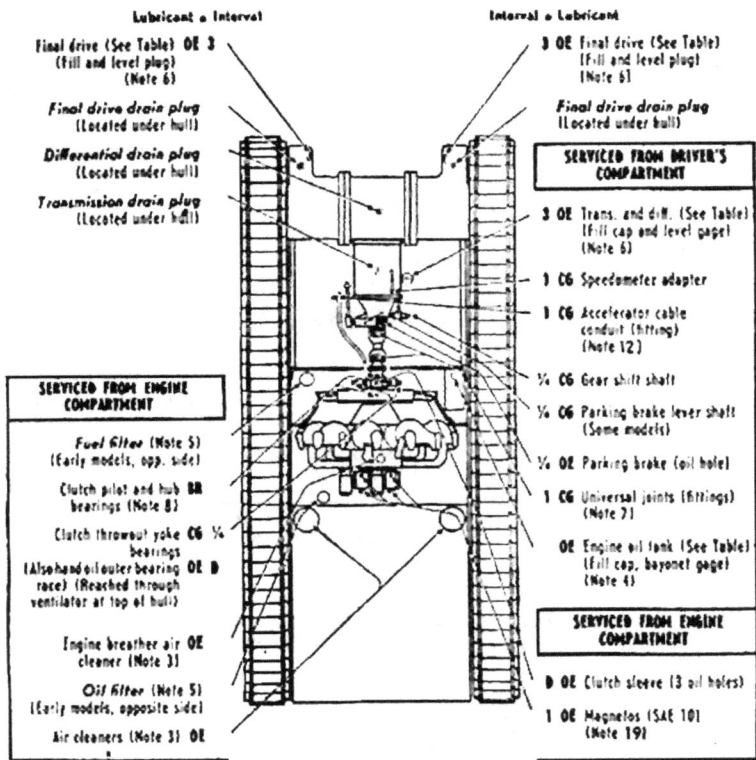

Figure 18. Lubrication guide, Cargo Carrier M30.

inspection, replacement, or overhaul, clean and repack bearings with grease, ball and roller bearing.

9. Winch Gears. Every 250 miles, clean and reoil with oil, engine, seasonal grade.

10. Winch Cable. Every 1,000 miles, clean and reoil with oil, engine, seasonal grade.

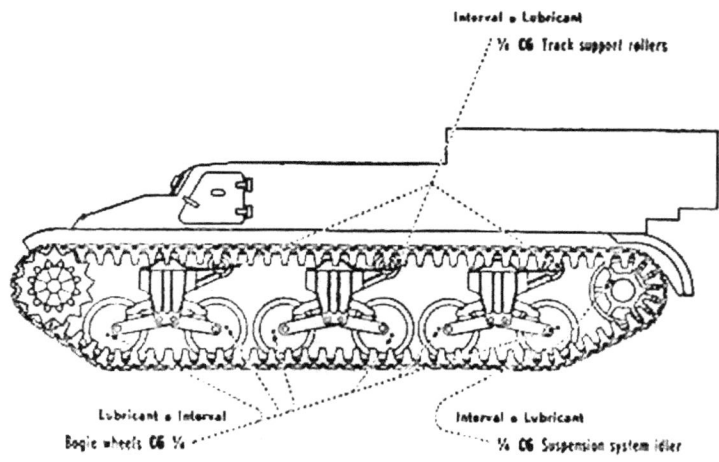

SUSPENSION SYSTEM

CAUTION—Lubricate SUSPENSION SYSTEM Points on BOTH SIDES of CARRIER

NOTE—See Figure 18 for lubrication of FINAL DRIVE, ENGINE COMPARTMENT and DRIVER'S COMPARTMENT points

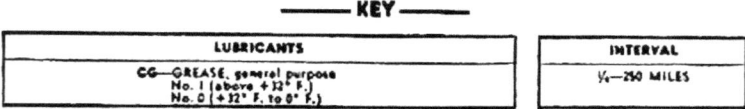

TABLE OF CAPACITIES AND LUBRICANTS TO BE USED

UNIT	CAPACITY (Approx.)	LOWEST EXPECTED AIR TEMPERATURE		
		32° F. and above	32° F. to 0° F.	Below 0° F.
Engine Oil Tank	36 qt.	OE SAE 50	OE SAE 30	Refer to OFSB 6-11
Transmission and Differential	64 qt.			
Final Drives (each)	36 qt.			

RA PD 301909

Figure 19. Lubrication guide, Cargo Carrier, M30.

11. Elevating and Traversing Worm and Sector. Clean and oil traversing gear daily with oil, engine, seasonal grade. Operate traversing gear to extreme left and right positions to reach sector and also oil top carriage thrust bearing surfaces.

12. Accelerator Cable Conduit. Every 1,000 miles, also lubricate other end of conduit, located under engine, by removing inspection plate under hull.

13. Recoil Slides and Chamber. Clean and oil exposed metal surfaces daily and before firing. Keep exposed surfaces covered with a thin film of oil, engine, seasonal grade. Recoil chamber may be filled using an oil screw filler when oil pump is not available.

14. Breech and Firing Mechanism. Clean and oil all moving parts and exposed metal surfaces daily with oil, engine, seasonal grade before and after firing. *Caution:* To insure easy breech operation and to avoid misfiring in cold weather, clean with dry-cleaning solvent, dry and lubricate with oil, lubricating, preservative, light. To clean firing mechanism, remove and operate pin in dry cleaning solvent.

15. Gun Bore. Daily and after firing, clean and coat with oil, engine, seasonal grade.

16. Oil Can Points.

(a) **Motor carriage.** Every 250 miles, lubricate door and shield hinges, peep hole protector slides, door latches, control rod pins, lever bushings, etc., with oil, engine, seasonal grade.

(b) **Gun Mount.** Weekly, lubricate traversing and elevating handwheel handles with oil, engine, seasonal grade.

17. Points Requiring No Lubrication Service. Bogie wheel suspension linkage and slides and final drive sprocket bearings require no lubrication service.

18. Cold Weather. For cold weather lubrication and service below zero, refer to OFSB 6-11 and OFSB 6-5.

19. Points To Be Serviced and/or Lubricated by Ordnance Maintenance Personnel.

(a) **Starter (direct electric and inertia).** At least every 6 months clean and repack all ball bearings with grease, special, high temperature.

(b) **Generator.** At least every 6 months, clean and repack ball bearings with grease, special, high temperature. Coat pole and exposed armature shaft surfaces with oil, engine, SAE 30, to prevent rust.

(c) **Magneto.** At least every 6 months, remove bearings, clean and repack with grease, special, high temperature. Apply grease, special, high temperature, to distributor gear. Lubricate felt wicks and pole pieces with oil, engine, SAE 30.

20. REPORTS AND RECORDS.

a. Reports. If lubrication instructions are closely followed, proper lubricants used, and satisfactory results are not obtained, a report will be made to the ordnance officer responsible for the maintenance of the matériel.

b. Records. Record of seasonal changes of lubricants and recoil oil will be kept in the Artillery Gun Book.

SECTION V

TOOLS AND EQUIPMENT STOWAGE ON 155-MM GUN MOTOR CARRIAGE

21. VEHICLE TOOLS.

a. Pioneer Tools.

Item	Number (carried)	Where carried
Ax, (chopping, single bit, 5-pound)	1	In bracket on engine compartment cover.
Crowbar, pinch point, 5-foot	1	In bracket on left sponson wall.
Handle, mattock	1	In bracket on engine compartment cover.
Pick, mattock, M1 (w/o handle)	1	Do.
Shovel, short handle	1	Do.
Sledge, blacksmith, double-face, 10-pound.	1	In bracket on driving compartment top plate.

b. Vehicular Tools.

Item	Number (carried)	Where carried
Adapter, button head to bayonet type.	1	In left rear floor compartment.
Adapter, button head to hydraulic type fitting.	1	Do.
Chisel, cold, ½-inch	1	
Cross bar	1	
Extension, handy grip ½-inch, square-drive.	1	Do.
Extension, ½-inch, square-drive, 10 inches long.	1	Do.
File, 3-inch, square, smooth, 6-inch.	1	
File, hand, smooth, 8-inch	1	
Fixture, track connecting	1	In center rear floor compartment.
Gun, grease, hand	1	In left rear floor compartment.
Hammer, machinist, ball peen, 32-ounce.	1	Do.
Handle, flexible, ½-inch, square-drive, 12 inches long.	1	Do.
Handle, combination tee, ½-inch, square-drive, 11 inches long.	1	Do.

Item	Number (carried)	Where carried
Handle, combination tee, 3/4-inch, square-drive, 17 inches long.	1	In left rear floor compartment.
Handle, speeder, 1/2-inch, square-drive, 17 inches long.	1	Do.
Hose, lubricating, heavy duty, 15-inch, button head fitting.	1	Do.
Joint, universal, 1/2-inch, square-drive.	1	Do.
Pliers, combination, slip joint, 8-inch.	1	Do.
Pliers, side cutting, 8-inch	1	Do.
Ratchet, reversible, 1/2 square-drive, 9-inch.	1	Do.
Screw driver, machinist, 5-inch blade.	1	Do.
Screw driver, special purpose, 1 3/4-inch blade.	1	Do.
Screw driver, special purpose, 1 1/2-inch blade.	1	Do.
Wrench, adjustable, single-end, 8-inch.	1	Do.
Wrench, adjustable, single-end, 12-inch.	1	Do.
Wrench, engine, 5/16 x 3/8 inch	1	Do.
Wrench, engine, 7/16 x 1/2 inch	1	Do.
Wrench, engine 9/16 x 11/16 inch	1	Do.
Wrench, engine, 5/8 x 3/4 inch	1	Do.
Wrench, engine, 13/16 x 7/8 inch	1	Do.
Wrench, engine 15/16 x 1 inch	1	Do.
Wrench, plug, 9/16-inch, hexagon (for transmission and oil drain plug).	1	Do.
Wrench, plug, 3/4-inch, hexagon (differential filler and drain plug).	1	Do.
Wrench, safety screw, 5/32-inch, hexagon.	1	Do.
Wrench, safety screw, 1/8-inch, hexagon.	1	Do.
Wrench, safety screw, 3/16-inch, hexagon.	1	Do.
Wrench, safety screw, 1/4-inch, hexagon.	1	Do.

Item	Number (carried)	Where carried
Wrench, safety screw, 5/16-inch, hexagon.	1	In left rear floor compartment.
Wrench, safety screw, 3/8-inch, hexagon.	1	Do.
Wrench, safety screw, 5/8-inch, hexagon.	1	Do
Wrench, socket, 1/2-inch, square-drive, 3/8-inch, square.	1	Do.
Wrench, socket, 1/2-inch, square-drive, 5/16-inch, hexagon.	1	Do.
Wrench, socket, 1/2-inch square-drive, 1/2-inch hexagon.	1	Do.
Wrench, socket, 1/2-inch square-drive, 9/16-inch hexagon.	1	Do.
Wrench, socket, 1/2-inch square-drive, 5/8-inch hexagon.	1	Do.
Wrench, socket, 1/2-inch square-drive, 3/4-inch hexagon.	1	Do.
Wrench, socket, 1/2-inch square-drive, 7/8-inch hexagon.	2	Do.
Wrench, socket, 1/2-inch square-drive, 15/16-inch hexagon.	2	Do.
Wrench, socket, 1/2-inch square-drive, 1-inch hexagon.	1	Do.
Wrench, socket, 1/2-inch square-drive, 1 1/16-inch hexagon.	1	Do.
Wrench, socket, 1/2-inch square-drive, 1 1/8-inch hexagon.	1	Do.
Wrench, socket, 3/4-inch square-drive, 1 1/2-inch hexagon.	1	Do.
Wrench, track adjusting	1	On brackets on under side of spade seat.

c. 155-mm Gun M1917A, M1918M1 Tools.

Item	Number (carried)	Where carried
Gear, opening, 1 1/8-inch square	1	
Wrench, face spanner, 68-mm x 99-mm.	1	In right rear floor compartment.
Wrench, face spanner, 111-mm	1	Do.
Wrench, firing mechanism	1	Do.
Wrench, open, 17-mm and 20-mm	1	Do.
Wrench, pin, 32-mm	1	Do.

Item	Number (carried)	Where carried
Wrench, piston rod, nut w/handle 81-mm x 108-mm.	1	In right rear floor compartment.
Wrench, safety, set screw, ⅛-inch	1	Do.
Wrench, screw adjusting, 12-inch	1	Do.
Wrench, screw adjusting, 18-inch	1	Do.
Wrench, ratchet, 24-inch	1	In center rear floor compartment.
Wrench, socket, 35-mm	1	In right rear floor compartment.

22. VEHICLE EQUIPMENT.

a. Miscellaneous Accessories and Equipment.

Item	Number (carried)	Where carried
Apparatus, decontaminating, 1½-quart M2.	2	1 on transfer case in driving compartment. 1 in left rear sponson compartment.
Bag, tool	1	In safety belt holders.
Belt, safety	6	Do.
Book, O. O. Form 7255	1	In right front sponson compartment.
Bucket, canvas, folding, 18-quart	1	
Bucket, galvanized iron, 14-quart	1	In bracket on spade seat.
Cable, towing	1	On left fuel tank cover on roof of driving compartment on right fuel tank cover.
Can, oil, ¼-gallon (stencil "oil engine" in black letters ½-inch high on can).	8	1 in bracket in driving compartment. 7 in left rear floor compartment.
Container, water, 5-gallon (Q. M. C. standard A–353).	4	In right rear sponson compartment.
Crank	1	In bracket on right sponson wall.
Extinguisher, fire, 4-pounds CO_2	2	1 driving in compartment. 1 on right rear fender.
Flashlight TL–122A	3	2 in driving compartment. 1 in left rear sponson compartment.

Item	Number (carried)	Where carried
Grouser	26	13 in box on left fuel tank cover.
		13 in center front floor compartment.
Guide, lubrication, laminated	1	On rack on right wall of driving compartment.
Helmet, tank	2	On personnel in driving compartment.
Kit, first-aid (24-unit)	1	On bracket on tranfer case in driving compartment.
Lamp, inspection light	1	In light inspection A213666.
Light, inspection	1	In left rear floor compartment.
Manual, field for hand grenades (FM 23-30)	1	In right front sponson compartment.
Manual, technical, 155-mm (Q. M. C. M12 TM 9-751)	1	Do.
Manual, technical, 155-mm gun matériel (TM 9-345)	1	In right front sponson compartment.
Manual, training (for engine)	1	Do.
Manual, spare parts, illustrated (for vehicle)	1	Do.
Mittens, asbestos, pair	1	In center rear floor compartment.
Net, camouflage, 45 x 45 feet	1	In right front sponson compartment.
Oil, engine, 1 quart	8	In bracket in driving compartment.
		In left rear floor compartment.
Oiler (trigger type, 1 pint)	1	In bracket on transfer case in driving compartment.
Padlock, with two keys	1	In left rear floor compartment.
Paulin, 12 x 12 feet	1	In strap holder on under side of spade seat.
Stove, cooking, gasoline M1941 (Coleman military burner #520 w/accessory cups)	1	Installed by troops.
Tape, adhesive, 4 inches wide, 15 yd OD	1	In right rear floor compartment.

Item	Number (carried)	Where carried
Tape, friction, ¾ inch wide, 30-foot roll.	1	In left rear floor compartment.
Tube, flexible, nozzle	2	In center rear floor compartment.

b. Rations.

Item	Number (carried)	Where carried
Type "C" 2-day rations for six men.	72 cans	In right rear sponson compartment.
Type "D" 1-day ration for six men.	1 box	In box in right front sponson compartment.

c. Signaling Equipment.

Item	Number (carried)	Where carried
Flag set, M113, complete composed of— 1 bag, canvas 2 flags, semaphore	1	On right rear sponson compartment cover.

d. Vehicle Spare Parts.

Item	Number (carried)	Where carried
Connector, end	12	In rear box on left front fender.
Headlight, service	1	In bracket on floor of driving compartment.
Lamp, 3 cp, 24–28 volts	4	In left rear floor compartment.
Link	6	In forward box on left front fender.
Nut, safety	16	In rear box on left front fender.
Pin, cotter, ¼ x 2¼-inches (for tow shackle pin).	2	Do.
Pin, locking (for tow shackle pin)	2	Do.
Wedge	12	Do.

e. 155-mm Gun M1917A, M1918M1 Equipment.

(1) Ammunition.

Item	Number (carried)	Where carried
Charge, propelling NH powder 155-mm M1917, M1917A1, and M1918M1.	10	6 in shelf under left rear seat. 2 in wells in lower carriage support. 2 on center rear floor compartment cover.
Fuse, P. D., M51 (w/booster M21)	25	In box on left rear seat.

Item	Number (carried)	Where carried
Grenade, hand	12	
Fragmentation, MK. II	4	2 in box on left fuel tank cover.
		2 in box in driving compartment.
Offensive, Mk. III (w/fuse detonation hand grenade M6).	2	1 in box on left fuel tank cover.
		1 in box in driving compartment.
Smoke	4	2 in box on left fuel tank cover.
		2 in box in driving compartment.
Thermite, incendiary	2	1 in box on left fuel tank compartment.
		1 in box in driving compartment.
Grenade, rifle, M9A1	10	In box in left rear sponson compartment.
Shell, HE 155-mm Gun Mk. III A1 or M101.	10	6 on floor to left of gun base.
		4 on floor to right of gun base.
Primer, percussion, 21-grain, Mk. 11A1.	50	In right front sponson compartment.
(2) Armament.		
Carbine, caliber .30, M1	5	1 in bracket in driving compartment.
		4 in spade rifle box.
Gun, 155-mm M1917A1, M1918M1.	1	On top carriage.
Launcher, grenade	1	In rifle grenade box.
(3) Accessories.		
Bit, vent cleaning	1	In right rear floor compartment.
Book, artillery gun, O. O. Form No. 5825.	1	In right front sponson compartment.
Brush, slush, M2	1	In right rear floor compartment.

Item	Number (carried)	Where carried
Case, carrying gunner's quadrant, M1.	1	In bracket in left rear sponson compartment.
Cover, muzzle, M32	1	On breech.
Cover, breech, M25	1	On muzzle.
Cover, quadrant sight, M1918	1	On quadrant sight M1918A1.
Cover, sponge	1	On sponge.
Lanyard M1	1	In right rear floor compartment.
Oiler, ½-pint, trade capacity	1	Do.
Opener, container	1	Do.
Quadrant, gunner's, M1	1	In case carrying gunner's quadrant M1.
Reamer, cleaning primer seat	1	In right rear floor compartment.
Setter, fuze, M14	1	In center rear floor compartment.
Sight, bore, breech	1	
Sight, bore, muzzle	1	
Sponge and rammer complete, composed of—	1	In center rear floor compartment.
1 Head, rammer		In center rear floor compartment.
1 Head, sponge, M10		On bracket on right fuel tank cover.
1 Staff end (short) assembly		
2 Staff, middle assembly		
Target, testing (set of four)	1	
Tray, loading	1	On bracket on right rear sponson compartment cover.
Wire, copper, 0.032 diameter 80-foot spool.	1	In right rear floor compartment.
Wrench, fuze, M1	1	Do.
(4) Sighting Equipment.		
Binocular M3 complete, composed of—	1	On bracket in driving compartment.
1 Binocular M3		
1 Case, carrying		
1 Strap, neck		

Item	Number (carried)	Where carried
Light, aiming post M14, complete composed of—	1	In right rear floor compartment.
8 Cell, flashlight BA-30		
1 Chest M14		
2 Lamp, electric 3v		
2 Light, aiming post, M14		
Light, instrument, M9	1	
Light, instrument, M17	1	
Post, aiming, M1, complete composed of—	1	On bracket on right fuel tank cover.
1 Cover, aiming post		
2 Post, aiming, M1		
Sight, quadrant, M1918A1	1	On gun.
Table, firing, FT-U-1	1	In right front sponson compartment.
Telescope, panoramic, M6	1	In box panoramic telescope M6.
Wrench, quadrant sight socket	1	In right rear floor compartment.
Wrench, wing teat, pin face	2	In box, panoramic telescope M6.
Prism, protectoscope	8	In box on floor driving compartment.
Telescope M53	1	In mount M40.
Extension, panoramic telescope, 14-inch.	1	

f. 155-mm Gun M1917A1, M1918M1 Spare Parts.

Item	Number	Where carried
Key, firing mechanism housing assembly.	1	In right rear floor compartment.
Key, woodruff, 6 x 22 x 9.7 mm	2	Do.
Lanyard, assembly, M12	2	In right rear floor compartment.
Mechanism, firing, M1918 assembly.	1	Do.
Pin, cotter, ⅛ x 1½-inch	5	Do.
Pin, cotter, ³⁄₃₂ x 1½-inch	3	Do.
Pin, cotter, ³⁄₃₂ x 1¾-inch	5	Do.
Pin, cotter, ¼ x 2¾-inch	3	Do.
Pin, firing	2	Do.

Item	Number (carried)	Where carried
Plug, filling and drain	1	In right rear floor compartment.
Screw, firing mechanism housing	1	Do.
Screw, operating lever catch	1	Do.
Spring compressor 0.072 diameter 7 coils 50D.	1	Do.
Spring, counterbalance	1	In center rear floor compartment.
Spring, firing mechanism block latch.	1	In right rear floor compartment.
Spring, obturator spindle	1	Do.
Spring, rack lock	1	Do.

SECTION VI

TOOLS AND EQUIPMENT STORAGE ON CARGO CARRIER

23. VEHICLE TOOLS.

a. Pioneer Tools.

Item	Number (carried)	Where carried
Ax, chopping, single bit 5-pound	1	In bracket on engine compartment top plate.
Crowbar, pinch point, 5-foot	1	In bracket on left rear propelling charge compartment cover.
Handle, mattock	1	In bracket on engine compartment top plate.
Mattock, pick, M1 (w/o handle)	1	Do.
Shovel, short handle	1	Do.
Sledge, blacksmith, double-face, 10-pound.	1	Do.

b. Vehicular Tools.

Item	Number (carried)	Where carried
Adapter, button head to bayonet type.	1	In tool box between rear engine bulkhead and ammunition rack.
Adapter, button head to hydraulic type.	1	Do.

Item	Number (carried)	Where carried
Cross bar	1	In tool box between engine bulkhead and ammunition rack.
Extension, handy grip, ½-inch square-drive, 5 inches long.	1	In tool box between rear engine bulkhead.
Extension, ½-inch square-drive, 10 inches long.	1	Do.
Fixture, set track connecting	1	Do.
Gun, grease, hand type	1	Do.
Hammer, machinist, ball peen, 32-ounce.	1	Do.
Handle, combination tee, ½-inch square-drive, 11 inches long.	1	Do.
Handle, combination, tee, ¾-inch square-drive, 17 inches long.	1	In tool box between rear engine bulkhead and ammunition rack.
Handle, flexible, ½-inch square-drive, 12 inches long.	1	Do.
Hose, (lubricating heavy duty 15 B. H. fitting).	1	Do.
Joint, universal, ½-inch square-drive.	1	Do.
Pliers, combination slip joint, 8-inch.	1	Do.
Pliers, side cutting, 8-inch	1	Do.
Ratchet, reversible, ½-inch square-drive, 9-inch.	1	Do.
Screw driver, machinist, 5-inch blade.	1	Do.
Screw driver, special purpose, 1¾-inch blade.	1	Do.
Screw driver, special purpose, 1½-inch blade.	1	Do.
Speeder, ½-inch square-drive, 17-inch.	1	Do.
Wrench, socket, ½-inch square-drive, ⅜-inch square.	1	Do.
Wrench, socket, ½-inch square-drive, ⁷⁄₁₆-inch hexagon.	1	Do.
Wrench, socket, ½-inch square-drive, ½-inch hexagon.	1	Do

Item	*Number (carried)*	*Where carried*
Wrench, socket, 1/2-inch square-drive, 9/16-inch hexagon.	1	In tool box between rear engine bulkhead and ammunition rack.
Wrench, socket, 1/2-inch square-drive, 3/4-inch hexagon.	1	Do.
Wrench, socket, 1/2-inch square-drive, 7/8-inch hexagon.	2	Do.
Wrench, socket, 1/2-inch square-drive, 15/16-inch hexagon.	2	Do.
Wrench, socket, 1/2-inch square-drive, 1-inch hexagon.	1	Do.
Wrench, socket, 1/2-inch square-drive, 1 1/16-inch hexagon.	1	Do.
Wrench, socket, 1/2-inch square-drive, 1 1/8-inch hexagon.	1	Do.
Wrench, socket, 3/4-inch square-drive, 1 1/2-inch hexagon.	1	Do.
Wrench, adjustable, single-end, 8-inch.	1	Do.
Wrench, adjustable, single-end, 12-inch.	1	Do.
Wrench, engine, 5/16 x 3/8-inch	1	Do.
Wrench, engine, 7/16 x 1/2-inch	1	Do.
Wrench, engine, 9/16 x 11/16-inch	1	Do.
Wrench, engine, 5/8 x 3/4-inch	1	Do.
Wrench, engine, 13/16 x 7/8-inch	1	Do.
Wrench, engine, 15/16 x 1-inch	1	Do.
Wrench, plug, 9/16-inch hexagon (for transmission and oil drain plug)	1	Do.
Wrench, plug, 3/4-inch hexagon (differential filler and drain plug)	1	Do.
Wrench, safety screw, 5/32-inch hexagon.	1	Do.
Wrench, safety screw, 1/8-inch hexagon.	1	Do.
Wrench, safety screw, 3/16-inch hexagon.	1	Do.
Wrench, safety screw, 1/4-inch hexagon.	1	Do.

Item	Number (carried)	Where carried
Wrench, safety screw, ⁵⁄₁₆-inch hexagon.	1	In tool box between rear engine bulkhead and ammunition rack.
Wrench, safety screw, ⅜-inch hexagon.	1	Do.
Wrench, safety screw, ¼-inch hexagon.	1	Do.
Wrench, track adjusting	1	In bracket on plate beneath caliber .50 machine-gun platform.

c. Caliber .50 Machine Gun Tools.

Wrench, combination, M2	1	In steel chest M5.

24. VEHICLE EQUIPMENT.
a. Miscellaneous Accessories and Equipment.

Item	Number (carried)	Where carried
Apparatus, decontaminating, 1½-quart, M11.	2	1 on transfer case in driving compartment. 1 on plate under caliber .50 machine-gun platform.
Bag, tool	1	
Belt, safety	6	In safety belt holders.
Book, O. O. Form 7255	1	Do.
Bucket, canvas, folding, 18-quart	1	
Cable, towing	1	On left fuel tank cover, on roof of driving compartment and on right fuel tank cover.
Can, oil, ¼-gallon (stencil "oil engine" in black letters ½-inch high on can).	8	1-inch bracket in driving compartment. 7 in box between engine bulkhead and ammunition rack.
Container, water, 5-gallon (Q. M. C. standard A 353).	3	In compartment on left sponson adjacent to ammunition rack.
Crank	1	On bracket on right sponson wall between engine bulkhead and ammunition rack.
Extinguisher, fire, CO_2 4-pound	2	1 in driving compartment. 1 on right fuel tank cover.

Item	Number (carried)	Where carried
Flashlight TL-122A	3	2 in driving compartment. 1 in caliber .50 ammunition compartment.
Grouser (when track D37889, D48067, D48076 is used).	26	13 in box on left fuel tank cover. 13 in box on right fuel tank cover.
Guide, lubrication, laminated	1	On rack on right wall of driving compartment.
Helmet, tank (see Q. M. C. head size chart).	2	On personnel in driving compartment.
Kit, first-aid (24-unit)	1	On bracket on transfer case in driving compartment.
Lamp, inspection light	1	In lamp inspection A213666.
Light, inspection	1	In box between engine bulkhead and ammunition rack.
Manual, field, for caliber .30 machine gun, M2, HB.	1	In right front sponson compartment.
Manual, field (for hand grenades)	1	Do.
Manual, training (for engine)	1	Do.
Manual, technical (for Cargo Carrier M30).	1	Do.
Manual, spare parts (illustrated) for vehicle.	1	Do.
Mittens, asbestos, pair	1	In box between engine bulkhead and ammunition rack.
Net, camouflage, 45 x 45 feet	1	
Oil, engine, 1-quart (in can B-101420).	8	
Oiler, (trigger type) 1-pint	1	In bracket on transfer case in driving compartment.
Padlock, 1½-inch w/2 keys	1	In box between engine bulkhead and ammunition rack.
Stove, cooking, gasoline, M1A41 (Coleman military burner #520 w/accessory cups).	1	

Item	Number (carried)	Where carried
Tape, friction, ¾-inch wide, 30-foot roll.	1	In box between engine bulkhead and ammunition rack.
Paulin, 12 x 12 feet	1	
Tube, flexible, nozzle	2	Do.

b. Signaling equipment.

Item	Number (carried)	Where carried
Flag set, M113, complete, composed of— 2 Flag, semaphore. 1 Bag, canvas.	1	In bracket on right fuel tank cover.

c. Rations.

Item	Number (carried)	Where carried
Type "C" 2-day rations for six men	72 cans	36 under right caliber .50 ammunition compartment. 36 under left caliber .50 ammunition compartment.
Type "D" 1-day ration for six men	2 boxes	In box in right front sponson compartment.

d. Vehicle Spare Parts.

Item	Number (carried)	Where carried
Lamp, 3 cp, 24–28 v	4	In tool box between rear engine bulkhead and ammunition rack.
Connector, end	12	In rear box on left front fender.
Headlight, service (spare)	1	In bracket on floor of driving compartment.
Link	6	In forward box on left front fender.
Nut, safety	16	In rear box on left front fender.
Pin, cotter, ¼ inch x 2¼ inches (for tow shackle pin).	2	Do.
Pin, locking, for tow shackle pin	2	Do.
Wedge	12	Do.

e. Gun Equipment.

(1) Caliber .50 Machine Gun M2 HB (Flexible) Accessories.

Item	Number (carried)	Where carried
Bag, metallic belt link	1	In steel chest M5.
Box, ammuniton, caliber .50, M2	10	In right caliber .50 ammunition compartment.
Brush, cleaning, caliber .50, M4	4	In steel chest M5.
Case, cleaning rod, M15	1	Do.
Chest, steel M5 (w/o contents)	1	On lower shelf right caliber .50 ammuniton compartment.
Chute, metallic belt link, M1	1	
Cover, spare barrel, M13, 45-inch	1	On spare barrel assembly.
Envelope, spare parts, M1 (w/o contents).	2	In steel chest M5.
Extractor, ruptured cartridge	1	Do.
Oiler, filling, oil buffer	1	Do.
Rod, jointed, cleaning, M7	1	Do.
Cover, gun and cradle, caliber .30 or caliber .50.	1	

(2) 155-mm Gun Accessories.

Item	Number (carried)	Where carried
Chest, oil pump, M2 w/contents, composed of—	1	In compartment between engine bulkhead and ammunition rack.

    1 Filler, oil screw.
    1 Funnel, ½-pint.
    2 Release, filling and drain valve.
    1 Pump, oil, M2A1.
    1 Rule, steel flexible, 25-cm.
    1 Wrench, open, 21-mm and 35-mm.

Item	Number	Where carried
Oil, recoil, gallon	2	

(3) Ammunition.

Item	Number	Where carried
Caliber .50 round (caliber .50 machine gun, M2HB).	1,000	In ten 100-round ammunition box, caliber .50 M2.
Charge, propelling, NH powder, 155-mm gun M1917, M1917A1, and M1918M1.	40	In compartment in rear of vehicle.
Fuze, PD M51 w/booster M21	50	25 in left upper sponson compartment.

Item	Number (carried)	Where carried
Grenade, hand	12 *	
4 Fragmentation Mk. II		2 in box on ammunition rack.
		2 in box in driving compartment.
2 Offensive Mk. III (w/fuze detonation hand grenades, M6).		1 in box on ammunition rack.
		1 in box in driving compartment.
4 Smoke		2 in box on ammunition rack.
		2 in box in driving compartment.
2 Thermite, incendiary		1 in box on ammunition rack.
		1 in box in driving compartment.
Grenade, rifle, M9A1	10	In box on ammunition rack.
Shell, HE 155-mm gun, Mk. III A1 or M101.	40	In ammunition rack.
Carbine, caliber .30, M1	4	1 in bracket in driving compartment.
		3 in brackets on rear compartment shelves.
Launcher, grenade (rifle)	1	In rifle grenade box.
Machine gun, flexible, caliber .50, M2HB.	1	In mount on rear of vehicle.
Rifle, caliber .30, M1903	1	In bracket on rear compartment shelf.

(4) **Sighting Equipment.**

Prisms, protectoscope (spare)	8	

f. Caliber .50 Machine Gun M2 HB (Flexible) Spare Parts.

Barrel, assembly	1	In brackets on left caliber .50-mm ammunition compartment cover.
Disk, buffer	1	In steel chest M5.
Extension, firing pin assembly	1	Do.
Extractor, assembly	1	Do.

f. Caliber .50 Machine Gun M2 HB (Flexible) Spare Parts—Continued.

Item	Number (carried)	Where carried
Lever, cocking	1	In steel chest M5.
Pin, cotter, belt feed lever pivot stud.	1	Do.
Pin, cotter pin	1	Do.
Pin, cotter, switch pivot	2	Do.
Pin, firing	1	Do.
Plunger, belt feed lever	1	Do.
Rod, driving spring w/spring assembly.	1	Do.
Slide, belt feed group, consisting of—	1	Do.
1 Arm, belt feed pawl, B8961		
1 Pawl, feed belt assembly B8961		
1 Pin, belt feed pawl assembly, B8962		
1 Slide, belt feed, assembly, B261110		
1 Spring, belt feed pawl, A9351		
Slide, sear	1	Do.
Spring, belt holding pawl	1	Do.
Spring, cover extractor	1	Do.
Spring, locking barrel	1	Do.
Spring, sear	1	Do.
Spring, belt feed lever plunger	1	Do.
Stud, bolt	1	Do.

SECTION VII

OPERATION UNDER UNUSUAL CONDITIONS

25. GENERAL COLD WEATHER PRECAUTIONS.

   a. Observe special precautions in starting and operating the vehicle at temperatures of zero and below. Greatest dangers come from lubrication failures due to thickened oil, slow or blocked circulation of oil, and improper or too frequent use of oil dilution valve. (See note, par. 11c.)

   b. Close engine compartment as tightly as possible during a halt or overnight stop. Always use the protection of a shed or other inclosed space if available.

   c. In extremely cold weather, remove oil from vehicle during overnight stops and keep oil in a warm place. When doing this, however, be sure to mark vehicle to indicate that oil has been removed. This will reduce the possibility of the engine being started without the oil in the tanks. Before attempting to start the engine in subzero weather, it is particularly important to warm up the engine oil. Otherwise, so little oil may be circulated that the engine may be seriously damaged during an attempted warm-up.

   d. To maintain the engine oil temperature at 140° F., cover the oil cooler with a curtain or piece of cardboard. This will cut off the flow of air through the cooler, thus keeping the circulating oil from cooling. Control the temperature of the engine oil by varying the oil cooler area covered. The extent of covering and uncovering will depend upon the atmospheric temperature.

   e. Be careful in operating oil dilution valve. Make sure that it is functioning properly and follow directions given on instruction plate located on the instrument panel to the left of dilution valve toggle switch (figs. 13 and 15). (See note, par. 11c.)

   f. Watch for any undue drop in oil pressure during engine warm-up. Stop engine at once if pressure drops sharply or if it fails to develop within 10 seconds.

   g. Due to condensation of moisture in the air, water will accumulate in gasoline tanks, drums, and containers. At low temperatures, this water will form ice crystals that will clog fuel lines and carburetor jets unless the following precautions are taken:

(1) Strain the fuel through a chamois skin or any other type of strainer that will prevent the passage of water. *Caution:* Gasoline flowing over a surface generates static electricity that will result in a spark unless means are provided to ground the electricity. A metallic contact between the container and the tank must be provided to assure an effective ground.

(2) Keep tank full, if possible. The more fuel there is in the tank, the smaller the volume of air will be from which moisture can be condensed.

(3) Be sure that all containers are thoroughly clean and free from rust before storing fuel in them.

(4) If possible, after filling or moving a container, allow the fuel to settle before filling the vehicle tank from it.

(5) Keep all closures of containers tight to prevent snow, ice, dirt, and other foreign matter from entering.

h. Cover the engine and transmission oil coolers with a curtain or piece of cardboard to cut off the flow of air through the coolers. When the desired temperatures have been reached, vary the extent of covering and uncovering of the coolers so as to maintain the oil temperatures within the desired range.

i. If transmission does not already contain SAE 30, engine oil, drain transmission immediately after use and while lubricant is still warm. Refill to proper level with SAE 30 engine oil. After standing overnight at subzero temperatures, warm up transmission as follows: After the engine has been warmed up, as outlined in paragraph 26, engage clutch and maintain engine speed of 1,600 revolutions per minute for 2 minutes or until gears can be engaged. Put transmission in low (first) gear and drive vehicle 100 yards, being careful not to stall engine.

j. All places requiring general purpose grease, at normal temperatures shall be lubricated with SAE 30 engine oil.

26. STARTING ENGINE IN COLD WEATHER.

a. Turn the engine over by hand until all cylinders have been proved free of hydrostatic lock. Flip booster switch on and off several times to remove any ice. Do not attempt to start until booster is working. If necessary, have another man listen for the buzz. This is important, because the booster switch has a tendency to ice-up and become inoperative. The same is true of the magneto switch, and it should also be switched back and forth several times before attempting to start.

b. Operate primer five strokes.

c. Depress clutch, open throttle one-quarter and operate starting controls in usual manner (par. 8**b**).

d. Continue priming until carburetor cuts in. (This will be evidenced by consistent firing.)

e. Booster switch should not be released until engine speed of 1,000 revolutions per minute is reached and firing is smooth. If engine does not start in three or four attempts, investigate to determine cause for failure to fire. With use of primer and addition of heat, engine should start to fire within 30 seconds. At no time operate starter for more than 30 seconds without allowing it to cool for 2 minutes. Accelerator should not be pumped during starting, except at temperatures of minus 30° F. to minus 40° F., and should be used only to control the speed of the engine until carburetor cuts in.

f. Check oil pressure. If gauge shows no oil pressure after engine has started, shut down engine and determine cause.

g. Idle engine at 1,600 revolutions per minute until oil temperature reaches 60° F. The cooling air outlet flap should then be opened and locked in its open position.

h. Clutch should be engaged with gears in neutral as soon as possible without stalling engine.

i. Vehicle is now ready for operation.

j. When oil temperature goes above 180° F., open cooling air intake flap to maintain proper oil temperature. *Caution:* Overheating of engine will cause detonation. When this occurs, open air inlet control and oil cooler covers immediately.

k. If vehicle is operated 4 hours or more at operating temperature, redilution is necessary to start if it is anticipated vehicle will be left standing for over 3 hours. This should be done in accordance with directions given on instruction plate located on the instrument panel. If operation has been less than 3 hours, reduce specified dilution by one-half.

27. INSPECTING ELECTRICAL SYSTEM IN COLD WEATHER.

a. Generator and Starter. Check the brushes, commutators, and bearings. See that the commutators are clean. The large surges of current which occur when starting a cold engine require good contact between brushes and commutators.

b. Wiring. Check, clean and tighten all connections, especially the battery terminals. Care should be taken that no short circuits are present.

c. Coil. Check coil for proper functioning.

d. Distributor. Clean thoroughly, and clean, adjust or replace points. Check the points frequently. In cold weather, slightly pitted points can prevent engine from starting.

e. Spark Plugs. Clean, test, and replace if necessary. If it is difficult to make the engine fire, reduce the gap 0.005 inch less than that recommended by the manufacturer. This will make ignition more effective at the reduced voltages likely to prevail. If engine is not so equipped, replace spark plugs with Champion 635 plugs with gap set at 0.018 to 0.020 inch.

f. Timing. Check carefully. Care should be taken that the spark is not unduly advanced or retarded.

g. Batteries.

(1) The efficiency of batteries decreases sharply with decreasing temperatures, and becomes practically nil at minus 40° F. Do not try to start the engine with the battery, when it has been chilled to temperatures below minus 30° F., until battery has been heated. See that the battery is always fully charged, with the hydrometer reading between 1.275 and 1.300. A fully charged battery will not freeze at temperatures likely to be encountered even in Arctic climates, but a fully discharged battery will freeze and rupture at 5° F.

(2) Do not add water to batteries when they have been exposed to subzero temperatures unless the battery is to be charged immediately afterward. If water is added and the battery not put on charge, the layer of water will stay at the top and freeze before it gets a chance to mix with the acid.

h. Lights.
Inspect the lights carefully. Check for short circuits and presence of moisture around sockets.

i.
Before every start, see that spark plugs, wiring, or other electrical equipment are free from ice.

28. INSPECTING ENGINE AND CHASSIS IN COLD WEATHER.

a. Inspecting Engine.

(1) Be sure that no heavy grease or dirt has been left on the starter throw-out mechanism. Heavy grease or dirt may keep the gears from being meshed or cause them to remain in mesh after the engine starts running. The latter will ruin the starter and necessitate repairs.

(2) A full choke is necessary to secure the air-fuel ratio required for cold weather starting. Check the butterfly valve to see that it closes all the way and otherwise functions properly.

(3) Carburetors which give no appreciable trouble at normal temperatures may not operate satisfactorily at low temperatures. Be sure the fuel pump has no leaky valves or diaphragm, as this will prevent the fuel pump from delivering the amount of fuel required to start the engine at low temperatures, when turning speeds are reduced to 30 to 60 revolutions per minute.

(4) Do not use oil in air cleaners at temperatures below 0° F. The oil will congeal and prevent the easy flow of air. Remove the air cleaner screens at temperatures below 0° F. Ice and frost formations on the air cleaner screens may cause an abnormally high intake vacuum in the carburetor air horn hose, resulting in collapse.

(5) Full flow oil filters have a bypass valve. These filters must be bypassed, below minus 30° F. because the viscous oil will not flow freely through them.

(6) Remove and clean sediment bulb, strainers, etc., from fuel system daily. Also drain fuel tank pump daily to remove water and dirt.

b. Inspecting Chassis.

(1) Brake bands, particularly on new vehicles, have a tendency to bind when they are very cold. Parking the vehicle with the brake released will eliminate most of the binding.

(2) Inspect the vehicle frequently. Shock resistance of metals, or resistance against breaking, is greatly reduced at extremely low temperatures. Operation of vehicles on hard, frozen ground causes strain and jolting, which will result in screws breaking or nuts jarring loose.

(3) Disconnect oil-lubricated speedometer cables at the drive end for operating vehicles at temperatures of minus 30° F. and below. These cables often fail to work properly at these temperatures, and sometimes break, due to the excessive drag caused by the high viscosity of the oil which is used to lubricate them.

29. OPERATION AT HIGH TEMPERATURES.

a. Make sure that the cooling system is functioning efficiently (pars. 101 and 102).

b. Watch Temperature Gauges (fig. 13).

30. DESERT OPERATION.

a. Desert operation and operation under other extremely sandy road conditions may necessitate cleaning the air cleaners as often as every 4 hours (par. 62).

b. Limit the vehicle to second and third gears, and the engine speeds from 1,500 to 2,200 revolutions per minute, when operating on loose or shifting sand, or desert terrain. Do not operate vehicle at high loads below 1,500 revolutions per minute for continuous duty.

31. OPERATION UNDER SLIPPERY CONDITIONS.

a. For operating the vehicle in mountainous terrain, in mud, or over ice and snow, where sufficient traction is not normally possible, grousers are provided for attaching to the track blocks (par. 147).

b. Keep tracks and bogies free of mud, snow, or ice.

SECTION VIII
MATERIEL AFFECTED BY CHEMICALS

32. MATÉRIEL AFFECTED BY CHEMICALS. For information on this subject, refer to the following publications:

   a. TM 3-220, Chemical decontamination materials and equipment.
   b. FM 17-59, Decontamination of armored force vehicles.
   c. FM 21-40, Defense against chemical attack.

CHAPTER 2
VEHICLE MAINTENANCE INSTRUCTIONS

SECTION I
GENERAL

33. SCOPE.

a. The scope of maintenance and repair by the crew, and other units of the using arms is determined by the availability of suitable tools, availability of necessary parts, capabilities of the mechanics, time available, and the tactical situation. All of these are variable and no exact system of procedure can be prescribed.

b. Indicated below are the maintenance duties for which tools and parts have been provided for the using arm personnel. Other replacements and repairs are the responsibility of ordnance maintenance personnel but may be performed by using arm personnel when circumstances permit, within the discretion of the commander concerned. Echelons and words as used in this list of maintenance allocations are defined as follows:

Second echelon: Line organization regiments, battalions, companies, detachments, and separate companies (first and second echelons).

Third echelon: Ordnance light maintenance companies ordnance medium maintenance companies, ordnance divisional maintenance battalions, and ordnance post shops.

Fourth echelon: Ordnance heavy maintenance companies, and service command shops.

Fifth echelon: Ordnance base regiments, ordnance bases, arsenals, and manufacturer's plants.

Service: (Including preventive maintenance) par. 23a(1) and (2), AR 850–15. Consists of servicing, cleaning, lubricating, tightening bolts and nuts, and making external adjustment of subassemblies or assemblies and controls.

Replace: Par. 23a(4), AR 850–15. Consists of removing the part, subassembly or assembly from the vehicles and replacing it with a new or reconditioned or rebuilt part, subassembly or assembly, whichever the case may be.

Repair: Par. 23a(3) and (5) in part, AR 850-15.—Consists of making repairs to, or replacement of the part, subassembly or assembly that can be accomplished without completely disassembling the subassembly or assemblies, and does not require heavy welding, or riveting, machining, fitting and/or aligning or balancing.

Rebuild: Par. 23a(5) in part and (6), AR 850-15. Consists of completely reconditioning and replacing in serviceable condition any unserviceable part, subassembly, or assembly of the vehicle, including welding, riveting, machining, fitting, aligning, balancing, assembling, and testing.

c. The following maintenance allocation list is applicable to both the 155-mm gun Motor Carriage M12 and the Cargo Carrier M30, except where noted and identified in note at end of list:

	Echelons			
	Second	Third	Fourth	Fifth
Boxes and racks, ammunition				
Boxes, ammunition—replace	X			
Boxes, ammunition—repair		X		
Racks, ammunition—replace	X			
Racks, ammunition—repair		X		
Controls, brackets, and levers				
Bearing, clutch release—replace	X			
Brackets and levers—replace	X			
Brackets and levers—repair		X		
Controls and linkage (all)—replace	X			
Controls and linkage (all)—repair		X		
Drive, final (gear train assembly) (three piece)				
*Drive, final, gear train assembly—replace	(*)	X		
Drive, final, gear train assembly—repair		X		
Drive, final, gear train assembly—rebuild			E	X
Differential assembly, controlled				
*Differential assembly, controlled—replace	(*)	X		
Differential assembly, controlled—repair		X		
Differential assembly, controlled—rebuild			E	X
Differential subassembly, controlled				
*Differential subassembly, controlled—replace	(*)	X		
Differential subassembly, controlled—repair		X		
Differential subassembly, controlled—rebuild			E	X
Drums, steering brake—replace or repair		X		
Shoes, steering brake—service and/or replace	X			
Shoes, steering brake—repair (reline)		X		
Reduction, final				
Hubs, sprocket—replace	X			
Hubs, sprocket—repair		X		
Hubs, sprocket—rebuild			E	X

	Echelons			
	Second	Third	Fourth	Fifth
Reduction, final—Continued				
Reduction assembly, final—replace	X			
Reduction assembly, final—repair		X		
Reduction assembly, final—rebuild			E	X
Sprockets—replace	X			
Sprockets—rebuild			E	X
Transmission assembly				
Brake, parking—service and/or replace	X			
Brake, parking—repair (reline)		X		
*Transmission assembly—replace	(*)	X		
Transmission assembly—repair		X		
Transmission assembly—rebuild			E	X
Electrical group				
Battery—service, recharge and/or replace	X			
Battery—repair		X		
Battery—rebuild			E	X
Box, terminal—replace	X			
Box, terminal—repair		X		
Brackets, mounting and support—replace	X			
Brackets, mounting and support—repair		X		
Breaker, circuit—replace	X			
Breaker, circuit—repair		X		
Breaker, circuit—rebuild			X	
Cables, battery—replace	X			
Cables, battery—repair		X		
Conduit—replace		X		
Conduit—repair		X		
Filters—replace	X			
Filters—repair		X		
Lamps (all)—service and/or replace	X			
Lamps (all)—repair		X		
Regulator, current and voltage—replace	X			
Regulator, current and voltage—service and/or repair		X		
Regulator, current and voltage—rebuild			X	
Siren—replace	X			
Siren—repair		X		
Siren—rebuild			X	
Solenoids—replace	X			
Solenoids—repair		X		
Switches—replace	X			
Switches—repair		X		
Switches—rebuild			X	
Wiring—replace	X			
Wiring—repair		X		
Engine, radial, Wright Whirlwind, Model R975-EC2				
Baffles and cowling—replace	X			
Baffles and cowling—repair		X		
Carburetor assembly—service and/or replace	X			
Carburetor assembly—repair		X		
Carburetor assembly—rebuild			X	
Clutch assembly—replace	E	X		
Clutch assembly—repair		X		
Clutch assembly—rebuild			E	X

	Echelons			
	Second	Third	Fourth	Fifth
Engine, radial, Wright Whirlwind, Model R975-EC2—Continued				
Cylinder assembly—replace		E	E	X
Cylinder assembly—repair		E	X	
Cylinder assembly—rebuild (recondition)			E	X
*Engine assembly—replace		X		
Engine assembly—repair		X		
Engine assembly—rebuild			E	X
Flywheel and fan assembly—replace		X		
Flywheel and fan assembly—repair		X		
Flywheel and fan assembly—rebuild			E	X
Generator assembly—replace	X			
Generator assembly—repair		X		
Generator assembly—rebuild			X	
Governor assembly—service and/or replace		X		
Governor assembly—rebuild			E	X
Harness, ignition wiring—replace		X		
Harness, ignition wiring—repair		X		
Harness, ignition wiring—rebuild			X	
Magneto assembly—replace	X			
Magneto assembly—repair		X		
Magneto assembly—rebuild			X	
Manifold, exhaust—replace	X			
Manifold, exhaust—rebuild			X	
Motor assembly, starting—replace	X			
Motor assembly, starting—repair		X		
Motor assembly, starting—rebuild			X	
Pipe, intake—replace and/or repair		X		
Pistons and rings—replace		E	E	X
Plugs, spark—service and/or replace	X			
Plugs, spark (two piece)—repair		X		
Pump assembly, fuel—service and/or replace	X			
Pump assembly, fuel—repair		X		
Pump assembly, fuel—rebuild			X	
Pump assembly, oil pressure and scavenger—replace.	X			
Pump assembly, oil pressure and scavenger—repair.		X		
Pump assembly, oil pressure and scavenger—rebuild.			E	X
Rocker assembly, valve—replace	X			
Rocker assembly, valve—repair		X		
Rocker assembly, valve—rebuild			E	X
Rod, valve push—replace	X			
Strainer, oil—service and/or replace	X			
Strainer, oil—repair		X		
Exhaust group				
Brackets—replace	X			
Mufflers and connections—replace	X			
Fire extinguishing system				
Control, remote—replace	X			
Control, remote—repair		X		
Cylinders, CO_2—replace	X			
Cylinders, CO_2—repair and/or recharge		X		
Extinguisher assembly, fire, CO_2—repair and/or recharge.		X		
Extinguisher assembly, fire, CO_2—rebuild			E	X
Lines and nozzles—replace	X			
Lines and nozzles—repair		X		

	Echelons			
	Second	Third	Fourth	Fifth
Fuel group				
Cleaners, air—service and/or replace	X			
Cleaners, air—repair		X		
Filter, fuel—service and/or replace	X			
Filter, fuel—repair		X		
Lines, valves and fittings—replace	X			
Lines, valves and fittings—repair		X		
Pump, priming—replace	X			
Pump, priming—repair		X		
Pump, priming—rebuild			X	
Pump assemblies, fuel (autopulse—3) (auxiliary)—replace	X			
Pump assemblies, fuel (autopulse—3) (auxiliary)—repair		X		
Pump assemblies, fuel (autopulse—3) (auxiliary)—rebuild			X	
Strainer assembly—service and/or replace	X			
Strainer assembly—repair		X		
Tanks, fuel—service and/or replace	X			
Tanks, fuel—repair		X		
Hull				
Arrester, flame—replace	X			
Arrester, flame—repair		X		
Brackets, engine support—replace			E	X
Door, rear compartment—replace	X			
Door, rear compartment—repair		X		
Doors and cover plates—replace	X			
Doors and cover plates—repair		X		
Guards, mud—replace	X			
Guards, mud—repair		X		
Hull—repair		X		
Hull—rebuild			E	X
Insulation and padding—replace	X			
Protectoscopes—service and/or replace	X			
Ring assembly, antiaircraft—replace	X			
Ring assembly, antiaircraft—repair		X		
Seats—replace	X			
Seats—repair		X		
Instruments and panels				
Instruments—replace	X			
Instruments—repair		X		
Instruments—rebuild			E	X
Panels and connections—replace	X			
Panels and connections—repair		X		
Lubrication group				
Cooler, oil, engine and transmission—replace	X			
Cooler, oil, engine and transmission—repair		X		
Cooler, oil, engine and transmission—rebuild			E	X
Filter, engine oil—service and/or replace	X			
Filter, engine oil—repair		X		
Lines, oil, engine and transmission—replace	X			
Lines, oil, engine and transmission—repair		X		
Tanks, oil—service and/or replace	X			
Tanks, oil—repair		X		

	Echelons			
	Second	Third	Fourth	Fifth
Shaft, propeller				
Shaft assembly, propeller, w/universal joints—replace	X			
Shaft assembly, propeller, w/universal joints—repair		X		
Shaft assembly, propeller, w/universal joints—rebuild			E	X
Track suspension group				
Bearings and seals, bogie and idler wheel—replace	X			
Bogie components—replace	X			
Bogie components—repair		X		
Bogie components—rebuild			E	X
Brackets, idler—replace	E	X		
Brackets, idler—repair		X		
Brackets, idler—rebuild			E	X
Roller and bracket assembly, track supporting—replace	X			
Roller and bracket assembly, track supporting—repair		X		
Roller and bracket assembly, track supporting—rebuild			E	X
Track assembly—replace or repair	X			
Track assembly—rebuild			E	X
Wheels, bogie—replace	X			
Wheels, bogie—repair (replace tire)		X		
Wheels, idler—replace	X			
Wheels, idler—repair		X		
Wheels, idler—rebuild			E	X
Vehicle assembly				
Carrier, Cargo—M30—service and preventive maintenance	X			
Carrier, Cargo—M30—rebuild (with serviceable unit assemblies)			X	E

NOTE: Operations allocated will normally be performed in the echelon indicated by "X."

Operations allocated to the echelons as indicated by "E" may be accomplished by the respective echelons in emergencies only.

*The second echelon is authorized to remove and reinstall items marked by an asterisk. However, when it is necessary to replace an item marked by an asterisk with a new or rebuilt part, subassembly or unit assembly, the assembly marked by an asterisk may be removed from the vehicle by the second echelon only after authority has been obtained from a higher echelon of maintenance.

SECTION II
ORGANIZATION PREVENTIVE MAINTENANCE

34. SECOND ECHELON PREVENTIVE MAINTENANCE SERVICES. Regular scheduled maintenance inspections and services are a preventive maintenance function of the using arms, and are the responsibility of commanders of operating organizations.

  a. Frequency. The frequencies of the preventive maintenance services outlined herein are considered a minimum requirement for normal operation of vehicles. Under unusual operating conditions it may be necessary to perform certain maintenance services more frequently.

  b. First Echelon Participation. The drivers should accompany their vehicles and assist the mechanics while periodic second echelon preventive maintenance services are performed. Ordinarily the driver should present the vehicle for a scheduled preventive maintenance service in a reasonably clean condition; that is, it should be dry and not caked with mud or grease to such an extent that inspection and servicing will be seriously hampered. However, the vehicle should not be washed or wiped thoroughly clean, since certain types of defects, such as cracks, leaks, and loose or shifted parts or assemblies are more evident if the surfaces are slightly soiled or dusty.

  c. If instructions other than those contained in the general procedures to follow, or the specific procedures in the chart are required for the correct performance of a preventive maintenance service or for correction of a deficiency, the motor sergeant or the vehicle operator's manual should be consulted.

  d. General Procedures. These general procedures are basic instructions which are to be followed when performing the services on the vehicle items listed in the specific procedures.

  Note: The second echelon personnel must be so thoroughly trained in these procedures that they will apply them automatically.

(1) When new or overhauled subassemblies are installed to correct deficiencies, care should be taken to see that they are clean and properly lubricated and adjusted.

(2) When installing new lubricant retainer seals, a coating of the lubricant should be wiped over the sealing surface of the lip of the seal. When the new seal is a leather seal, it should be soaked in SAE 10 engine oil (warm if practicable) for at least 30 minutes. Then, the leather lip should be worked carefully by hand before installing the seal. The lip must not be scratched or marred.

(3) The general inspection of each item applies also to any supporting member or connection, and usually includes a check to see whether the item is in good condition, correctly assembled, secure, or exces-

sively worn. The mechanics must be thoroughly trained in the following explanations of these terms.

(a) The inspection for good condition is usually an external visual inspection to determine whether the unit is damaged beyond safe or satisfactory limits. The term "good condition," is explained further by such terms as the following: not bent or twisted, not chafed or burned, not broken or cracked, not bare or frayed, not dented or collapsed, not torn or cut, not deteriorated.

(b) The inspection of a unit to see that it is correctly assembled is usually an external visual inspection to determine whether it is in its normal assembled position in the vehicle.

(c) The check of a unit to determine if it is secure is usually an external visual inspection, a hand-feel, or a pry-bar check for looseness in the unit. Such an inspection should include any brackets, and all lock washers, lock nuts, locking wires, or cotter pins used to secure the tightening.

(d) "Excessively worn," which is a frequently used term, will be understood to mean worn close to or beyond satisfactory limits, and likely to result in a failure if not replaced before the next scheduled inspection.

(4) **Special Services.** These are indicated by added item numbers in the 50-hour or 100-hour column or both, and indicate that the part or assembly is to receive certain mandatory services. For example, an item number in one or both columns opposite a *Tighten* procedure, means that the actual tightening of the object must be performed. The special services include—

(a) **Adjust.** Make all necessary adjustments in accordance with the vehicle operators' manual, special bulletins, or other current directives.

(b) **Clean.** Clean units of the vehicle to remove excess lubricant, dirt, etc., using dry-cleaning solvent. After the parts are cleaned, rinse them in clean fluid and dry them well. Take care to keep the parts clean until reassembled, and to keep cleaning fluid away from rubber or other material which it will damage. Clean the protective grease coating from new parts. This material is usually not a good lubricant. Clean hydraulic brake cylinder parts in clean brake fluid. Do not use dry-cleaning solvent fluids on such parts.

(c) **Special Lubrication.** This applies either to lubrication operations that do not usually appear on the vehicle lubrication chart, and to items that do appear on such charts but should be performed in connection with the maintenance operations if parts have to be disassembled for inspection.

(d) **Serve.** Serving a part usually consists of performing special operations, such as replenishing batterywater, brake fluid, and shock absorber fluid; draining and refilling units with oil; and changing the oil filter cartridge.

(e) **Tighten.** All tightening operations should be performed with sufficient wrench torque (force on the wrench handle) to tighten the unit according to good mechanical practice. Use torque-indicating wrench where specified. Do not overtighten, as this may strip threads or cause distortion. Tightening will always be understood to include the correct installation of lock washers, lock nuts, and cotter pins provided to secure the tightening.

(5) When conditions make it difficult to perform the complete preventive maintenance service at one time, it can sometimes be handled in sections, planning to complete all operations within the week if possible. All available time at halts, rest periods, and in bivouac areas must be utilized if necessary to assure that maintenance operations are completed. When limited by the tactical situation, items with special services in the columns, should be given first consideration.

(6) The numbers of the preventive maintenance procedures that follow are identical with those outlined on W. D. A. G. O. Form No. 462, which is the Preventive Maintenance Service Work Sheet for Wheeled and Half-track Vehicles. Certain procedures that do not apply to this vehicle are deleted. The numerical sequence in general is followed, but in some cases there is deviation for conservation of the mechanic's time and effort.

c. **Specific Procedures.** These procedures for performing each item in the 50-hour and 100-hour preventive maintenance inspections and services are described on the following chart. Each of these pages has two columns at its left edge corresponding to the 50-hour and the 100-hour service, respectively. Very often it will be found that a particular procedure does not apply to both services. In order to determine which procedure to follow, look down the column corresponding to the service due, and wherever an item number appears perform the inspection and/or service indicated opposite the number.

ROAD TEST

NOTE: If the tactical situation does not permit a full road test, perform items 2, 3, 4, 6, 9, 12, 13, 14, and 15, which require slight or no movement of the vehicle. When a road test is possible it should be for 3 miles preferably, but not over 5 miles.

100-hour maintenance	50-hour maintenance	
1	1	*Before Operation Service.* Perform as outlined in paragraph 14.
2	2	*Dash Instruments and Gauges.*
		Oil Pressure Gauge. Normal operating pressure is 50 to 80 pounds at operating speed.

ROAD TEST—Continued

100-hour maintenance	50-hour maintenance	
		Dash Instruments and Gauges—Continued. *Engine Hour Meter.* See that it indicates accumulating engine running hours. Observe that meter runs at all times when engine is running. *Ammeter.* Reading should show high charge for short time after starting (45 to 55 amperes); then return to zero or slight positive charge. *Oil Temperature Gauge.* Normal range 165° F. to 190° F. **Caution:** Investigate cause if temperature gauge indicates more. *Tachometer.* Reading should indicate engine speed with no excessive fluctuation or noise. *Fuel Gauge.* Reading should indicate approximate amount of fuel in tanks. *Compass.* Reading should react with change of direction. *Clock.* Make sure clock is present and running. *Fuel Shut-off Valve.* See that valve turns freely. *Speedometer.* The speedometer should indicate without excessive fluctuation or noise.
3	3	*Siren.* If tactical situation permits test siren for normal signal
5	5	*Steering Brakes.* In second speed pull back on both steering levers. Brakes should offer resistance just before coming to a vertical position, and come back evenly. Test for quick stop, side pull, and independently for right and left steering. Parking brake must hold vehicle securely in applied position, and lock should operate freely.
6	6	*Clutch.* Free travel of clutch is 1 inch and must operate smoothly without slipping, chattering, or squeaking.
7	7	*Transmission.* Gears must shift smoothly, operate quietly, and not slip out of mesh during operation. Note excessive foaming from breather.
9	9	*Engine.* Engine must idle smoothly at 800 revolutons per minute. Note tendency to stall or misfire. Accelerate engine, observing for adequate pulling power in all speeds. Note excessive exhaust smoke. Listen for unusual noises that may indicate damaged, excessively worn, or inadequately lubricated accessories. Governed speed is 2,400 revolutions per minute
10	10	*Unusual Noises.* Listen for unusual noises in propeller shaft, universal joints, final drive, sprockets, idlers, wheels, support rollers, and tracks that might indicate damaged, loose, or inadequately lubricated units.
11	11	*Temperatures.* Stop vehicle; check by hand-feel, transmission, final drive, hubs, sprockets, idlers, wheels, and rollers for abnormal heat.
13	13	*Leaks.* Look for indication of fuel or lubricant leaks in drivers and engine compartment under vehicle.
12	12	*Gun Elevating and Traversing Mechanism.* Elevate and traverse the gun through its full range observing that mechanism operates freely with no binding, excessive lash, or erratic action. See that "carrying position" locking device operates freely.
14	14	*Noise and Vibrations.* Accelerate and decelerate engine, noting unusual noise and virbration in clutch, engine, accessories, drives, and exhaust.
15	15	*Track Tension.* See that track has sag of 1½ to 2 inches.

MAINTENANCE OPERATION

100-hour maintenance	30-hour maintenance	
16	16	*Fuel Pump Test.* Normal pressure 3 to 3½ pounds.
17	17	*Crankcase.* Check crankcase for indications of oil leaks. Examine organizational records for indication of excessive oil consumption. Two quarts per hour is normal. Drain and refill with correct grade of engine oil. (Refer to sec. IV, ch. 1.) **Caution:** In cases where engine will be removed, do not refill until engine has been reinstalled.
18	18	*Side Armor.* Examine for loose mounting and damage to fenders, towing shackles, cable, siren, paint, and markings.
19	19	*Bottom.* See that all bottom drain plugs are secure and not damaged.
20	20	*Differential and Final Drive.* Examine housings for damage, loose mounting, and lubricant leaks.

NOTE: If organizational records indicate a lubrication change is due, drain and refill to correct level with correct seasonal grade of gear oil. Tighten all assembly and mounting bolts.

21	21	*Track.* Examine track, connectors, and wedges for damage, looseness and excessive wear. Inspect for broken, bent, or cracked guides.

NOTE: Reverse rubber tracks when they are worn to a point where rubber is even with edge of connectors. If they have been reversed previously, replace when a majority of the cross tubes are exposed. Tighten all track connector wedge nuts.

22	22	*Idler.* Examine wheels, arm, adjustment nuts, idler bearings, and springs for damage, looseness, and excessive wear. Inspect oil seals for leaks. Tighten all assembly bolts, and lubricate idler hub.
23	23	*Bogie.* Examine levers, arms, springs and seats and brackets for damage, looseness or excessive wear. Tighten assembly and mounting bolts. Replace missing cotters and locking wires.
24	24	*Wheels.* See that tires are not separating from wheels. Inspect wheels and rollers for lubricant leaks.
24	----	*Test.* Track wheel and support rollers should be tested for damaged or loose bearings. Tighten assembly and mounting bolts securely.
25	25	*Sprockets.* Look for damage, looseness, and excessive wear. If badly worn, reverse sprockets or replace.
26	26	*Track Tension.* Adjust track tension to 1½ to 2 inches sag. Tighten locking devices securely.
27	27	*Top Armor.* Inspect for damage and loose mounting. Hatch covers and latches must operate freely. Repaint rusted or polished surfaces. Inspect vehicle markings for legibility.
28	28	*Caps and Gaskets.* Inspect for damage and serviceability. Test vent operation by shaking. Clean vent.
63	63	*Battery.* Inspect for leaks, loose mounting, corroded posts, terminals, and connections. Tighten hold-downs and terminal connections. If corroded, clean and apply fresh coat of grease. Gravity, after being tested and recorded, should not be below 1.225. Reddish brown color of electrolyte may indicate overcharging.
63	----	*Test.* Make high rate discharge test and record. Report if variation between cells is over 30 percent.
63	63	*Servc.* Clean exterior with water or soda wash, and dry. Fill with distilled water to ½ inch above plates.
82	82	*Handcrank, Ratchet, and Lever.* Inspect for damage, loose mounting, and excessive wear.
30	----	*Engine Removal.* Engine will be removed on 100-hour service if inspection made in items 6, 9, 13, and 14, and a check of organizational records on oil consumption indicate need (2 quarts per hour is normal.) Clean exterior with dry-cleaning solvent and dry thoroughly. Keep dry-cleaning solvent away from wiring and equipment.

NOTE: The above cleaning, and the following services, in items 31 to 60, should be performed in the best possible manner on the engine when it is not removed.

100-hour maintenance	50-hour maintenance	
31		*Valve Mechanism.* Observe that oil is being delivered to valves and rocker arms properly. Inspect push rod housing for damage, looseness, and oil leaks. Adjust valve clearance to 0.006 inch.
32		*Spark Plugs.* Inspect for burned, corroded, or coated electrodes. If unserviceable, replace with new or reconditioned plugs. Use new gaskets in installing.
34	34	*Generator and Starter.* Examine for damage, looseness, and loose wiring connections.
34		*Serve.* Remove commutator inspection cover and inspect for damage and dirt, brushes free and securely connected, and wiring broken or chafing. Tighten starter mounting bolts securely.
37	37	*Magnetos.* Inspect for damage and loose mounting. Observe evidence of oil leaks at mounting gaskets. Observe that points are clean and well aligned.
37		*Adjust.*—Adjust breaker point gaps; Scintilla Magneto to 0.012 inch; Bosch, 0.009 inch.
38	38	*Ignition Wiring and Conduits.* Examine for loose connections, chafing, and breaks. Clean all exposed ignition wiring.
		Note: Do not disturb connections unless loose.
39	39	*Coil.* Examine for loose mounting and connections. Clean terminals and tighten mounting bolts.
40	40	*Radial Engine.* Inspect oil pump, sump pump, screens and lines, accessory plate, fuel screens and line and control linkage for wear, dirt, loose mounting or connections, and leakage. Clean oil, fuel, and scavenger pump screens. Drain old oil from sump. Tighten assembly mountings accurately.
43	43	*Air Cleaner.* Disassemble air cleaner and clean element and reservoirs in dry-cleaning solvent and drain. Refill reservoir to correct level with seasonal grade engine oil. Apply engine oil to element and drain. Replace all gaskets. Tighten all mounting bolts.
44	44	*Carburetor.* Inspect carburetor for loose mounting and correct assembly. See that fittings and lines of primer system are not damaged or leaking.
45	45	*Manifold.* Inspect manifold for cracking, warping, looseness, and leakage. Examine stacks and intake pipes for looseness. Tighten all mounting and assembly bolts.
46	46	*Cylinders.* Clean cooling fins of dirt and grease.
47	47	*Radial Engine.* Inspect cowling, air deflectors, flywheel, fan and guard, and support beam for damage or looseness. Tighten all mounting bolts.
48	48	*Clutch Assembly.* If engine is removed, disassemble clutch, clean parts thoroughly and inspect for excessive wear.
51	51	*Engine Compartment.* Inspect all control linkage and bulkheads for damage or looseness.
51		*Clean.* Compartment should be cleaned as thoroughly as possible.
52	52	*Engine Oil.* Examine tank, lines, and fittings for breaks, looseness, and leakage. Check oil level in tank. Tighten all mounting bolts.
53	53	*Fuel.* Inspect tanks, lines, and pump for damage, loose mountings, and leaks.
54	54	*Engine Oil Filter.* Inspect filter for damage, loose mounting, and leakage.
54		*Clean.* Clean filter element and reinstall.
55	55	*Fuel Filters and Screens.* Inspect main and auxiliary filters for damage, loose connections, and leaks.
		Note: On vehicles of later manufacture a new type of fuel filter is installed in the gas tank.

100-hour maintenance	50-hour maintenance	
	55	*Fuel Filters and Screens*—Continued. *Clean.* Turn handle on main filter one complete turn, remove plugs, and drain filters.
55		*Clean.* Clean elements in dry-cleaning solvent. Do not scrape disks or dry with compressed air.
56	56	*Oil Coolers.* Inspect for damage, loose mounting and leakage. See that core air passages are not clogged. Clean if necessary. Flush out oil tank.
57	57	*Exhaust Pipe.* Inspect for cracks, loose mounting, and exhaust leaks. Tighten all mounting bolts.
58	58	*Engine Mountings.* Inspect for damage and looseness.
58		*Tighten.* Tighten all accessible mounting bolts.
60	60	*Fire Extinguisher System.* Inspect tanks, lines, valves, and nozzles for damage and loose mounting. See that tanks are fully charged. Disconnect main feed line at tank control valve and apply compressed air *cautiously* to blow out dirt and corrosion in lines and nozzles. See that nozzles are properly aimed and valves in locked position. Replace extinguishers if weight is less than 9½ pounds.
60	60	*Special Lubrication.* Apply few drops of engine oil to control cable pulleys and guides.
60		*Tighten.* Tighten mounting bolts.
88	88	*Radio Bonding.* See that suppressors, filters, condensors, and shielding are clean, secure, and not damaged. Note: Any irregularities other than cleaning and tightening must be reported to Signal Corps personnel.
59	59	*Clutch Release.* Inspect yoke, rollers, linkage, and mounting for damage, correct assembly, and excessive wear. See that yoke rollers do not have any flat spots on their outside diameter. Tighten accessible mounting bolts. Adjust free travel between the yoke and release bearing to ⅛ inch.
61	61	*Engine.* Install engine if it has been removed. Install mountings, lines and fitting, wiring, control linkage, and oil supply. Tighten mountings and connections securely. Refill oil tank.
64	64	*Accelerators.* Depress accelerator and see that carburetor throttle opens fully. Inspect connecting linkage for security.
65	65	*Starter, Primer, and Instruments.* Start engine observing operation of primer, cranking speed of starter, and if engine starts readily. Check instruments immediately to see that they are indicating properly.
66	66	*Leaks.* Inspect for fuel or oil leaks in drivers and engine compartment.
68	68	*Regulator Unit.* Inspect for damage, loose mounting, and loose connections.
68		*Test.* Make low voltage test after regulator has reached normal operating temperature, according to instructions on lid of test set.
69	69	*Engine Idle.* Engine must idle smoothly and not stall. Must have normal acceleration. Check for unusual noises that might indicate loose or inadequately lubricated units.
69		*Adjust.* Idle speed should be adjusted as required.
71	71	*Drivers Compartment.* See that seat and adjusting mechanism are in good condition. Examine crash pads, stowage boxes, and mounting brackets for damage and looseness. See that ammunition is safely stowed in compartments on gun deck.
73	73	*Protectoscopes.* See that they are in good condition and secure. See that holders are secure and lever and locking device operates freely and is not excessively worn.

100-hour maintenance	50-hour maintenance	
74	74	*Clutch Pedal.* Inspect for free travel, approximately 1 inch. See that linkage is secure and not excessively worn, and that return spring operates properly and is not weak.
75	75	*Steering Brakes.* Inspect levers, linkage, and shafts for loose mounting, connections, and excessive wear. See that levers meet resistance just before reaching vertical position. See that parking brake holds vehicle securely in applied position, locks freely, and releases fully.
75		*Adjust.* Adjust brake to 5½ inches free travel.
77	77	*Differential and Breather.* Inspect accessible part of differential gear case for secure mounting. See that breathers are secure and not clogged.
77		*Clean.* Remove and clean breathers. Tighten all external assembly and mounting bolts.
78	78	*Transmission.* Inspect for cracks and loose mounting. See that breather is not clogged, and oil is not leaking from case or seals.
78		*Clean and Tighten.* Remove and clean breather. Tighten all assembly and mounting bolts.
81	81	*Propeller Shaft.* Inspect shaft for loose assembly or mounting. See that universal and slip joints are correctly alined and not excessively worn.
81		*Tighten.* Universal joint assembly and companion flange bolts should be tightened securely.
83	83	*Oil Dilution Valve and Lines.* Inspect lines and controls for looseness or damage. Disconnect fuel line at engine inlet, operate valve and see that fuel is delivered properly and does not leak when valve is closed. (See note, par. 11c.)
84	84	*Compass.* Check accuracy against a compass of known accuracy. Inspect for low level or bubbles in fluid bowl. See that light switch operates properly.
85	85	*Lamps and Switches.* If tactical situation permits, test all lights and switches. Inspect lights for loose mountings and broken lenses. Adjust aim of headlight beam.
86	86	*Wiring.* Inspect exposed wiring, plugs, blocks and boxes for damage and loose mounting, and loose connections. See that fuses and spares are serviceable.
126	126	*Gun.* See that gun is not damaged, and is clean, well lubricated and correctly and securely assembled. See that firing control operates freely. Elevate and traverse gun through complete range and observe excessive lash or binding. Tighten all assembly and mounting bolts.
130	130	*Tools.* See that vehicle and pioneer tools are in good condition and properly stowed or mounted.
131	131	*Equipment.* Check against vehicle stowage list in section VI, chapter 1. Inspect ground spade for damage and loose mounting. Inspect winch mounting, ratchet, and brake for damage and serviceability. Examine cable for worn or broken strands and spade pulleys for serviceability.
132	132	*Grousers and Spare Track Blocks.* Inspect grousers and blocks for presence and proper stowage.
134	134	*Decontaminator.* Inspect decontaminator for full charge by removing filler plug. Examine for damage and loose mounting.
		Note: Solution must be renewed every 30 days.
135	135	*Fire Extinguishers.* Inspect portable extinguishers for damage and loose mounting. Clean nozzles. Refill if extinguisher weight is less than 3½ pounds. See that red safety blow-off valve is intact. **Caution:** Any cylinder containing gas under pressure is dangerous. Do not drop, handle roughly, or expose to unnecessary heat.

100-hour maintenance	50-hour maintenance	
136	136	*Publications and Form No. 26.* See that maintenance manual, parts list, and Accident Form No. 26 are present and legible.
137	137	*Vehicle Lubrication.* Lubricate all points in accordance with instructions in section IV, chapter 1, and the following instructions: Use only clean lubricant; replace missing or damaged lubrication fittings; wipe off excess lubricant; parts or assemblies that have already been lubricated while disassembled for inspection, gear cases that have been drained and refilled as mandatory items in the procedures, and those parts that have been indicated in the procedures for Special Lubrication may be omitted from the general lubrication of the vehicle.
139	139	*Final Road Test.* Recheck items 2 to 15 inclusive. Check transmission, differential, and final drive lubricant levels and examine for leaks. Confine this test to the minimum distance necessary to make satisfactory observations.

Note: Correct or report all defects found during road test.

SECTION III
ORGANIZATION TOOLS AND EQUIPMENT

35. ALLOCATION OF TOOLS AND EQUIPMENT.

a. The tools and equipment included in this section, together with the vehicle tools listed in section VI, provide the using arms with necessary tools and equipment for servicing the vehicle.

b. Organization Maintenance Tools, Standard Sets.

Standard tool set	Federal stock numbers
Tool Set, Motor Vehicle Mechanics	41-T-3538
Tool Set, Welders'	41-T-3555
Tool Set, Unit Equipment, Second Echelon No. 1.	41-T-3545-10
Tool Set, Unit Equipment, Second Echelon No. 2.	41-T-3545-11
Tool Set, Unit Equipment, Second Echelon No. 3.	41-T-3545-12
Tool Set, Unit Equipment, Second Echelon No. 4.	41-T-3545-13
Tool Set, Unit Equipment, Second Echelon No. 5.	41-T-3545-14
Tool Set, Unit Equipment, Second Echelon No. 6.	41-T-3545-15
Tool Set, Unit Equipment, Second Echelon No. 7.	41-T-3545-16
Tool Set, Unit Equipment, Second Echelon No. 9.	41-T-3545-18

e. Regimental Maintenance Platoon Tools.

(1) 155-MM GUN MOTOR CARRIAGE M12.

Name	Mfrs. symbol	Mfrs. tool No.	Federal stock No.	Vehicle set	Mech. set	Company set	Bn. or regt. crew set	Regt. or bn. maint. pltl set
Bolt, eye, lifting engine compartment top plate	MTM	M3-197	41-B-1586-200			4	2	4
Bolt, eye 1 inch 8NC-2, transmission lifting compressor, suspension volute spring.	MTM	M3-3	41-B-1586-350 41-C-2556			2	2	4 2
Drift, bogie wheel, installing	MTM	M3-13	41-C-1463				1	1
Drift, clutch, oil seal, inner, installing	MTM	M3-503	41-D-1540-50				1	1
Drift, clutch, spindle bearing, inner, installing	MTM	M3-502	41-D-1540-150				1	1
Drift, clutch, spindle bearing, outer, installing	MTM	M3-504	41-D-1540-200				1	1
Drift, engine clutch, oil retainer, installing	MTM	M3-496	41-D-1540-800			1	1	1
Drift, idler wheel, inner, bearing w/MTM3 8-inch puller.	MTM	M3-41						
Drift, idler wheel, outer bearing	MTM	M3-14	41-D-1540-550			1	1	1
Drift, idler wheel, inner bearing	MTM	M3-15	41-D-1540-500			1	1	1
Fixture, track connecting w/simplex jack	TK	7278	41-F-2997-85			1	1	1
Gauge, pr brake linkage adjusting, pr	MTM	M3-158			1			
Gauge, thickness, special, 0.010- and 0.070-inch	MTM	M3-211	41-G-412-75			1	1	1
Guide, bogie wheel gudgeon, installing	MTM	M3-5	41-G-2500			1	1	1
Handle, tubular 36 inches long, ⅜-inch inside diameter 1⁷⁄₁₆-inch outside diameter.	MTM	M3-16C	41-H-1498-50			1	1	1
Head, square 1-inch male	MTM	M3-16E	41-H-1779-50			1	1	1
Lift, bogie wheel, medium tank	MTM	M3-813	41-L-1375			1	1	1
Plate, lifting, clutch spindle and cover assembly	MTM	M3-499	41-P-1542-200			1	1	1
Protector, bogie wheel gudgeon, driving	MTM	M3-133	41-P-2838			1	1	1
Puller, idler wheel outer bearing	MTM	M3-40						
Puller, idler wheel	MTM	M3-8	41-P-2940-800			1	1	1
Puller, slide hammer type, bogie gudgeon			41-P-2957-33			1	1	1

Description		Part No.	Federal Stock No.			
Screw driver, socket, ½ inch square-drive for ½ inch armor plate bolts.			41–S–3869–150		2	2
Screw driver, ½ inch square-drive for ¾ inch armor plate bolts.			41–S–3867–157	1	2	2
Screw driver, valve clearance adjusting	MTM	M3–239	41–S–1725		1	1
Sling, engine			41–S–3832			4
Sling, final drive and transmission	MTM	M3–136	41–S–3832			4
Stand, engine, staley, inspection		15036	41–S–4942–23			4
Stand, engine, staley, transport		15040	41–S–4972–23			4
Timer, engine com. w/fan flywheel and cowl attachment.	MTM	M3–501	41–T–1598			1
Tool, clutch housing, oil retainer, guide	MTM	M3–498	41–T–3083–200		1	1
Tool, idler wheel installing	MTM	M3–9	41–T–3216–150		1	1
Wrench, box, flare nut, 1⅜ inch oil line fitting	SN	44	41–W–636–590		1	1
Wrench, box, generator attaching, ½ inch special	MTM	M3–506	41–W–636–550		1	1
Wrench, box, oil relief valve body check nut and cap.			41–W–636–620		1	1
Wrench, box, socket, suspension spring compressor special 1½ inch hexagon.	MTM	M3–5A	41–W–640–200		1	1
Wrench, box, special 3 inch hexagon 44⅝ inches long	MTM	M3–7	41–W–640–100		1	1
Wrench, crowfoot, starter attaching, ⁹⁄₁₆ inch, special.	MTM	M3–505	41–W–871–45		1	1
Wrench, cylinder hold down nut, ½ inch hexagon (EC1 and EC2 only).	MTM	M3–290	41–W–871–37		. 1	1
Wrench, cylinder hold down nut, ⁹⁄₁₆ inch hexagon (EC3 only).	MTM	M3–287	41–W–872–375		1	1
Wrench, drain plug, final drive, differential, ¾-inch, hexagon.	MTM	M3–130	41–W–877		1	1
Wrench, drain plug, transmission and oil tank, ⅞₁₆-inch, hexagon.	MTM	M3–131	41–W–878	1		1
Wrench, engine support nut, 1⁹⁄₁₆-inch, hexagon	MTM	M3–254	41–W–906–25		1	1
Wrench, engineers, double head, ⁹⁄₁₆-inch opening, each, 45° and 90°.			41–W–1465	2	1	1
Wrench, exhaust elbow to cylinder stud nut	MTM	M3–300	41–W–1471		1	1
Wrench, head ratchet socket, 1-inch, square-drive	MTM	M3–16M	41–H–1838		1	1
Wrench, intake pipe packing nut	MTM	M3–210	41–W–1537		1	1
Wrench, magneto, w/gauge, ½-inch	SCI	11–490	41–W–1555		2	2
Wrench, oil pump to crankcase rear section attaching nut, ⁷⁄₁₆-inch, hexagon.	MTM	M3–299	41–W–1577–500		1	1

75

Name	Mfrs. symbol	Mfrs. tool No.	Federal stock No.	Vehicle set	Mech. set	Company set	Bn. or regt. crew set	Regt. or bn. maint. just set
Wrench, oil pressure relief valve body, 7/16-inch by 9/16-inch, hexagon.	MTM	M3-341	41-W-1577-400			1	1	1
Wrench, plug, male, 3/4-inch, hexagon	MTM	M3-10	41-W-1960				1	1
Wrench, push rod housing adapter			41-W-1985				1	1
Wrench, crowfoot, 1 1/8-inch	MTM	M3-263	41-H-1543-85				1	1
Handle, wrench, crowfoot	MTM	M3-259	11-E-515				1	1
Extension, wrench, crowfoot	MTM	M3-214	41-W-2622				1	1
Wrench, set socket, special, 1-inch, square-drive, extra heavy duty.	MTM	M3-16A						
Wrench, socket, spark plug type, 1-inch, double hexagon.	MTM	M-3-460	41-W-3328				1	1
Wrench, socket, 1/2-inch, square-drive, double hexagon, 1 5/8-inch (for brake adjusting).	MTM	M 3-129	41-W-2573				1	1
Wrench, socket, bogie wheel gudgeon nut, 2 3/4-inch, hexagon.	MTM	M 3-137	41-W-2573-150				1	1
Wrench, socket idler wheel, shaft locknut	MTM	M-3-21						
Wrench, socket, 1-inch, square-drive, 2 1/8-inch	MTM	M-3-16F	41-W-3058-430			1	1	1
Wrench, socket, 1-inch, square-drive, 2 1/8-inch, hexagon.	MTM	M-3-16H	41-W-3058-450			1	1	1
Wrench, socket, 1-inch, square-drive, 2 3/8-inch, hexagon.	MTM	M-3-16K	41-W-3058-480			1	1	1
Wrench, socket, 1-inch, square-drive, 1 1/2-inch, hexagon.	MTM	M-3-16J	41-W-3058-200			1	1	1
Wrench, socket, 1-inch, square-drive, 1 13/16-inch, hexagon.	MTM	M 3-16X	41-W-3058-300			1	1	1
Wrench, valve adjusting and clamp screw	MTM	M3-266	41-W-3412-190			1	1	1

(2) Cargo Carrier M30.

Description							
Bolt, eye, 1-8NC-2, transmission lifting	MTM	M3-497	41-B-1586-350			4	1
Bolt, eye lifting engine compartment top plate	MTM	M3-8	41-B-1586-200		2	4	2
Compressor, suspension volute spring			41-C-2556	2	1	1	1
Drift, bogie wheel, installing	MTM	M3-13	41-D-1463			1	1
Drift, clutch, oil seal, inner, installing	MTM	M3-503	41-D-1510-50	1	1	1	1
Drift, clutch, spindle bearing, inner, installing	MTM	M3-502	41-D-1540-150		1	1	1
Drift, clutch spindle bearing, outer, installing	MTM	M3-504	41-D-1540-200		1	1	1
Drift, engine clutch, oil retainer, installing	MTM	M3-196	41-D-1510-800		1	1	1
Drift, idler wheel, inner bearing (used W MTM M3-8 puller)	MTM	M3-41					
Drift, idler wheel outer bearing	MTM	M3-14	41-D-1540-550	1	1	1	1
Drift, idler wheel inner bearing	MTM	M3-15	41-D-1540-500		1	1	1
Fixture, track connecting, with simplex jack and lever bar	TK	7278	41-F-2997-85			1	1
Gauge, thickness, special 0.010 and 0.070 inch	MTM	M3-211	41-G-412-75				
Gauge, brake, linkage, adjusting, pr	MTM	M3-158				1	1
Guide, bogie wheel gudgeon, installing	MTM	M3-5	41-G-2500			1	1
Lift, bogie wheel, M TK	MTM	M3-813	41-L-1375			1	1
Plate, lifting, clutch, spindle and cover assembly	MTM	M3-499	41-P-1542-200		1	1	1
Protector, bogie wheel gudgeon, driving	MTM	M3-153	41-P-2838	1	1	1	1
Puller, bogie gudgeon, screw type with adapter No. A1 61884	MTM	M3-6A	41-P-2905-65			1	1
Puller, slide hammer type, bogie gudgeon	MTM	M3-40	41-P-2957-33		1	1	1
Puller, idler wheel, outer bearing	MTM	M3-8	41-P-2940-800		1	1	1
Puller, idler wheel	MTM	M3-239	41-S-1725			1	1
Screw driver, valve clearance, adjusting			41-S-3832			1	1
Sling, engine			41-S-3832-72				
Sling, final drive and transmission	MTM	M3-136	41-S-3867-150	1	2	2	2
Socket, screw driver, ½-inch square drive (for ½-inch armor plate bolts)			41-S-3867-157	1		2	2
Socket, screw driver, ½-inch square drive, for ¾-inch armor plate bolts							
Stand, engine, staley inspection		15036	41-S-1972			1	1
Stand, engine, staley transport		15040	41-S-1942			1	1
Timer, engine, w/fan, flywheel, and cowl	MTM	M3-501	41-T-1598			1	1
Tool, clutch housing, oil retainer	MTM	M3-498	41-T-3083-200			1	1

Name	Mfrs. symbol	Mfrs. tool No.	Federal stock No.	Vehicle set	Mech. set	Company set	Bn. or regt. crew set	Regt. or bn. maint. plat set
Tool, idler wheel, installing	MTM	M3-9	41-T-3216-150			1	1	1
Wrench, box, flare nut, 1⅜-inch, oil line fitting	SN	44	41-W-636-590			1	1	1
Wrench, box, generator attaching, ½-inch, special	MTM	M3-506	41-W-636-550			1	1	1
Wrench, box, 3-inch hexagon, 44⅝ inches long	MTM	M3-7	41-W-640-400			1	1	1
Wrench, box socket, suspension spring compressor, special 1½-inch hexagon.	MTM	M3-2A	41-W-640-200					
Wrench, crowfoot, starter, attaching, ⅞₆-inch special.	MTM	M3-505	41-W-871-45			1	1	1
Wrench, cylinder hold-down nut, ½-inch hexagon (EC1 and EC2 only).	MTM	M3-290	41-W-871-37			1	1	1
Wrench, cylinder hold-down nut, ⅞₆-inch hexagon (EC3 only).	MTM	M3-287	41-W-872-375			1	1	1
Wrench, drain plug, final drive differential, ¼-inch hexagon.	MTM	M3-130	41-W-877		1	1	1	1
Wrench, drain plug, transmission and oil tank, ⅞₆-inch hexagon.	MTM	M3-131	41-W-878		1	1	1	1
Wrench, engine support nut, 1⅝₁₆-inch hexagon	MTM	M3-254	41-W-906-25			1	1	1
Wrench, engineer's double-head, ⅞₁₆-inch opening, 45° and 90° angle.			41-W-1465			1	1	1
Wrench, exhaust elbow to cylinder stud nut	MTM	M3-300	41-W-1471			1	1	1
Wrench, intake pipe packing nut	MTM	M3-210	41-W-1537			1	1	1
Wrench, magneto w/gauge, ¼-inch	SCI	11-490	41-W-1555		2	2	2	2
Wrench, oil pump to crankcase rear section attaching nut, ⅞₁₆-inch hexagon.	MTM	M3-299	41-W-1577-500			1	1	1
Wrench, oil pressure relief valve body, ⅝-inch x ⅞₁₆-inch hexagon.	MTM	M3-341	41-W-1577-400					
Wrench, plug, male ⅝-inch hexagon	MTM	M3-10	41-W-1960	1		1	1	1

78

Wrench, push rod housing adapter, consisting of—	MTM	M3-263	41-W-1985	1
Wrench, crowfoot, 1⅛-inch	MTM	M3-259	41-H-145385	1
Handle, wrench, crowfoot	MTM	M3-244	41-E-515	1
Extension, wrench, crowfoot	MTM	M3-460	41-W-3328	1
Wrench, socket, spark plug type, ½-inch square-drive				
Wrench, socket, bogie wheel gudgeon nut, 2⅜-inch hexagon	MTM	M3-137	41-W-2573-150	1
Wrench, socket, idler wheel shaft locknut	MTM	M3-21	41-W-3261	1
Wrench, spanner, track support roller retainer	MTM	M3-11	41-W-3260	1
Wrench, spanner, track support roller lock ring	MTM	M3-19	41-W-2622	1
Wrench, set, socket, 1 inch square-drive	MTM	M3-16A		1

SECTION IV
ENGINE AND ACCESSORIES

36. GENERAL DESCRIPTION (figs. 20, 21, 22 and 23). The Continental engines are of the single row, static radial, air-cooled type, operating on the conventional four-strokes cycle. The R975C-1 model has nine cylinders, a 5.00-inch bore, and a 5.50-inch stroke, giving a total piston displacement of 973 cubic inches. Throughout this book the flywheel end of the engine will be referred to as the "front" and the antiflywheel end will be referred to as the "rear." The terms "right" and "left" refer to the sides of the engine as viewed from the rear. Horizontal and vertical positions of the engine will be referred to with respect to the position of the crankshaft. Directions of rotation are determined by looking from the rear toward the front of the engine. The cylinders are numbered in the clockwise direction, the top cylinder being number one. Following this designation the firing order is 1, 3, 5, 7, 9, 2, 4, 6, 8.

a. Cylinders. Each cylinder is built up by screwing and shrinking a cast aluminum head onto a forged steel barrel. The rocker support boxes are cast integrally with the head. The head is finned for cooling, the fins being a part of the casting. The exhaust port faces to the right side and the intake port to the rear of the cylinder. Bronze valve guides are shrunk into bosses within the valve ports. Valve seat inserts are shrunk into the inside of the head.

b. Crankcase. The crankcase is composed of five aluminum alloy castings, secured together with studs through substantial flanges. The five sections are referred to as the front section, the front main bearing support, the main section, the diffuser section, and the rear section.

c. Crankshaft. The crankcase is a two-piece, single-throw, counterbalanced assembly. The front section of the shaft consists of the shaft proper, the front crankcheek with its counterweight, and the crankpin. The rear section of the shaft consists of the rear crankcheek with its counterweight and the rear main bearing journal. The dynamic damper consists of a steel counterweight, similar in shape to that of the conventional counterweight, hung on an extension of the rear crankcheek by two loose-fitting pins.

d. Master and Connecting Rods. The connecting rod assembly consists of the master rod and articulated or connecting rods. The master rod is of one-piece construction and operates in No. 1 cylinder.

e. Pistons. The aluminum alloy forged pistons are of the full-trunk type. There are five piston ring grooves in each piston.

f. Valve Operating Mechanism. A circular cam, riding on a steel sleeve which is screwed onto the crankshaft rear main bearing exten-

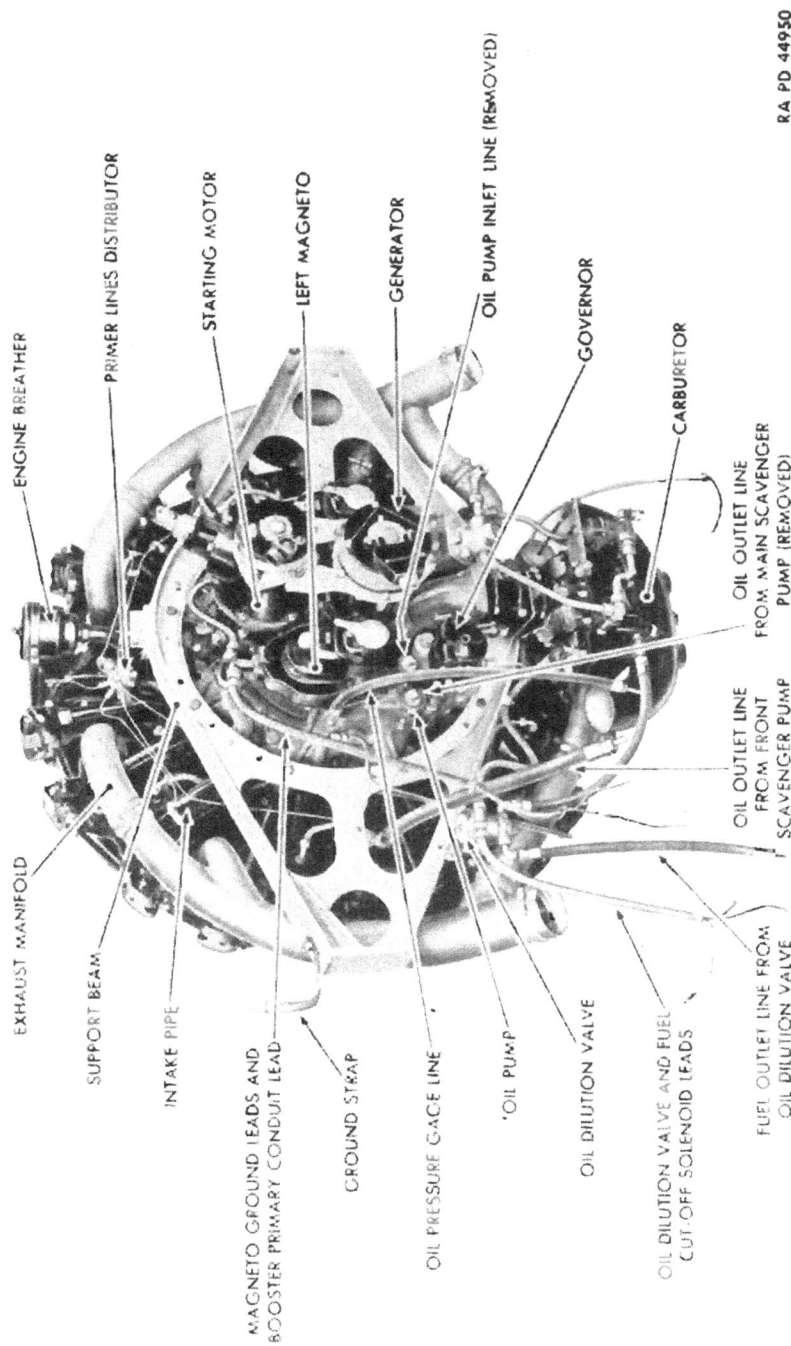

Figure 20. Left rear of engine with support beam.

81

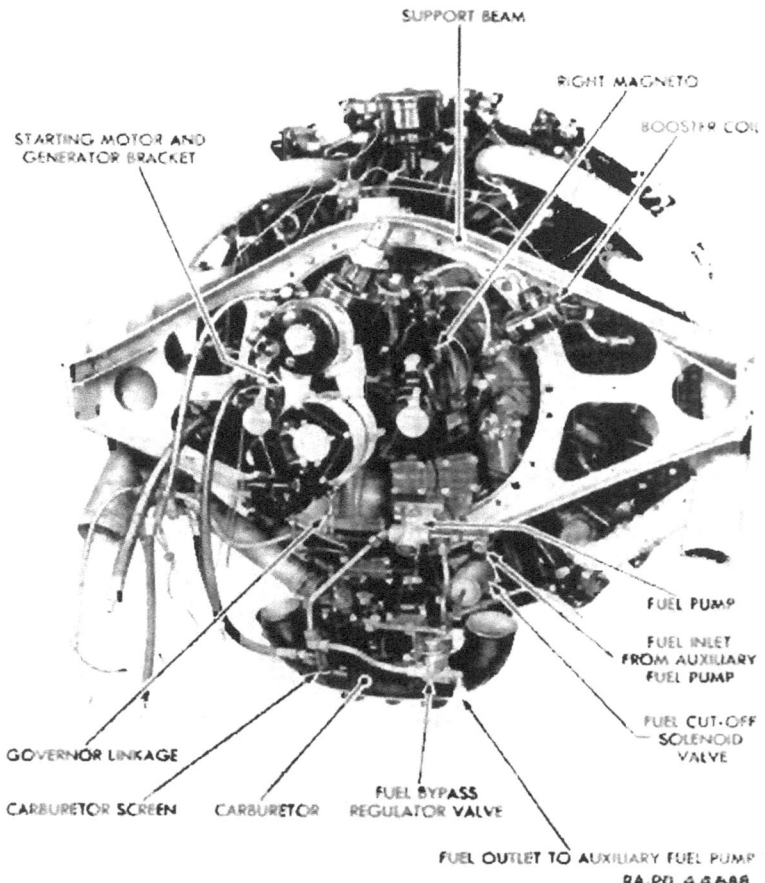

Figure 21. Right rear of engine with support beam.

sion, actuates the intake and exhaust valves through cam followers, push rods, and rocker arms.

g. Accessories. The two magnetos, starter, generator, governor, and oil pump, are mounted on the accessory case at rear of engine (figs. 20 and 21). The fuel pump is mounted at lower right rear of engine compartment. The carburetor is mounted on extreme lower end of rear section. The accessory case is attached to rear face of crankcase and houses the gear train which drives the various accessories, such as magnetos, starter, generator, etc.

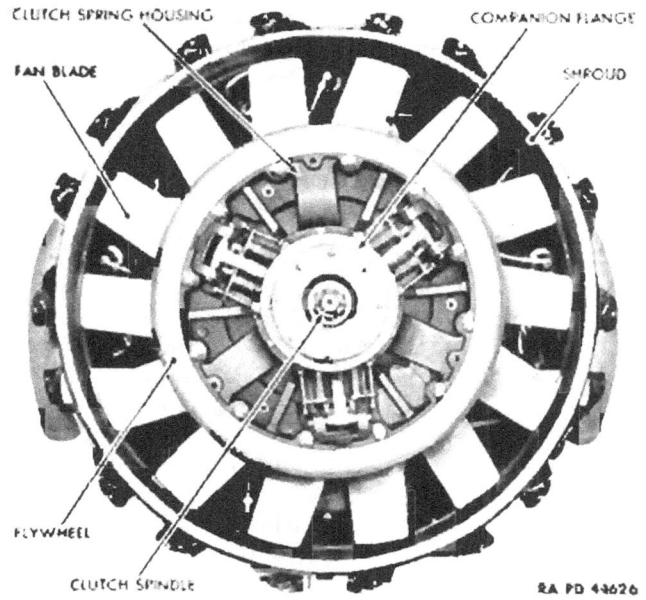

Figure 22. Front of engine with clutch and flywheel installed.

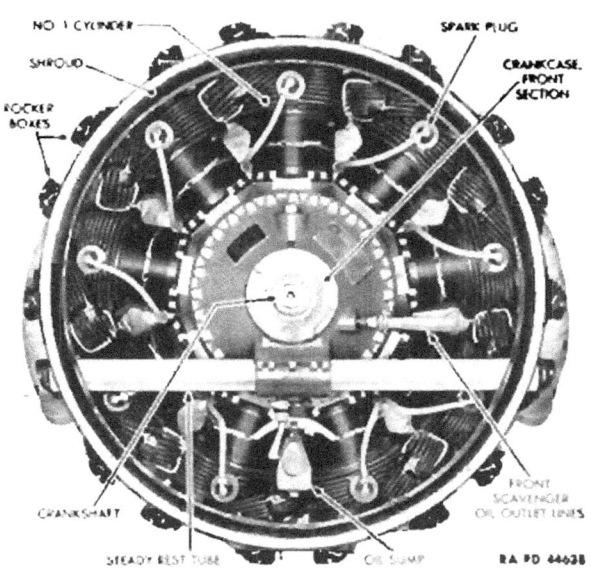

Figure 23. Front of engine with clutch and flywheel removed.

h. Engine Supports. The engine is supported in the engine compartment by support brackets and a support beam (banjo) at the rear (figs. 20 and 21) and a steady rest tube and support bearings at the front (fig. 23).

i. Engine Lubrication System.

(1) **Description.** In the engine lubrication system, the main oil supply is kept in a tank. Oil is drawn from the bottom of this tank by means of the engine oil pump, entering through a finger strainer and being forced through the engine lubrication system. Oil returns to a sump which is emptied by a main scavenging pump and a nose scavenging pump. Before returning to the engine oil tank, the oil is forced through a disk type filter and then through an engine oil cooler.

(2) **Breather Line.** To relieve pressure from the oil tank, and to vent foam from the expansion hopper, a breather line is provided from the tank to the crankcase of the engine (figs. 85 and 88).

(3) **Oil Tank.** The oil tank is filled through a filler pipe on the top of the oil tank (fig. 86). The filler cover is equipped with a bayonet gauge to indicate the amount of oil in the tank (fig. 86). A drain plug is located in the bottom of the oil tank. To drain the tank, it is necessary to remove the cover plate in the hull floor, directly below the tank, and then remove the drain plug.

(4) **Oil Filter.** The disk type oil filter is cleaned by turning the handle on the top of the cover (fig. 87). This should be done daily. Disks in automatic oil filters are turned and cleaned by a motor driven by oil pressure. No daily manual cleaning is necessary.

(5) **Oil Dilution Valve.** To assist in engine starting and warm-up in cold weather, an oil dilution valve permits engine oil to be thinned with gasoline. Dilution is performed before stopping the engine, when low temperatures are anticipated. The valve is solenoid operated, and is located on the left side of the engine support beam. When open, the valve allows gasoline to flow through a line from the carburetor into the main oil inlet line, thus diluting the oil. The solenoid is operated by a toggle switch on the instrument panel. Since excessive oil dilution can cause serious injury to the engine, the operating instructions on the instrument panel plate should be followed exactly. The oil dilution valve should be checked frequently, to be sure that it completely shuts off the gasoline line when the switch is in the OFF position. (See note, par. 11c.)

j. Tabulated Data.

(1) **Direction of Rotation** (from rear or magneto end).
        Crankshaft_____ _____ _____ Clockwise.
        Tachometer drive_____ Counterclockwise.

    Fuel pump.................................... Counterclockwise.
    Starter... Counterclockwise.
    Generator..................................... Counterclockwise.
    Magneto....................................... Counterclockwise.

(2) **Gaps and Clearances.**
    Magneto breaker points gap (Scintilla) 0.012 inch.
    Magneto breaker points gap (Bosch).... 0.009 inch.
    Spark plug gap, ceramic plugs.......... 0.017 inch to 0.019 inch.
    Spark plug gap, mica plugs............. 0.012 inch.
    Valve clearance (cold engine).......... 0.006 inch.

(3) **Engine Oil Pressure at Operating Speeds 60 pounds Minimum to 80 pounds Maximum.**
    Quantity..................................... 36 quarts.
    Temperature at inlet (maximum allowable). 190° F.
    Temperature at inlet (desired)......... 140° F.

(4) Firing Order 1, 3, 5, 7, 9, 2, 4, 6, 8 (No. 1 cylinder at top; cylinders numbered clockwise, from rear).

(5) **Compression and Fuel.**
    Compression ratio (Model R975C-1)... 5.7 to 1.
    Minimum fuel octane................... 80.

(6) **Displacement.**
    Bore... 5.00 inches.
    Stroke....................................... 5.50 inches.
    Displacement............................... 973 cubic inches.
    Rated horsepower.......................... 380 horsepower.

k. Accessories.

(1) The following accessories can be removed with the engine in the vehicle:

Booster coil	Magnetoes
Carburetor (first remove generator, fuel pump and governor)	Oil dilution valve
	Oil pump
Fuel pump	Spark plugs
Generator	Starter
Governor	Upper rocker arms and push rods
Ignition shielding harness	

(2) The following accessories cannot be removed while the engine is in the vehicle:

Baffles and cowling	Exhaust manifold
Clutch	Fan and flywheel

37. PERIODIC INSPECTION (50-hour). Periodic inspections given in full in paragraph 34 included inspections applying to the engine. These inspections have been grouped together for convenient reference and must be made regularly and completely. Periodic inspection, with proper adjustments or repairs being made immediately, is the surest way to protect the life and efficiency of the engine.

38. PERIODIC INSPECTION (100-hour). As indicated in paragraph 34, the engine is inspected and cleaned every 100 hours of operation. It is important at this time to inspect all lines for signs of leaks before the engine has been cleaned, since evidence of leakage may be entirely removed by cleaning. Spark plugs are not to be removed until after engine has been cleaned. Replace old plugs with dummy plugs to keep all dirt out of the engine interior until new plugs can be installed. Take particular care in disconnecting and connecting all lines and leads to avoid the necessity of replacement because of damage directly due to removal or installation of engine.

39. TROUBLE SHOOTING. The following trouble shooting table provides a list of common engine symptoms, their possible causes, and their possible remedies. Paragraph references to corrective procedure elsewhere in this manual are also included.

a. Engine Cannot be Cranked with Hand Crank or Cranks with Undue Resistance.

Possible cause	*Possible remedy*
Water or oil in cylinders.	Drain cylinders (par. 7**b**).
Improper seasonal grade of oil.	Use correct seasonal grade of oil (fig. 19).
Cold or thickened oil.	Warm engine oil (par. 25**h**).

b. Engine Fails to Start—Faulty Fuel System.

Lack of fuel.	Fill fuel tanks.
Fuel tank valves closed.	Open fuel tank valves.
Clogged fuel lines.	Clean fuel lines (par. 82**b**(2)).
Inoperative fuel pump.	Replace fuel pump (par. 56**d**(1) and (2)).
Clogged carburetor inlet screen.	Clean screen (par. 57**d**(1)).
Incorrect throttle opening.	Adjust throttle opening (par. 57**d**(2)).
Water in carburetor.	Drain water from carburetor (par. 57**b**).
Engine overprimed.	Wait 5 minutes for excess fuel to drain.
Raw fuel in air scoop.	Replace carburetor (pars. 58 and 59).

c. Engine Fails to Start—Faulty Ignition System.

Possible cause	Possible remedy
Inoperative booster coil.	Replace coil (par. 53c(1) and (2).)
Defective ignition harness.	Replace harness (par. 52).
Inoperative spark plugs.	Replace spark plugs (par. 51).
Faulty magneto breaker points.	Inspect points. Adjust point gap (pars. 49 and 50).
Defective magnetos.	Replace magnetos (par.49b(5)).

d. Engine Fails to Start—Mechanical Troubles.

Incorrect timing.	Check ignition timing (par. 49b(2)).
Mechanical failure.	Report to ordnance maintenance.

e. Engine Starts But Stops.

Faulty fuel system.	See trouble shooting procedure ((2) above).
Defective magneto (engine stops abruptly when switched from one magneto to another).	Replace defective magneto (par. 49b).

f. Engine Fails to Stop.

Inoperative solenoid fuel cut-off valve.	Replace valve (par. 84c and d).
Loose solenoid linkage.	Tighten solenoid linkage properly (par. 84d).
Carburetor flooding.	Replace carburetor (pars. 58 and 59).

g. Low Power.

Incorrect governor setting.	Check governor setting (par. 64). Report to ordnance maintenance if setting is incorrect.
Faulty fuel system.	See trouble shooting procedure ((2) above).
Faulty ignition system.	See trouble shooting procedure ((3) above).
Mechanical troubles.	See trouble shooting procedure ((4) above).

h. Low Power and Uneven Running with Black Smoke.

Mixture too rich.	Adjust idle mixture control (par. 57d(2)).
Defective fuel bypass regulator valve.	Replace fuel bypass regulator valve (par. 87).
Carburetor flooding.	Replace carburetor (pars. 58 and 59).

I. Low Power with Overheating and Backfiring.

Possible cause	Possible remedy
Mixture too lean.	Adjust idle mixture control (par. 57d(2)).
Defective fuel bypass regulator valve.	Replace fuel bypass regulator valve (par. 87).
Leaking intake pipe connections.	Replace intake pipe packing (par. 52f(4)).
Defective intake pipes.	Replace intake pipes (par. 52e(4)).
Defective carburetor.	Replace carburetor (pars. 58 and 59).
Loose or improperly adjusted solenoid linkage.	Tighten or adjust (par. 84d).
Damaged solenoid valve linkage.	Replace solenoid valve linkage (par. 84).
Insufficient oil supply.	Add oil to proper level (par. 19c(4)).
Improper seasonal grade of oil.	Drain oil tank and fill with proper seasonal grade of oil (par. 19).
Incorrect ignition timing.	Check ignition timing (par. 49b(4)).

J. Loss of Oil Pressure.

Possible cause	Possible remedy
Insufficient oil supply.	Add oil to proper level (par. 19c(4)).
Improper seasonal grade of oil.	Drain oil tank and fill with proper seasonal grade of oil (par. 19).
Clogged oil lines.	Disconnect and clean oil lines (par. 45).
Leaking oil lines and connections.	Replace oil lines and connections (par. 45).
Clogged oil filter.	Disassemble and clean filter (par. 95).
Defective oil pump.	Replace oil pump (pars. 45 and 46).
Clogged sump screen.	Clean screen (pars. 45, 46 and 47).
Clogged sump strainer.	Clean strainer (par. 47).
Oil pump needs priming.	Prime oil pump (par. 44).
Foam in oil system.	Drain oil tank and fill with proper seasonal oil (par. 19).
Improper bearing clearance.	Report to ordnance maintenance.
High oil temperature.	Check oil coolers (par. 96).
Cold-thickened oil.	Warm up engine properly (par. 7).
Worn piston rings or cylinders.	Report to ordnance maintenance.

k. High Oil Consumption.

Inadequate oil cooling.	Check oil coolers (par. 96).
Insufficient oil supply.	Add oil to proper level (par. 19c (4)).
Improper seasonal grade of oil.	Drain oil tank and fill with proper seasonal grade of oil (par. 19).
Defective oil pump.	Replace oil pump (pars. 46 and 47).
Overdiluted oil.	Check oil dilution system (par. 25e).
Dirty oil.	Drain oil tank and fill with proper seasonal grade of oil (par. 19).
Worn piston rings.	Report to ordnance maintenance.
Overheated bearings.	Report to ordnance maintenance.

40. ENGINE REMOVAL.
a. Equipment.

Cable, length	Wrench, open-end, 7/16-inch.
Cloth, wiping	Wrenches, open-end, 9/16-inch (2).
Extensions, long (2)	Wrench, open-end, 5/8-inch.
Handle, sliding T	Wrench, open-end, 3/4-inch.
Hoist	Wrench, open-end, 15/16-inch.
Pliers	Wrench, open-end, 1-inch.
Screw driver	Wrench, open-end, 1 1/4-inch.
Sling, engine lifting	Wrench, open-end, 1 5/8-inch.
Wire, binding	Wrench, socket, 7/16-inch.
Wrench, open-end, 5/16-inch	Wrench, socket, 9/16-inch.
Wrench, open-end, 13/32-inch	Wrench, socket, 5/8-inch.
Wrench, open-end, 3/8-inch	Wrench, socket, 1 1/4-inch.

b. Procedure. Operations involving the 155-mm gun are applicable only to the 155-mm Gun Motor Carriage M12. In all other respects, the removal of the engine from either the gun carriage or cargo carrier is identical.

(1) **Preliminary Steps** (fig. 14).

(a) Open battery switch and disconnect battery cables at the terminals.

(b) Close the two fuel shut-off valves by turning the two valve handles in the left and right side of the driving compartment clockwise.

(2) **Remove Engine Compartment Top Guards** (fig. 24).

Wrench, socket, 9/16-inch.

Remove cap screws which secure front and rear engine top compartment guards (9/16-inch socket wrench). Lift off guards.

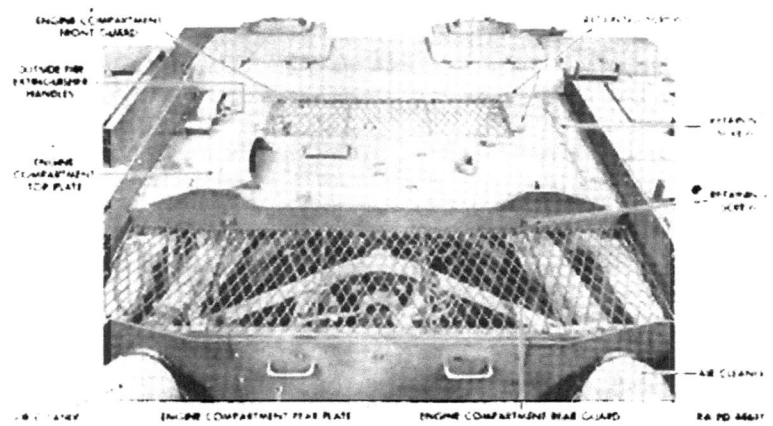

Figure 24. Engine compartment guards, top plate and rear plates.

(3) **Remove Engine Compartment Top Plate** (figs. 24 and 25).

    Cable, length   Screw driver
    Hoist   Wrench, open-end, ⁹⁄₁₆-inch
    Pliers   Wrench, socket, ⁹⁄₁₆-inch

(a) Elevate gun to its full rest position, then traverse to the left as far as possible.

(b) Remove cap screws which secure engine compartment top plate (⁹⁄₁₆-inch socket wrench). Do not remove top plate at this time.

(c) Remove cap screws from under side of center of top plate which secure upper engine shroud (⁹⁄₁₆-inch socket wrench).

(d) From within driving compartment, remove lock wire and

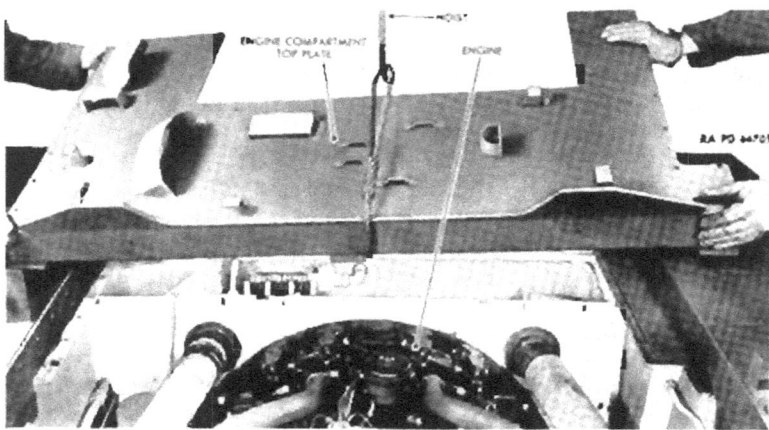

Figure 25. Engine compartment top plate removal.

three screws which secure cover of fixed fire extinguisher control head (pliers and screw driver) (fig. 146). Lift off cover.

(e) Loosen screws which secure outside fire extinguisher handle cable to block inside control head (screw driver). Pull cable out of block, then remove block.

(f) Repeat the operation on the rear fire extinguisher to disconnect the outside fire extinguisher cable.

(g) Disconnect the tubes through which the cables run, at the fire extinguisher control head (%6-inch open-end wrench). The opposite end of the tubes will be disconnected after the top plate has been partially lifted from the vehicle.

(h) From on top the engine compartment top plate, pull the two fire extinguisher outside handles, with attached cables, out of their mounting (fig. 24).

(i) Lift the top plate by hand enough to pass a length of cable completely under the top plate. Hook a hoist to the cable, and lift the top plate clear of the vehicle (fig. 25). *Caution:* Do not let the top plate slip. Disconnect and remove the two fire entinguisher outside handle cable tubes (%6-inch open-end wrench). Swing the top plate to one side and lower it to the ground.

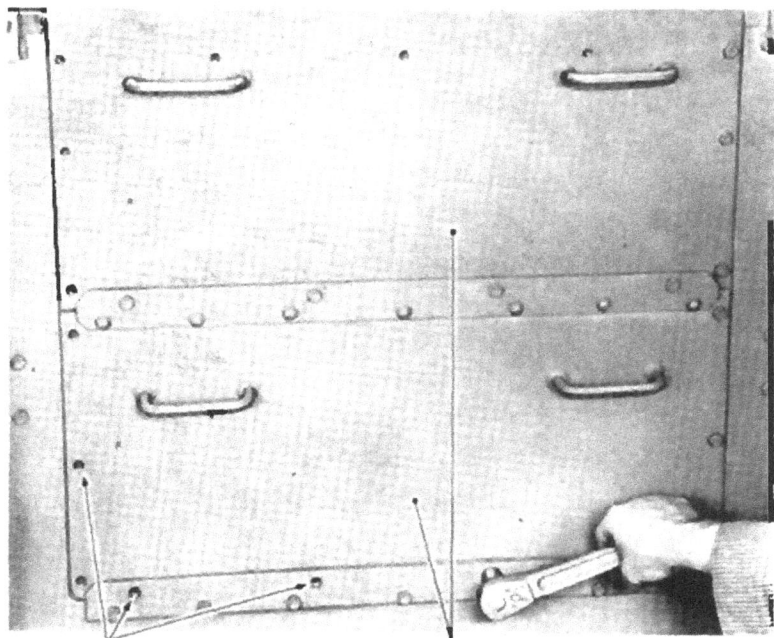

RETAINING SCREWS (REMOVED)   ENGINE COMPARTMENT REAR PLATES   RA PD 44618.

Figure 26. Engine compartment rear plates removal.

Figure 27. Engine through engine rear plates opening.

Figure 28. Engine installed—from upper rear of engine compartment.

Figure 29. Engine installed—from upper front of engine compartment.

(4) **Remove Engine Compartment Rear Plates** (fig. 26).
　　Wrench, socket, ⁹⁄₁₆-inch.
　Remove cap screws which secure engine compartment rear plates (⁹⁄₁₆-inch socket wrench). Lift off the plates by hand. Remove carburetor inspection plates.

(5) **Remove Upper Air Intake Tubes** (figs. 28 and 29).
　　Screw driver
　(a) Disconnect hose connections which clamp upper intake tubes (right and left) to air cleaners (screw driver).
　(b) Slide tubes forward and upward to free from air cleaners, then pull tubes out of upper half of engine compartment shroud.

(6) **Remove Upper Half of Engine Compartment Shroud** (figs. 28 and 29).
　　Wrench, open-end, ⁷⁄₁₆-inch.
　(a) Remove two cap screws which secure small bracket on right lower side of upper half of engine compartment shroud (⁷⁄₁₆-inch open-end wrench) (through which passes right rear fire extinguisher line).
　(b) Lift off upper half of engine compartment shroud.

(7) **Remove Lower Air Intake Tubes** (fig. 27).
　　Screw driver
　(a) Disconnect hose connections which clamp lower air intake tubes (right and left) to air cleaner and air scoop of carburetor (screw driver).
　(b) Slide tubes back from connections and remove.
　(c) Tie cloths over air scoop openings.

(8) **Disconnect Rear Terminal Box Electrical Connections** (figs. 30 and 31).
　　Pliers　　　　　　　　Wrench, open-end, ⁵⁄₁₆-inch
　　Screwdriver　　　　　Wrench, open-end, ⅜-inch
　(a) Remove four screws and lock washers which secure rear terminal box cover (screw driver). Lift off cover.
　(b) Disconnect knurled nut which secures magneto ground leads and booster primary conduit to top conduit inlet of rear terminal box (pliers).
　(c) Remove both magneto ground leads and the booster primary lead from their terminals in the rear terminal box, tagging each post and wire as disconnection is made to assure correct installation (⅜-inch open-end wrench).
　(d) Disconnect knurled nut which secures oil dilution valve and fuel cut-off solenoid leads conduit to bottom conduit inlet of rear terminal box (pliers). (See note, par. 11c.)
　(e) Remove oil dilution valve lead and fuel cut-off solenoid leads from their terminals in the rear terminal box, tagging each post and wire as they are removed to assure correct installation (⁵⁄₁₆-inch open-end wrench).

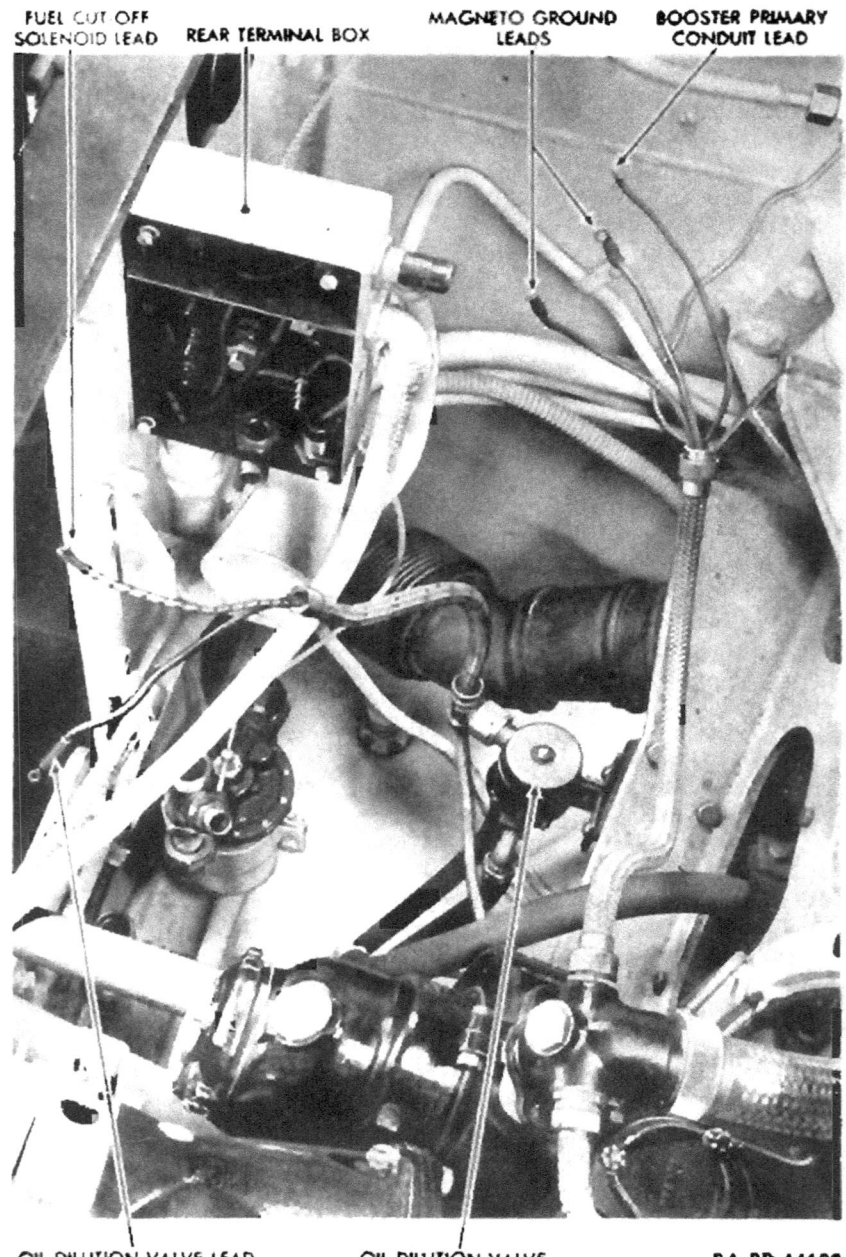

Figure 30. Rear terminal box and disconnected wires.

(9) **Disconnect Remaining Electrical Connections.**
　　Pliers　　　　　　　　　　Wrench, open-end, 1½₂-inch
　　Screw driver　　　　　　　Wrench, open-end, ⅝₆-inch

(a) Remove generator shield (screw driver), disconnect wires (⅝₆-inch open-end wrench), and remove generator cable from generator (pliers) (fig. 27).

(b) Remove starter shield (screw driver), disconnect conduit (pliers) and disconnect and remove starter cable at starter (1½₂-inch open-end wrench) (fig. 27).

(10) **Disconnect oil lines.**
　　Cloth, wiping　　　　　　Wrench, open-end, ¾-inch
　　Screw driver　　　　　　Wrench, open-end, 1-inch
　　Wire, binding　　　　　　Wrench, open-end, 1¾-inch

(a) Inspect all oil line connections for evidence of leaks before disconnecting. Disconnect main oil pump inlet connection and main scavenger outlet connection at the lower end of each hose (1¾-inch open-end wrench) (figs. 27 and 34). Avoid loosening the wrong part of the coupling (loosen the large lower nut). Attempting to turn the two upper hexes may shear the hose. To avoid necessity of priming oil pump when installing engine, tie up oil inlet line to generator, thus keeping oil from draining out.

(b) Disconnect front section scavenger line at Y-connection (above oil filter) with oil outlet line (1-inch open-end wrench) (fig. 34). As soon as oil has drained out of lines, cover ends with wiping cloth, and wire into place. *Caution:* When making disconnections at Y-fittings, care must be taken not to injure one connection while disconnecting other line.

(c) Disconnect oil dilution valve line at T-connection with oil inlet line (¾-inch open-end wrench) (fig. 34). As soon as oil has drained out of lines, cover ends with wiping cloth and wire into place.

(11) **Disconnect Fuel Lines.**
　　Cloth, wiping　　　　　　Wire, binding
　　Pliers　　　　　　　　　　Wrench, open-end, ¾-inch.

(a) Inspect all fuel connections for leaks before disconnecting.

(b) Disconnect fuel bypass line at auxiliary fuel pump (¾-inch open-end wrench) (fig. 32).

Note: The auxiliary pump on later vehicles has been replaced by an AC type fuel pump.

(c) Disconnect fuel inlet line at fuel pump (¾-inch open-end wrench) (figs. 27 and 32).

(d) Cover all fuel openings with wiping cloth wired in place (pliers).

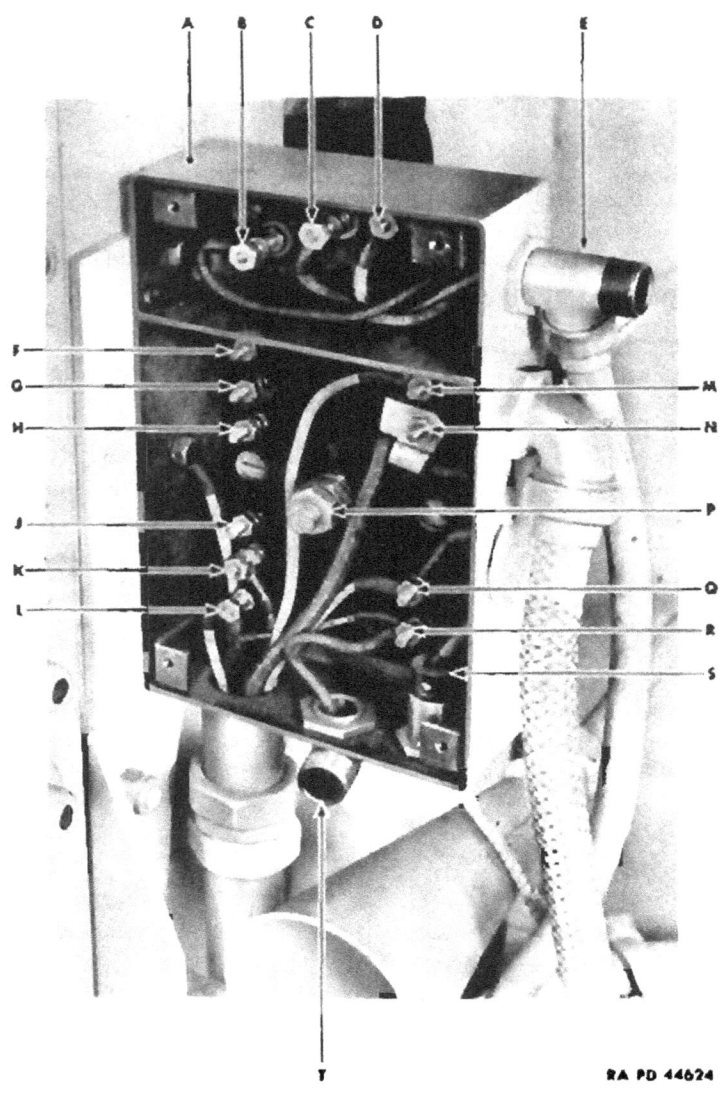

A. Rear terminal box.
B. Magneto ground wire terminal.
C. Magneto ground wire terminal.
D. Booster primary lead terminal.
E. Top conduit inlet.
F. Not used.
G. Not used.
H. Not used.
J. Not used.
K. Oil dilution valve terminal.
L. Fuel cut-off solenoid terminal.
M. Generator field terminal.
N. Generator armature terminal.
P. Not used (formerly used as starter terminal).
Q. Trouble light terminal.
R. Auto pulse fuel pump terminal.
S. Trouble light socket.
T. Bottom conduit inlet.

Figure 31. Rear terminal box and terminals.

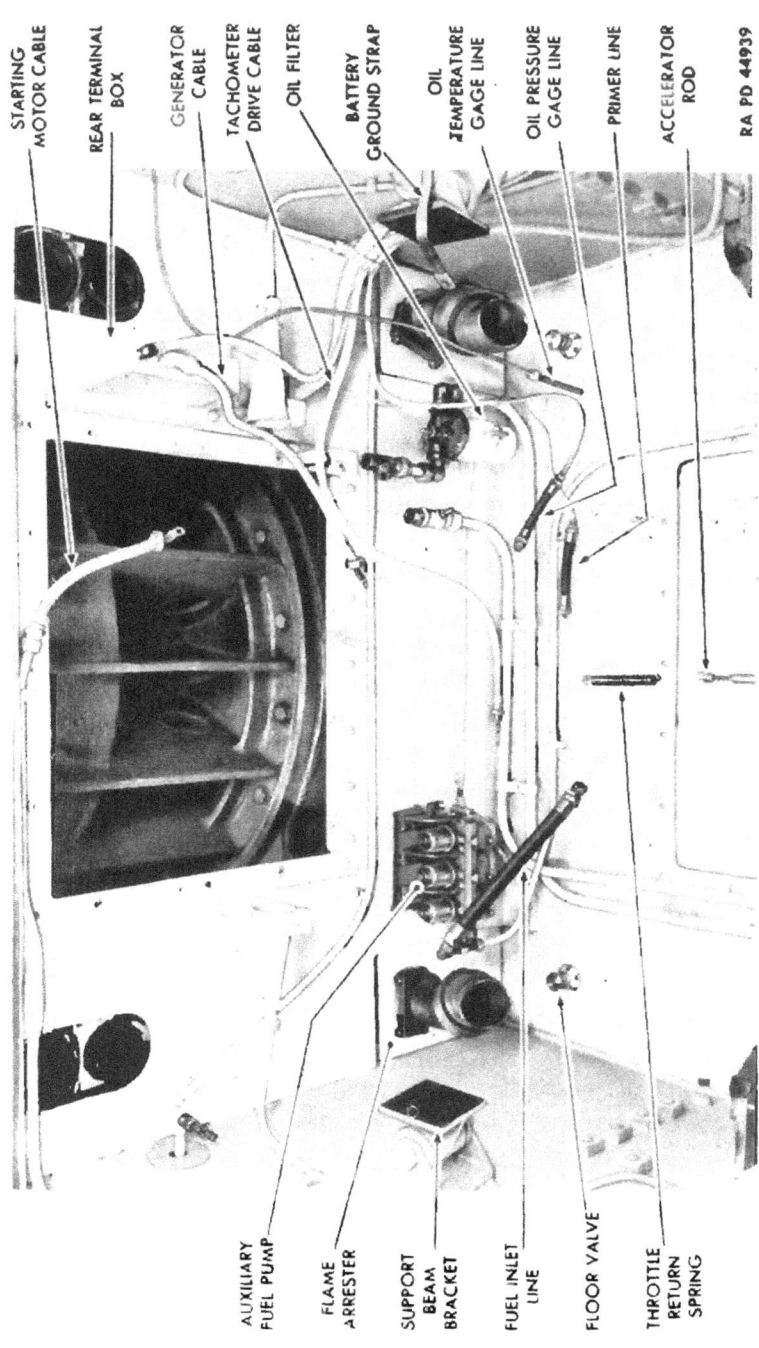

Figure 82. Engine compartment—looking toward rear of vehicle.

99

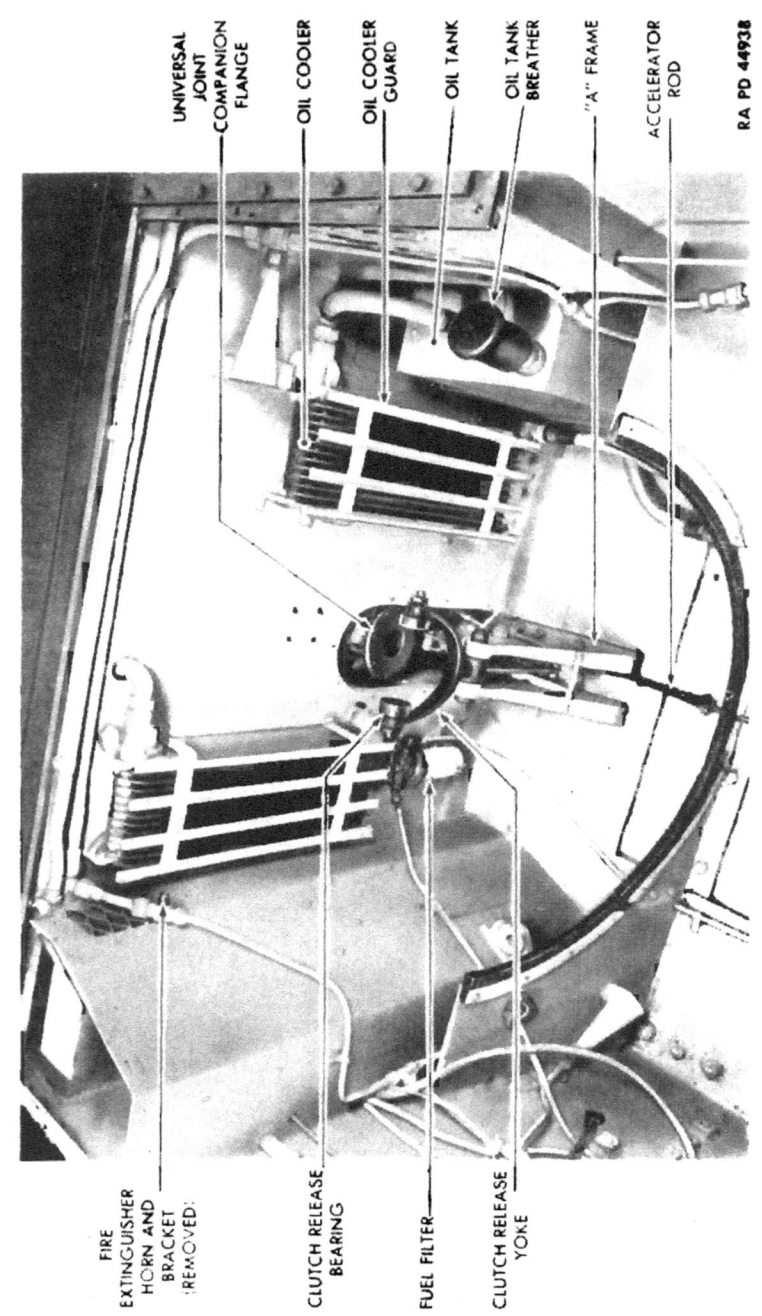

Figure 38. Engine compartment—looking toward front of vehicle.

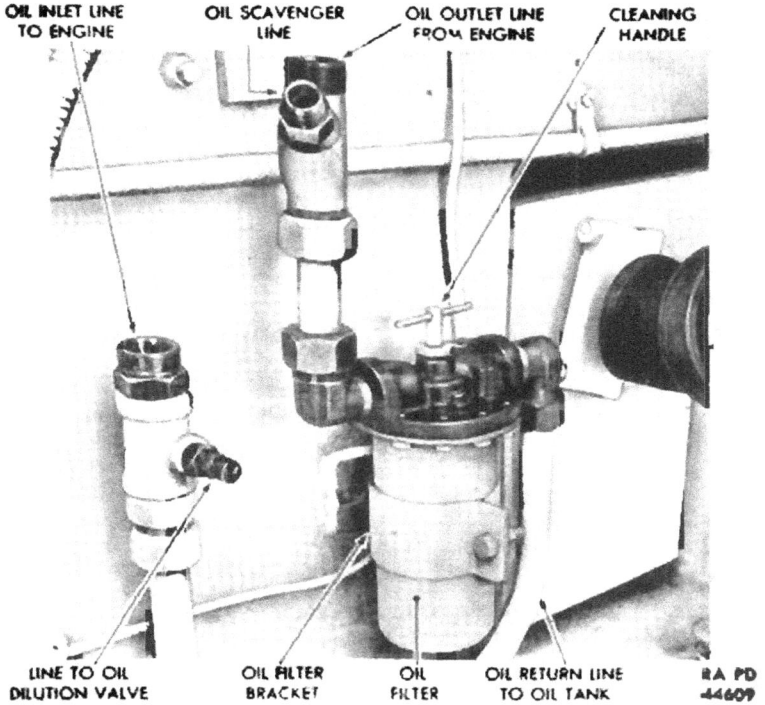

Figure 34. Oil filter installed.

(12) Disconnect Oil Pressure Gauge Line and Primer Distributor Line (fig. 32).

    Cloth, wiping   Wrench, open-end, $\frac{7}{16}$-inch
    Pliers   Wrench, open-end, $\frac{5}{8}$-inch
    Wire, binding

Disconnect oil pressure gauge line ($\frac{5}{8}$-inch open-end wrench), and primer distributor line ($\frac{7}{16}$-inch open-end wrench) under left side of engine support beam. Cover openings with wiping cloth wired in place (pliers).

(13) Disconnect Oil Temperature Gauge Line and Tachometer Drive Cable (fig. 32).

    Pliers   Wrench, open-end, $\frac{5}{8}$-inch

(a) Inspect oil temperature gauge connection for breaks **before** removing.

(b) Loosen retaining nut and remove oil temperature gauge bulb from oil pump finger strainer ($\frac{5}{8}$-inch open-end wrench). Plug opening (wiping cloth).

(c) Remove tachometer drive cable by unscrewing knurled nut (pliers).

(14) Disconnect Accelerator Linkage (fig. 32).

Pliers

(a) Unhook throttle return spring on floor of engine compartment which is attached to arm on clevis pin, holding accelerator rod clevis to carburetor throttle arm.

(b) Remove cotter pin from accelerator rod clevis pin and remove clevis pin, disconnecting linkage (pliers).

Note: The easiest way to disconnect the accelerator linkage, which is extremely difficult to get at, is to stand to the left of the engine, in the engine compartment facing toward the rear of the vehicle.

(15) Disconnect Exhaust Tubes (fig. 32).

Wrench, open-end, 7/16-inch Wrench, socket, 7/16-inch

Loosen clamps which secure exhaust tubes (right and left) to smaller section of exhaust tube clamped to flame arrester (7/16-inch open-end wrench and 7/16-inch socket wrench).

(16) Disconnect Engine From Propeller Shaft.

Wrenches, open-end, 9/16-inch (2)

Remove the eight bolts and nuts that secure the clutch companion flange to the universal joint companion flange (two 9/16-inch open-end wrenches) (fig. 33).

Note: Turn the engine with the crank, to rotate lower bolts into a position from which they may more easily be removed.

(17) Remove Left Front Fire Extinguisher Horn and Bracket.

Wrenches, open-end, 9/16-inch (2) Wrench, socket, 9/16-inch.

Wrench, open-end, 1-inch

(a) Remove two cap screws which secure left front fire extinguisher horn to bracket (9/16-inch socket wrench). Unscrew and remove horn (1-inch open-end wrench).

(b) Remove two bolts and nuts which secure fire extinguisher horn bracket to recessed compartment bulkhead (two 9/16-inch open-end wrenches). Lift off bracket.

(18) Remove Oil Tank Breather (fig. 33).

Wrench, open-end, 15/16-inch.

Unscrew and remove oil tank breather (15/16-inch open-end wrench). Tie cloths over exposed ends.

(19) Disconnect Engine From Engine Supports.

Extensions, long (2) Wrench, open-end, 9/16-inch.
Handle, sliding T   Wrench, open-end, 1¼-inch.
Pliers   Wrench, socket, 1¼-inch.

(a) Remove bolts which hold steady-rest adapters (which slide in steady-rest tube) to steady-rest brackets (1¼-inch open-end wrench, 1¼-inch socket wrench with extension and sliding T-handle) (figs. 23

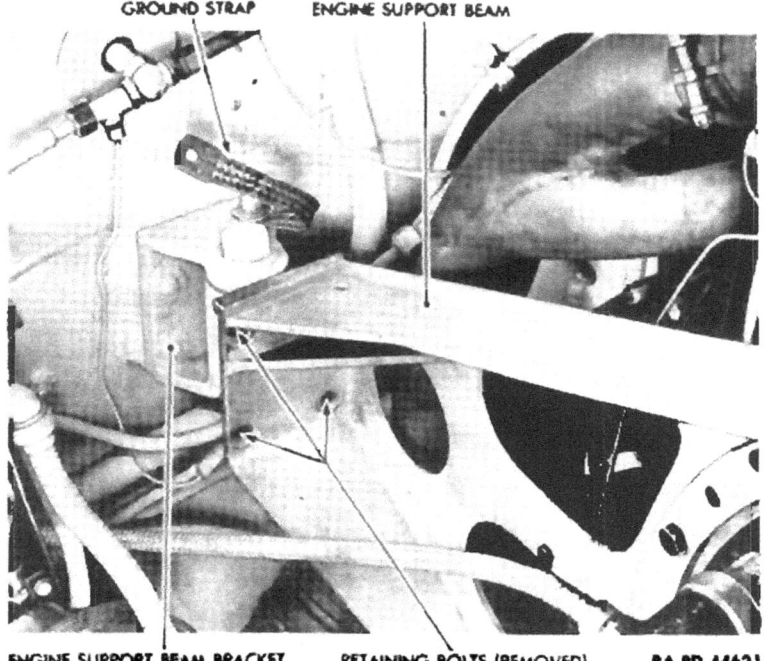

Figure 35. Engine support beam.

and 37). This will require two men; one to hold the bolt and one to turn the nut.

(b) Remove bolt, nut, and ground strap from engine support beam bracket (two ⁹⁄₁₆-inch open-end wrenches) (fig. 35).

(c) Remove cotter pins from nuts on engine support beam holding bolts (pliers). Then remove the eight nuts and bolts (four on each side) (two ⁹⁄₁₆-inch open-end wrenches) that hold the support beam to support brackets (fig. 35).

(20) **Attach Lifting Sling to Engine** (figs. 36 and 37).

Sling, engine lifting   Wrench, open-end, ⅞₆-inch.

(a) Remove bolts which secure bracket arms of engine breather (⁷⁄₁₆-inch open-end wrench) (figs. 20 and 37).

(b) Lay engine lifting sling on top engine. The two hooked cables at the rear pass down the rear of the engine inside the engine support beam, and hook to the engine support beam.

(c) Wrap the hooked cable on the front of the engine sling twice around the clutch spring housing hub, then hook the cable to the eye on the sling.

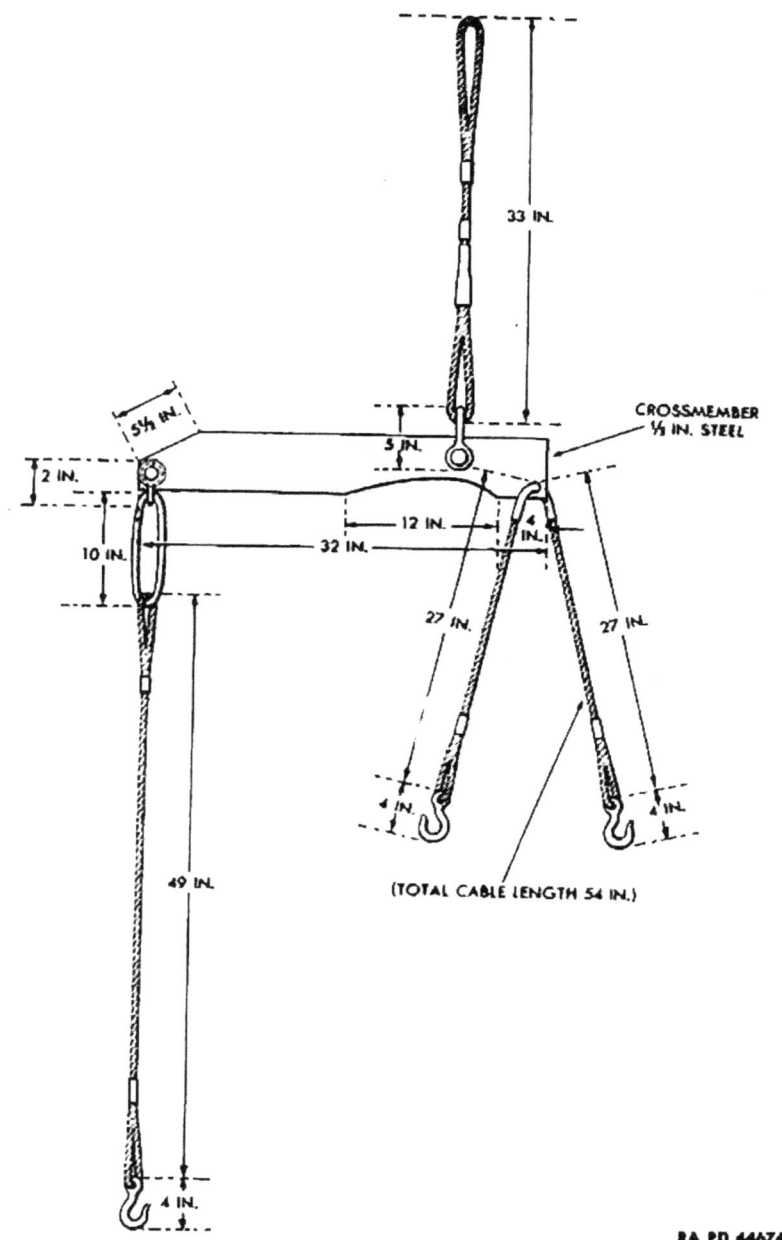

Figure 36. Engine lifting sling.

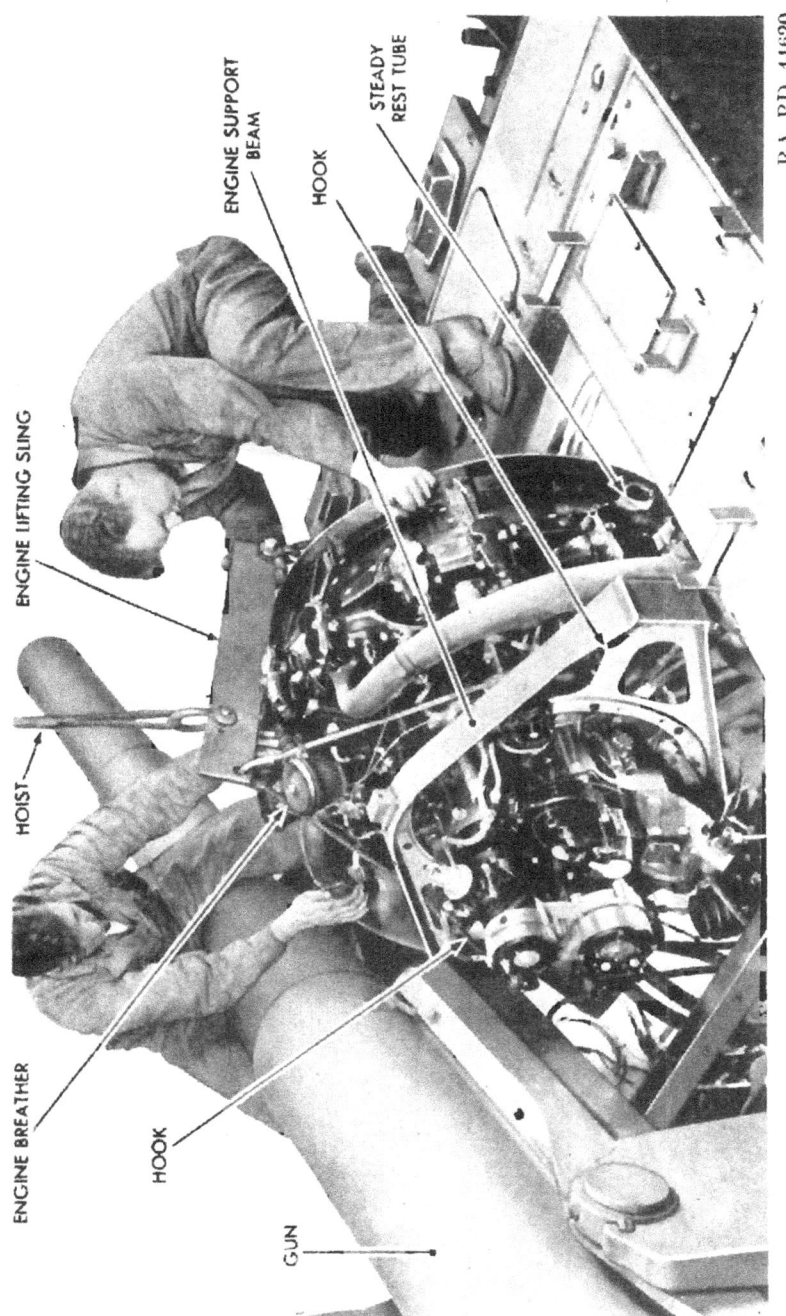

Figure 87. Engine partially lifted from engine compartment.

Figure 38. Engine lifted from engine compartment.

(21) **Lifting Out Engine** (figs. 37 and 38).
  Hoist   Wrench, socket, ¾-inch
  Stand, engine

(a) Elevate the gun fully.

(b) Remove the right-hand traversing stop screw from the pedestal mount and traverse the gun as far as it can go to the right.

(c) Attach hoist to lifting sling (fig. 37). Raise hoist sufficiently to lift engine vertically until the engine support beams are just free of their brackets, and the steady-rest adapters may be pulled out of the steady-rest brackets and steady-rest tube. *Caution:* When lifting the engine out of the compartment, it will be necessary to swing the engine slightly in a clockwise direction about its vertical axis, to prevent the accessories from striking the recoil mechanism.

(d) Carefully rock the front end of the engine downward, then turn the engine slightly clockwise so that the terminal shields on the starting motor and generator will clear the top of the rear engine compartment bulkhead when the engine is lifted out of the compartment.

(e) Lift the engine out very slowly. Be careful that the oil dilution valve does not strike against the recoil mechanism. (See note, par. 11c.) Be careful that the clutch companion flange does not catch on the four bolts projecting through the center of the front engine bulkhead. These hold the fire extinguisher bracket to the bulkhead and must be removed if they interfere with the removal of the engine. *Caution:* TAKE EXTREME CARE TO SEE THAT NO PART OF THE ENGINE CATCHES ON ANY PROJECTION IN THE ENGINE COMPARTMENT. In order to lift engine high enough to clear the hull, the boom of the wrecker can be raised.

(f) When the engine is clear of the hull, pull wrecker ahead and lower engine to the ground. Place pipe extensions into the steady-rest tube if needed, and secure properly. Remove lifting sling.

41. EXHAUST SYSTEM.

a. **Description.** The exhaust system consists of left and right exhaust manifold assemblies which lead down through elbows at the lower left and right side of the engine, just beneath the engine support beam. These each connect to flame arresters mounted in the floor of the fighting compartment, next to the rear engine bulkhead. Exhaust gas is expelled through two tail pipes connected to the flame arresters and projecting out the center side of the hull, one on each side of the vehicle.

b. **Maintenance.** If the exhaust manifolds or tail pipes become broken or damaged, replace with new ones.

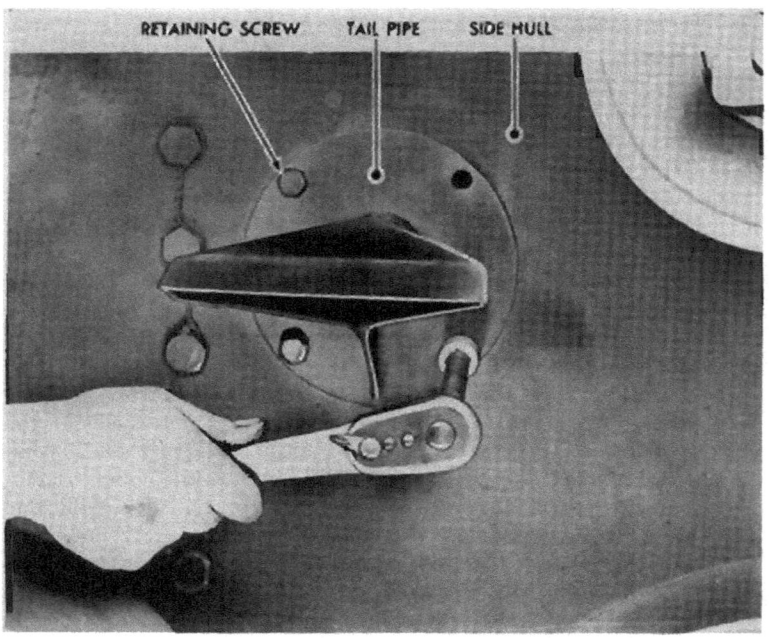

Figure 39. Tail pipe removal.

RA PD 44603

42. EXHAUST SYSTEM REMOVAL. The procedure for removal of either the left or right exhaust system is similar.

a. Equipment.

    Pliers   Wrenches, open-end, 7/16-inch (2)

    Pliers, diagonal cutting   Wrench, socket, 3/4-inch

b. Procedure.

(1) Remove Tail Pipe.

    Pliers, diagonal cutting   Wrench, socket, 3/4-inch

(a) Cut and remove wire which locks the four cap screws securing tail pipe to hull (diagonal cutting pliers).

(b) Remove the four cap screws (3/4-inch socket wrench) (fig. 39). Lift off tail pipe.

(2) Remove Exhaust Tube.

    Pliers   Wrenches, open-end, 7/16-inch (2)

(a) Remove cotter pin from nut which holds exhaust tube clamp to manifold and exhaust tube (pliers).

(b) Loosen exhaust tube clamp bolt (two 7/16-in open-end wrenches).

(c) Remove exhaust tube. Lift off exhaust tube clamp with bolt, nut and washers loosely installed.

(3) **Exhaust Manifold Removal.**
    Pliers, diagonal cutting   Wrench, socket, ⅝₆-inch
    (a) At each of the nine exhaust ports, cut and remove safety wires.
    (b) Remove two ⅝₆-inch castle nuts and washers from the exhaust port studs holding manifold flange at each exhaust port.
    (c) Slide the two sections of the exhaust manifold off the exhaust port studs and lift out between engine and support.
    (d) Remove gaskets.

43. EXHAUST SYSTEM INSTALLATION. The procedure for installation of either the left or right exhaust system is identical.
   a. Equipment.
       Pliers   Wrench, socket, ¾-inch
       Wrenches, open-end, ⅞₆-inch (2)
   b. Procedure.
(1) **Exhaust Manifold Installation.**
    Pliers, diagonal cutting   Wrench, socket, ⅝₆-inch
    (a) Put new copper gaskets in place on exhaust port studs.
    (b) Slide manifolds into place between engine and support and over exhaust port studs. Secure in place with castle nuts, and plain washers.
    (c) Install safety wire to secure nuts.
(2) **Install Exhaust Tube.**
    Pliers   Wrenches, open-end, ⅞₆-inch (2)
    (a) Place the exhaust tube clamp, with bolt, nut, and washer loosely installed, on the lower end of exhaust manifold.
    (b) Slip the exhaust tube inside the clamp. Do not tighten clamp until exhaust tube is also in position in flame arrester.
    (c) Tighten exhaust tube clamp bolt at manifold (two ⅞₆-inch open-end wrenches). Install cotter pin through bolt (pliers).
(3) **Install Tail Pipe.**
    Pliers   Wrench, socket, ¾-inch
    (a) Place tail pipe in position against side of hull. Line up holes in tail pipe with tapped holes in flame arrester.
    (b) Install four cap screws which secure tail pipe to hull and flame arrester (¾-inch socket wrench) (fig. 39). Lock wire screws securely in position (pliers).

44. OIL PUMPS (fig. 40). The supply of oil in the engine crankcase is maintained by an oil pump which draws oil from the engine oil tank located in the lower right front of the engine compartment. This is a gear type pump supported on the lower left side of the rear crankcase section. The pump is divided into two sections, a separate pressure pump and a main scavenger pump. In addition, there is a scavenger pump located in the front section of the engine. The pressure pump draws oil from the engine oil tank through an external line and a finger strainer into the rear section of the engine. After oil has lubricated internal engine parts, it drains off through passages into the engine oil sump located between the two bottom cylinders. The front section scavenger pump draws the oil from the lower end of the sump. The main scavenger pump draws oil from the upper section of the sump. Both return lines join together at Y-connection and continue on in a single line to the oil filter (fig. 85).

45. OIL PUMP REMOVAL (fig. 40).
a. Equipment.
Pliers   Wrench, open-end, 1⅛-inch.
Wrench, open-end, ⅞₆-inch.

b. Procedure.
(1) Disconnect inlet line at connector, and outlet line at Y-connection (1⅛-inch open-end wrench) Remove inlet and outlet lines from oil pump. Tie a cloth over the exposed end of each oil line.

(2) **Remove governor** (par. 64b(1)).

(3) Remove safety wire from nuts which hold oil pump to crankcase (pliers), and remove the nuts and plain washers (support pump with one hand while removing last nut) (⅞₆-inch open-end wrench). Remove oil pump and gasket from crankcase studs. Caution must be exercised to keep foreign substances out of crankcase during the operation. Cover opening in crankcase with a clean cloth.

46. OIL PUMP INSTALLATION (fig. 40).
a. Equipment.
Pliers   Wrench, open-end, ⅞₆-inch.
Wire, safety   Wrench, open-end, 1⅛-inch.

b. Procedure.
(1) Remove cloth from opening in crankcase.
(2) Install new gasket and oil pump on studs in crankcase.
(3) Install retaining nuts and plain washers on studs. Tighten and secure nuts with safety wire (⅞₆-inch, open-end wrench, pliers, and safety wire).
(4) Remove cloth covering exposed ends of oil lines. Connect inlet and outlet lines to oil pump (1⅛-inch open-end wrench).
(5) Install governor (par. 64 **b(2)**).
(6) Connect free end of inlet line at connector, and free end of outlet line at Y-connection (1⅛-inch open-end wrench).

47. OIL PUMP FINGERS TRAINER REMOVAL, CLEANING, AND INSTALLATION (fig. 40.) Perform this operation after every 15 hours of operation.
a. Equipment.
Solvent, dry-cleaning   Wrench, socket, ⅜-inch.
Wrench, open-end, ⅝-inch.

b. Procedure.

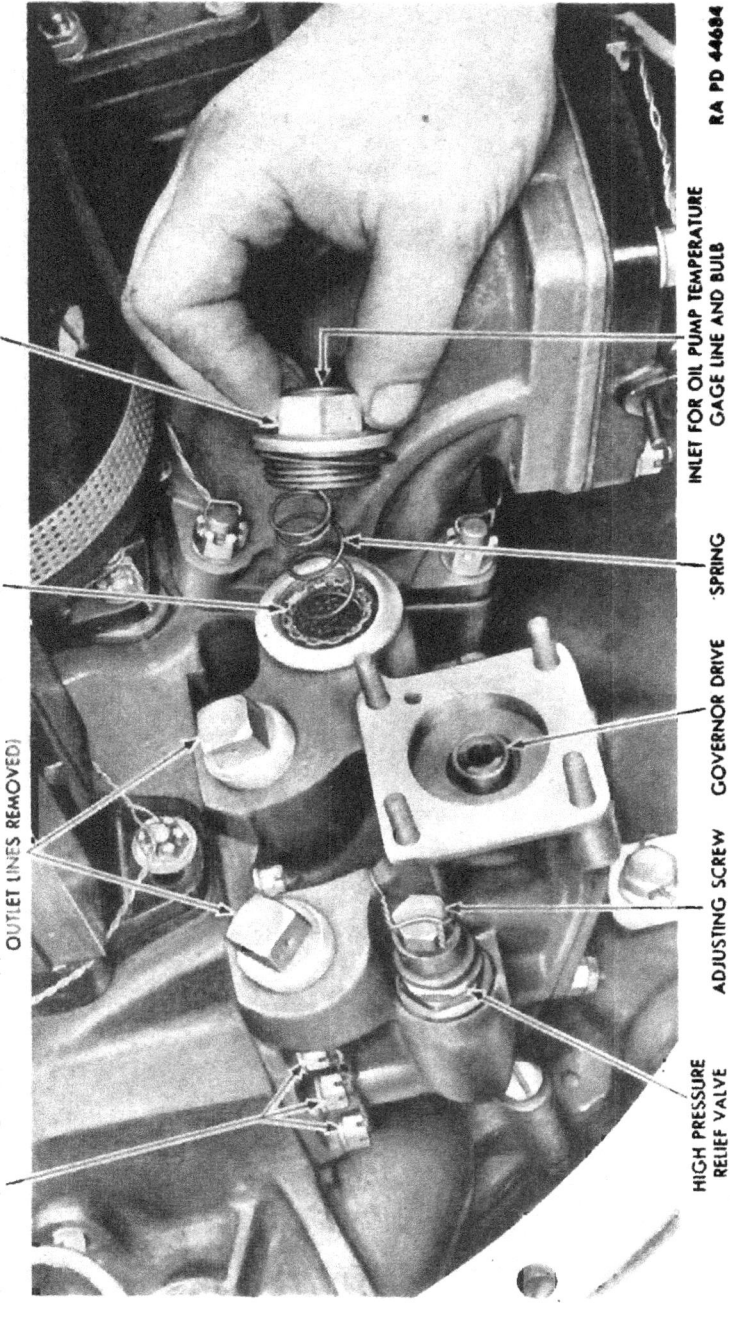

Figure 40. Engine oil pump and oil pump finger strainer (governor removed).

(1) **Removing Strainer.**
   Wrench, open-end, ⅜-inch   Wrench, socket, ⅜-inch.
   (a) Remove oil pump temperature gauge line and bulb (⅜-inch open-end wrench).
   (b) Remove reducing nipple and retainer plug together with copper gaskets for each, from finger strainer (⅜-inch socket wrench).
   (c) Remove spring and pull out strainer.

(2) **Cleaning Parts.**
   Solvent, dry-cleaning
   (a) Wash strainer in dry-cleaning solvent.
   (b) Clean spring and retaining plug.

(3) **Installing Strainer.**
   Wrench, open-end, ⅜-inch.
   (a) Install strainer, insert spring, and install copper gasket and retainer plug.
   (b) Install gasket and reducing nipple.
   (c) Install oil pump temperature gauge bulb and line. *Caution:* Be sure assembly is correctly seated and oiltight.

48. MAGNETOS. Dual ignition is furnished by two Bosch type MJT-9A-306 or Scintilla type VAG-9-DFA magnetos mounted on the crankcase (fig. 27). The right magneto fires front spark plugs and the left magneto fires rear plugs. The magnetos are equipped with automatic spark advance and shielded to prevent radio interference. Numbers on magneto ends of the ignition wires correspond to order in which they fire spark plugs, and not to the cylinder numbers. Thus the wire to cylinder No. 3 is stamped "2" as it is the second in the cylinder firing order (cylinder firing order, clockwise viewed at magneto end, is 1-3-5-7-9-2-4-6-8). Since Bosch and Scintilla magnetos require slightly different adjustments, they are treated separately (pars. 49 and 50).

49. MAINTENANCE OF SCINTILLA MAGNETO (figs. 41 and 42).

  a. **Equipment.**

  Disk, timing and pointer   Wrench, magneto
  Gauge, feeler   Wrench, open-end, ⅜-inch
  Indicator, top dead center   Wrench, open-end, ½-inch
  Pliers   Wrench, socket, ½-inch, with
  Pliers, diagonal cutting   universal extension
  Straightedge   Wrench, spark plug
  Tool KM-80292

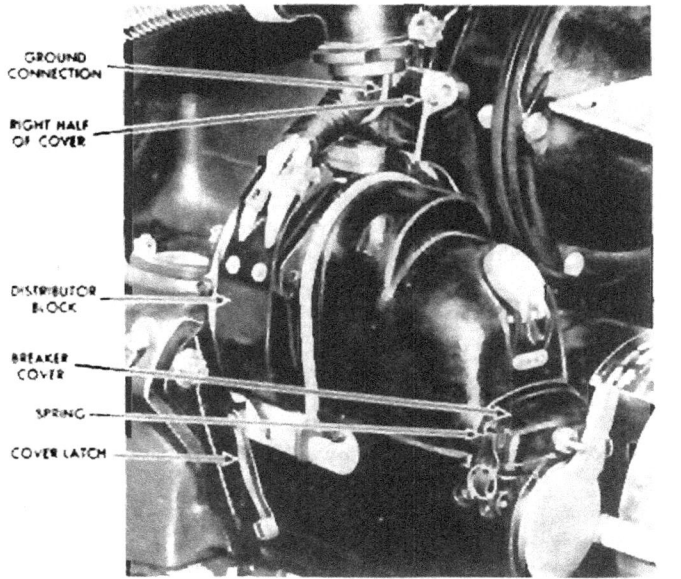

Figure 41. Magneto with left half of distributor cover removed.

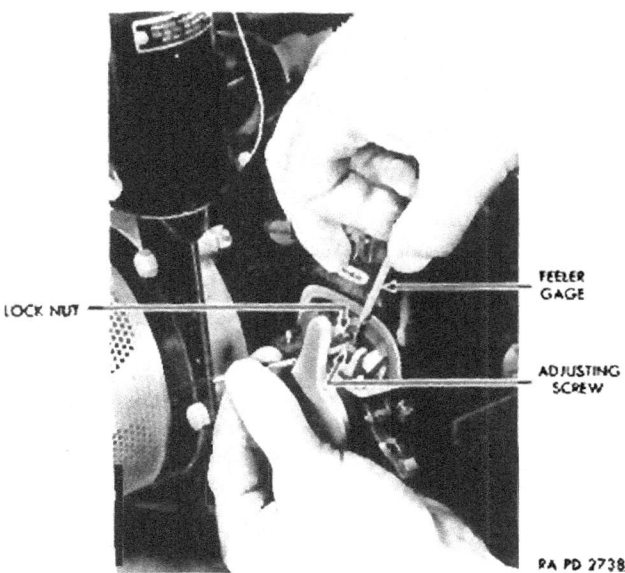

Figure 42. Breaker point gap adjustment on Scintilla magneto.

113

b. Procedure.
(1) Checking Magneto.

Pliers

(a) Check magneto and spark plug lead, by disconnecting lead from plug and holding it 1/4 inch from a positive ground while engine is running. If spark is evident, trouble is in plugs.

(b) If no spark is delivered, remove distributor blocks and check secondary by holding a grounded wire about 1/8 inch from distributor segments while rotating engine. Spark indicates magneto is satisfactory, and that fault is in wiring harness or plugs.

Note: Because of the inaccessibility of the front bank of spark plugs, the output of right magneto is checked at distributor segments only.

(2) Adjustment of Breaker Gap (fig. 42).

Gauge, feeler   Wrench, magneto

(a) Remove breaker cover by releasing safety ring on each side (fig. 39).

(b) Measure clearance of contacts when held wide open by the lobe of the cam (feeler gauge) (fig. 42). The engine will have to be turned over with a hand crank to open the magneto contacts to their widest position. Correct clearance is 0.012 inch. Adjust if necessary by following procedures:

  1. Loosen lock nut on adjusting screw (magneto wrench) (fig. 42).
  2. Adjust points by turning screw until there is 0.012-inch clearance between the contact points of the magneto.
  3. Tighten lock nut (magneto wrench).
  4. Install breaker cover.

(3) Replacing Breaker Mechanism (fig. 43).
With breaker cover off, breaker mechanism can be removed or replaced by squeezing breaker advance lever and latch lever together (fig. 43).

Note: Breaker points must be replaced when badly burned, when surfaces are rough or deeply pitted, or when points are worn down so close to the breaker arm that further dressing is impossible.

(4) Checking Scintilla Magneto Timing (engine out of vehicle).

Disk, timing and pointer   Tool KM-80292
Indicator, top dead center   Wrench, open-end, 3/4-inch
Pliers, diagonal cutting   Wrench, spark plug
Straightedge

(a) Remove spark plugs and rocker box covers from No. 1 cylinder (fig. 23).

(b) Install top dead center indicator in the rear No. 1 cylinder spark plug opening (fig. 44).

BREAKER ADVANCE LEVER   LATCH LEVER   RA PD 12503

Figure 43. Magneto breaker mechanism installation (Scintilla).

(c) Install timing disk (Tool KM-80292) and pointer on front end of crankshaft. A piece of wire, stiff enough to hold position, makes an excellent pointer (fig. 45).

(d) Find the dead center on No. 1 cylinder by the following steps (fig. 45):

1. Rotate crankshaft clockwise until top dead center indicator pointer comes up to one of the lines on dial of top dead center indicator (fig. 44). Mark line.
2. Mark location of wire pointer on timing disk at this point (fig. 45).
3. Continue rotation of crankshaft until pointer of top dead center indicator moves to its maximum point and returns to the line marked in step 2 above.
4. Mark location of wire pointer on timing disk (fig. 45) at this point.
5. Move crankshaft so that wire pointer points to the line on the timing disk midway between the two pencil marks on timing disk. No. 1 cylinder is now at top dead center. Move wire pointer until it points to zero on timing disk.

115

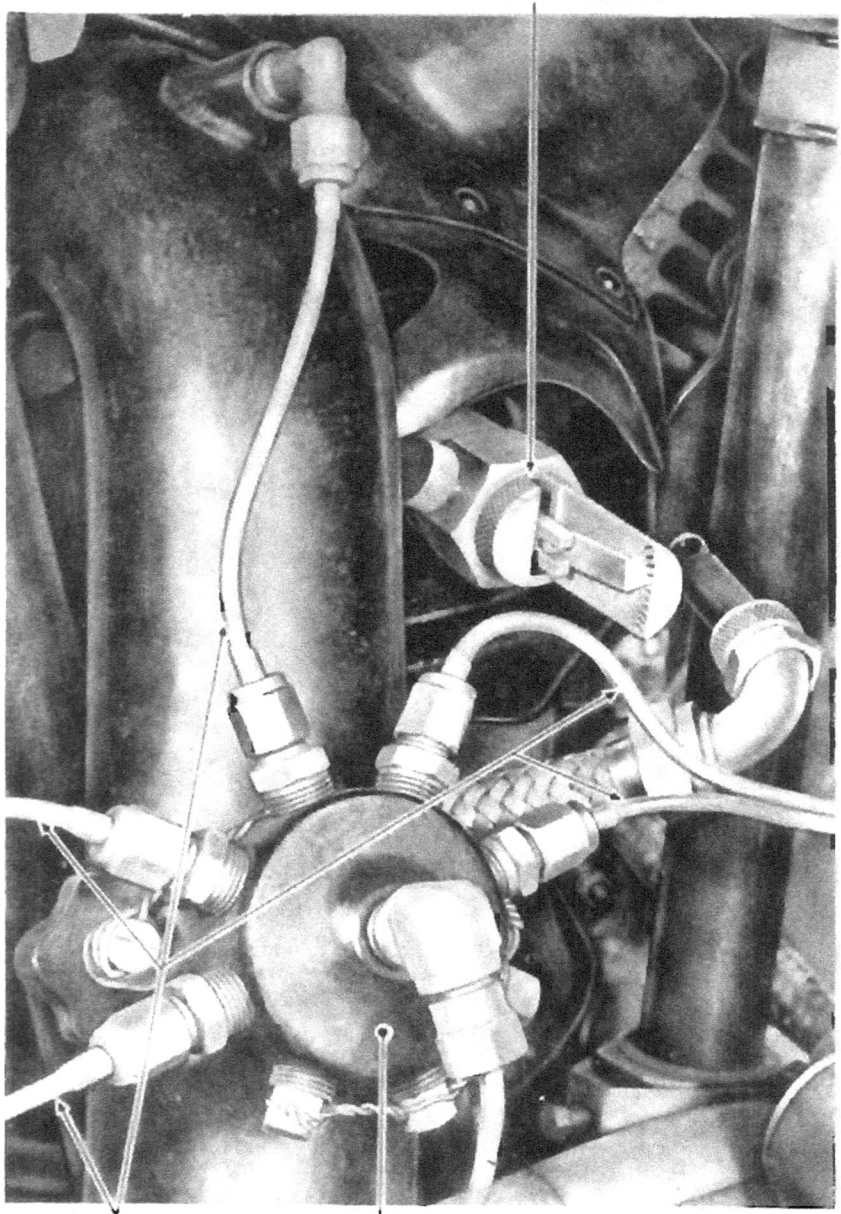

Figure 44. Checking piston location with top dead center indicator.

Figure 45. Use of timing disk.

(e) Move crankshaft until wire pointer points to 25°, before top dead center on compression stroke (fig. 45).

NOTE: Valve rocker arms will be free at 29° before top dead center on compression stroke.

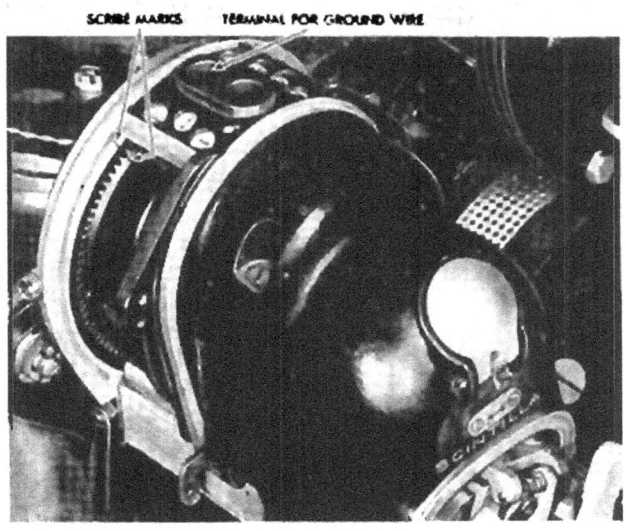

Figure 46. Scribe marks. 12448 RA PD

(f) Remove cover on magneto to be checked, by removing safety pins from cover latches at bottom, and by removing bolts and cap screws at top. Unscrew ground wire connection from top of magneto (fig. 47), also booster secondary wire if on right magneto. Remove the two distributor blocks from magneto.

(g) Check scribe marks on magneto housing and distributor gear. If magneto is properly timed, the single lines will be exactly in line, as illustrated (fig. 46). The double lines on the other side of the magneto will also be lined up. For accuracy, check with a straightedge. If marks are not exactly in line, loosen hold-down nuts slightly and tap magneto to right or left to secure perfect alinement.

NOTE: If timing operation is in connection with replacement of a magneto, line up scribe marks on magneto before installation, holding alignment during installation as illustrated (fig. 47).

(h) Tighten hold-down nuts securely, and wire nuts.

(i) Install magneto ground wire, distributor blocks, and covers, and install bolts, cap screws, and latch fastenings. Install booster secondary wire if on right magneto.

(5) **Field Method of Replacing One Scintilla Magneto** (fig. 47) (engine in vehicle when crankshaft has not been turned).

    Pliers, diagonal cutting   Wrench, socket, ½-inch, with universal extension
    Straightedge
    Wrench, open-end, ½-inch

(a) Before starting to remove a magneto, remove distributor blocks ((4) (f) above) and turn crankshaft until scribe marks on magneto housing and distributor gear line up. Cut locking wires.

(b) Remove hold-down nuts which secure magneto to crankcase.

(c) Hold magneto in place while removing the last nut, and then lift off magneto and gasket.

(d) Check magneto to be installed for proper 0.012-inch gap, cleanliness of points, and for timing ((1) (b) above).

(e) Line up scribe marks in magneto to be installed ((4) (g) above) and, holding marks securely in line with thumb, install gasket and magneto (fig. 47).

(f) Install hold-down nuts, and, before tightening them all the way down, check scribe marks with a straightedge. *Caution:* Lining scribe marks up by eye is not sufficiently accurate. To secure exact alignment, tap magneto to left or right. Then partially tighten hold-down nuts.

(g) Install distributor blocks and covers ((4) (i) above).

NOTE: If engine has been turned over while old magneto was off, the correct position of crankshaft can be determined by removing blocks of other magneto on engine, and turning engine over with crank until scribe marks line up.

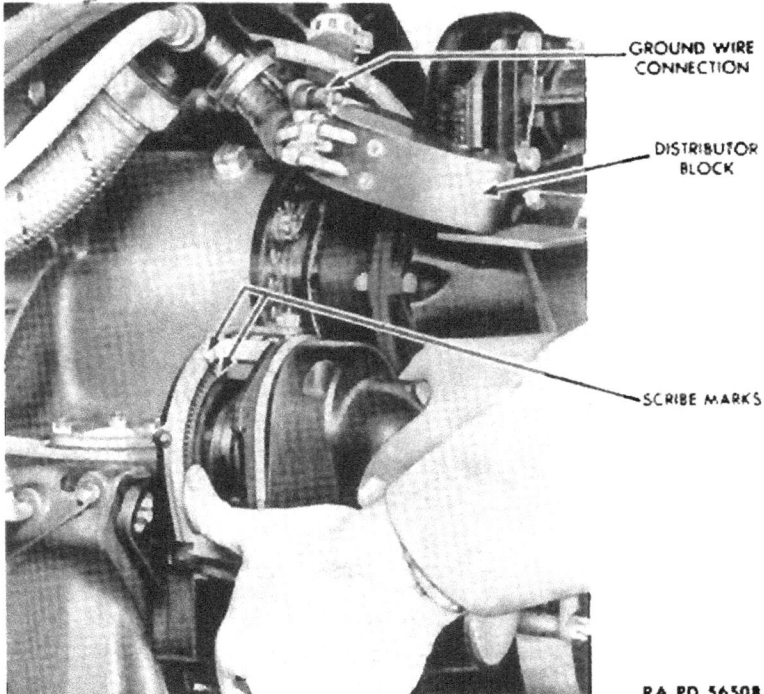

Figure 47. Holding alignment of scribe marks while installing magneto.

(6) **Emergency Method of Installing Scintilla Magnetos on Engine in Field When Position of No. 1 Piston is Unknown Due to Engine Having Been Moved With Both Magnetos Removed.**

(a) Remove rear spark plug in No. 1 cylinder and use top center indicator (or screw driver if no indicator is available). Remove rocker box covers from intake and exhaust rocker boxes on No. 1. Find top dead center of compression stroke on No. 1 (fig. 44) (both valves will be closed).

(b) Check magneto for proper point gap (0.012 inch), cleanliness of points.

(c) Line up scribe marks on magneto distributor gear and magneto housing and turn distributor gear in clockwise direction until points just start to break. This will be about 3½ to 4 teeth past alignment of scribe marks. (Distributor gear rotates in same direction as crankshaft, clockwise, when magneto is installed on accessory case.)

(d) Hold distributor gear with thumbs, with points just starting to open, and install magneto on accessory case. Tap magneto slightly to right or left after installing to obtain exact position with points just breaking.

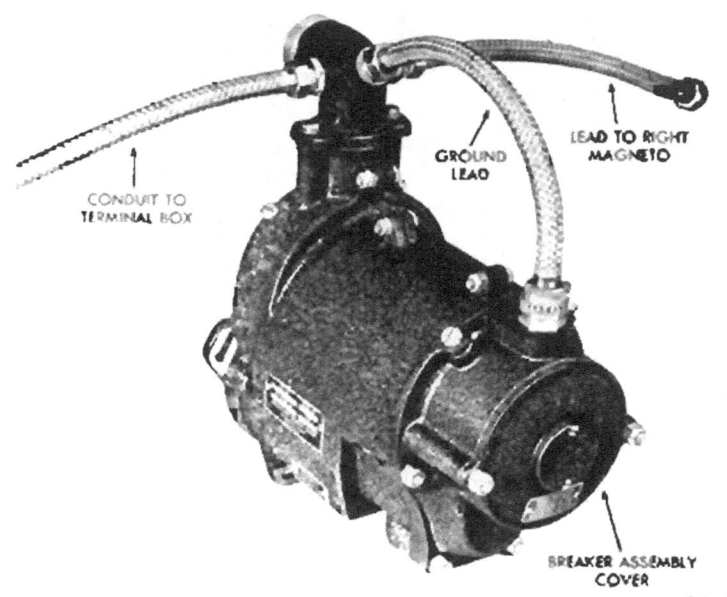

Figure 48. Bosch magneto.

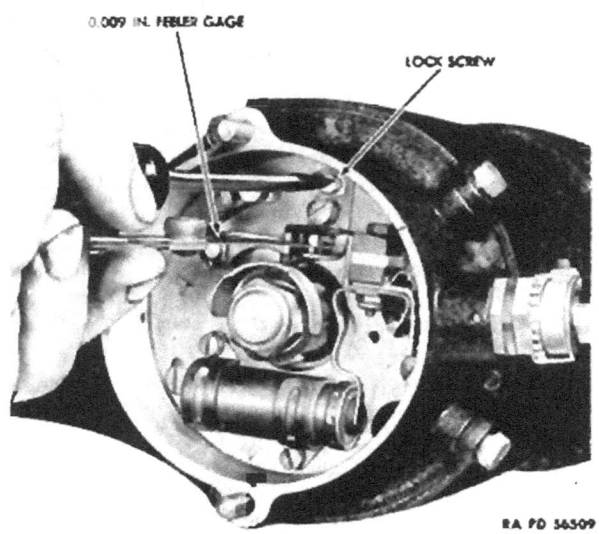

Figure 49. Breaker point gap adjustment on Bosch magneto (before installation).

(e) Tighten hold-down nuts and install lock wire; install blocks, covers, and clamps.

(f) Other magneto may be installed in same manner, or, before replacing blocks on first magneto, turn engine over until scribe marks line up (approximately two revolutions). Then line up scribe marks on other magneto and install it as in field method.

Note: Magnetos installed in this manner will be a degree or two out of time, but will serve to run engine until such time as engine can be removed and magnetos are properly timed with a timing disk.

50. MAINTENANCE OF BOSCH MAGNETO.

a. Equipment.

Disk, timing and pointer
Gauge, feeler, 0.009-inch
Gauge, feeler, 0.0015-inch
Indicator, top dead center
Pliers, cutting
Screw driver
Socket, universal, with extension and handle, ½-inch

Straightedge
Wrench, open-end, ½-inch
Wrench, open-end, ⅝-inch
Wrench, open-end, ½-inch
Wrench, spark plug

b. Procedure.

(1) **Description.** The Bosch magneto furnished differs from the Scintilla magneto in method of adjusting breaker gap and in procedure of checking and adjusting timing.

(2) **Adjusting Breaker Point Gap** (fig. 49).

Gauge, feeler, 0.009-inch Wrench, open-end, ½-inch
Screw driver

(a) Remove two self-locking nuts that hold breaker assembly cover to housing, and remove cover (fig. 48).

(b) Rotate the cam (by hand if magneto is off engine, or by using hand crank if magneto is installed on engine) until maximum breaker point opening is reached. Correct clearance is 0.009-inch. Adjust clearance if necessary by following procedure:

   1. Loosen screw that holds breaker point bracket to housing and, with 0.009-inch feeler gauge between points, set gap and tighten screw (fig. 49).
   2. Rotate cam to next maximum breaker point opening and check gap. Readjust if necessary.

(3) **Checking Bosch Magneto Timing.**

Disk, timing and pointer
Gauge, feeler, 0.012-inch
Gauge, feeler, 0.0015-inch
Indicator, top dead center
Pliers, cutting

Screw driver
Straightedge
Wrench, open-end, ⅝-inch
Wrench, spark plug

121

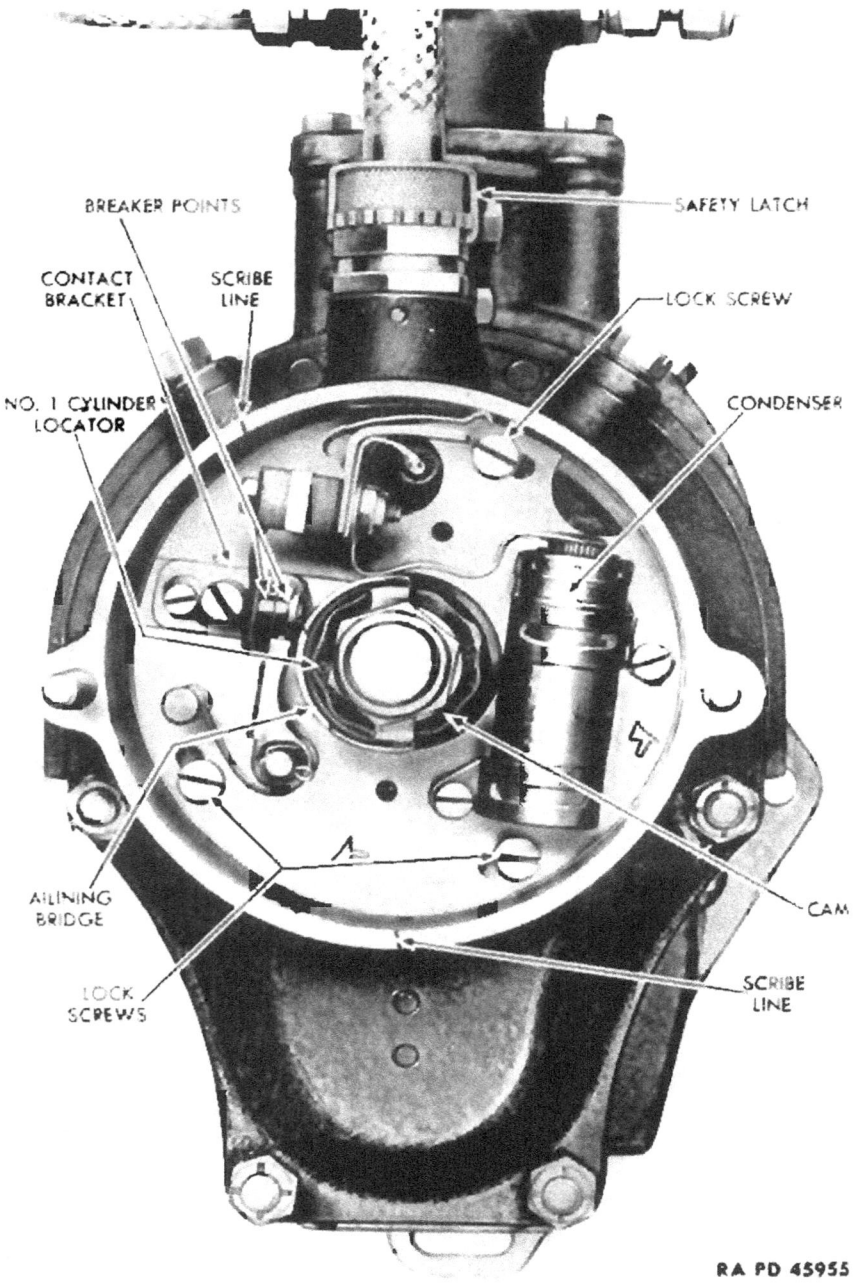

Figure 50. Breaker mechanism of Bosch magneto.

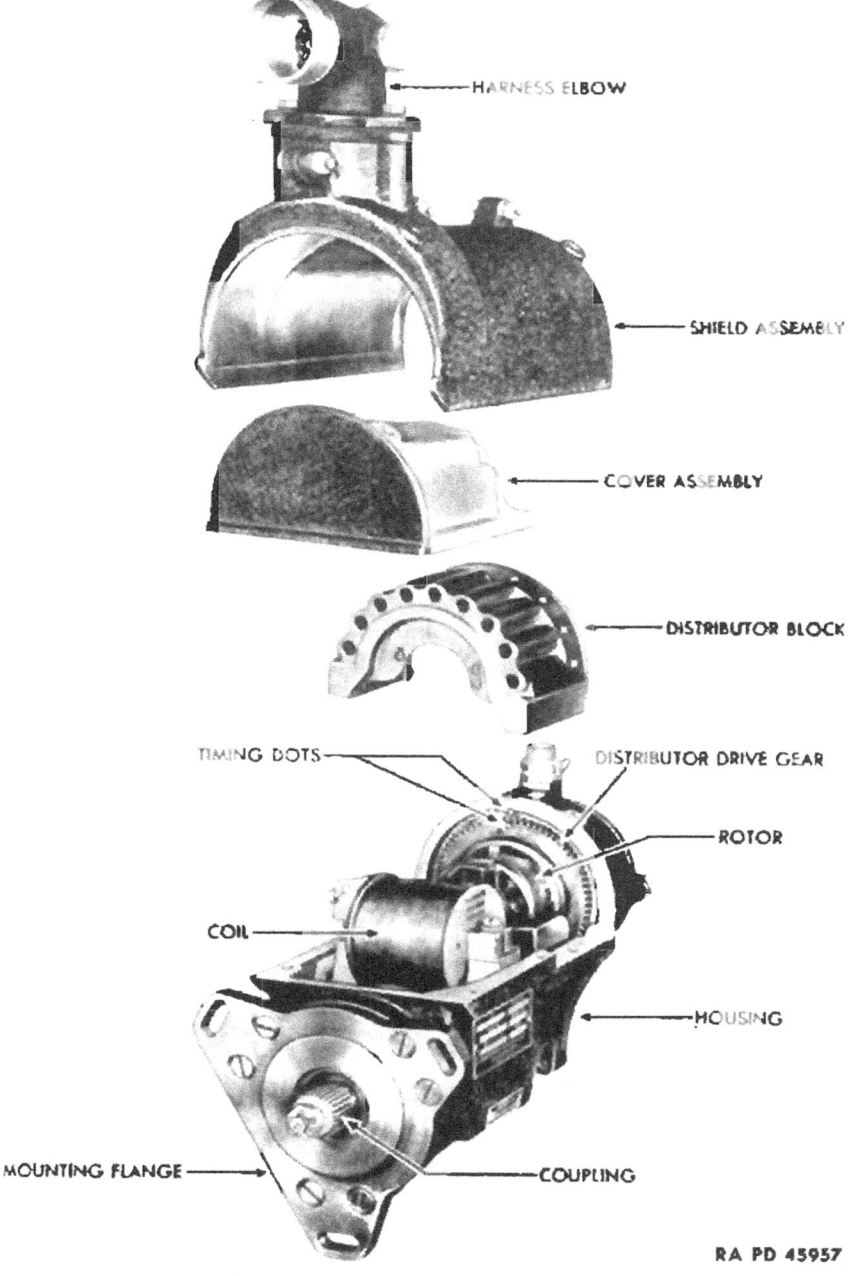

Figure 51. Parts of the Bosch magneto.

Crank engine until No. 1 piston is at 4° before top dead center on compression stroke. Procedure for timing Bosch magneto is the same as for Scintilla (par. 50).

NOTE: Valve rocker arms will be free at 5° before top dead center on compression stroke.

Caution: If magnetos have been changed from Scintilla to Bosch or vice versa, it will be necessary, after retiming the engine, to replace the engine name plate with a new one. New plates are available which will be completely stamped with the exception of the engine serial number, which must be stamped on the plate at the time the change is made. Care must be exercised to be certain that no errors are made in the serial number when stamping the name plate.

(4) **Installing Magneto on Engine** (fig. 51).

Screw driver
Socket, universal, with extension and handle, ½-inch
Wrench, open-end, ⁷⁄₁₆-inch
Wrench, open-end, ½-inch

(a) Remove four fillister head bolts that hold housing cover to housing, and remove cover.

(b) Turn rotor until white timing dot on distributor drive gear lines up with red dot on housing.

(c) Rotate engine until No. 1 piston is at top dead center as shown by screw driver or top dead center indicator installed through spark plug opening (fig. 42).

(d) Holding distributor gear from turning with fingers of left hand, install magneto in place, tipping it to right or left if necessary, to engage splines on magneto drive.

(e) Install and tighten magneto retaining nuts on studs by hand. Check alignment of timing dots and move magneto to right or left, if necessary, to correct alignment. Tighten outside nuts.

(f) Remove field coil cover, install distributor block in place, and replace field coil cover. The field coil cover must always be removed before removing or replacing distributor block.

(g) Separate housing cover into its two parts, and install halves in place on housing. Install three bolts joining two halves, and install and tighten harness elbow to cover.

(h) Check timing, as outlined in (3) above.

51. SPARK PLUGS.

a. General. BG or Champion 63S spark plugs are used, and are regular aircraft type, and radio shielded. When a spark plug is removed from an engine due to suspected trouble, it should be inspected thoroughly before disassembly. If the electrodes are heavily coated with carbon, the compression of the cylinder should be checked to determine whether the piston rings are worn or stuck, therefore allowing oil to pass. A coating of fresh oil also indicates that the spark plug may not have been firing. Check the interior of the intake pipes for possible oil leakage past the impeller shaft oil seal. If the spark plug electrodes are free from carbon but discolored or burned, and if they have been running hot, detonation from a poor grade of fuel, or operating under excessive loads may have been the cause. The spark plug gap should be checked. The core should be inspected for tightness, and the ignition wire and terminals should be inspected for failure. If the plug is disassembled, the core should be inspected for defects in the mica insulation. The plug should be replaced if the insulation is found to be broken, flaked, or dented, or if any mica laminations project beyond adjacent laminations, or if contacts are pitted or burned.

b. Locating Faulty Spark Plugs.

(1) To determine whether it is a front or rear plug, switch from one magneto to the other, noting uneven engine operation.

(2) Locate faulty plug by running engine at not less than 1,800 revolutions per minute on faulty bank long enough for stack on inoperative cylinder to cool (about 1 minute). With engine at normal operating temperature, locate cool stack by putting drops of oil on stacks and noting stack on which oil fails to disappear.

c. Spark Plug Removal.

    Wrench, open-end, $\frac{5}{8}$-inch   Wrench, socket, $\frac{11}{16}$-inch
    Wrench, open-end, $\frac{3}{4}$-inch   Wrench, socket, 1-inch

(1) At time of 50-hour check, or whenever spark plug replacement is found necessary, unserviceable plugs will be replaced by serviceable plugs.

(2) With engine compartment top guards and plate removed (par. 40b(2) and (3)), remove cable connection from spark plug and remove plug from cylinder recess.

(3) Remove carburetor inspection plate from bottom of hull under engine compartment.

(4) To facilitate removal of front spark plugs, remove inspection cover plates on fan cowling.

Note: Use of a socket wrench with a universal joint extension facilitates removal of certain plugs difficult to remove with an ordinary socket wrench. Five plugs may be removed from the front and five from the rear of the engine by reaching down from the top of the compartment. Remaining plugs can be reached through opening in the hull beneath engine.

d. Spark Plug Installation.

    Wrench, open-end, $\frac{5}{8}$-inch   Wrench, socket, $\frac{11}{16}$-inch
    Wrench, open-end, $\frac{3}{4}$-inch   Wrench, socket, 1-inch

(1) When installing spark plugs in engine, do not use wrenches with handles more than 10 inches long. Too much force used in tightening plugs in engine cylinders may distort certain sections of the plug. Solid copper gaskets 0.095 inch thick are to be used. Tighten spark plugs to 40 foot-pounds of torque if a tension wrench is available.

(2) Apply a thin coating of compound, antiseize, mica base, to all plug threads and install plugs in cylinder recess. Connect cable connections.

(3) Install carburetor inspection plate.

52. IGNITION HARNESS (fig. 52).

a. Description. All parts of the ignition wiring system are shielded. Shielding is constructed so that if any high-tension wire is damaged, the entire ignition harness must be replaced. Each conduit end is equipped with an elbow which attaches directly to the shielded spark plug and facilitates easy removal of the harness from the spark plugs.

b. Ignition Harness Removal.

(1) Remove Intake Pipes.

Wrench, lug, intake pipe packing nut   Wrench, open-end, $\frac{7}{16}$-inch

(a) Loosen the intake pipe gland nut and turn it out all the way (intake pipe packing nut lug wrench).

(b) Remove the three cap screws ($\frac{7}{16}$-inch open-end wrench), holding the intake pipe flange to the cylinder intake port. Pull pipe away from intake port and lift out.

(c) Remove packing from intake manifold port.

(2) Remove Air Deflector Disks.

Screwdriver

(a) Remove two screws which hold each air deflector disk to the intercylinder air deflector.

(b) Slide the air deflector disk and grommet back on each conduit.

(3) Disconnect Harness from Engine.

Wrench, open-end, $\frac{5}{8}$-inch   Wrench, socket, $\frac{1}{2}$-inch

(a) Disconnect each conduit at the cylinder ($\frac{5}{8}$-inch open-end wrench). Pull out the plug type connection.

(b) Remove two nuts on each of four clips which hold the harness shield to the crankshaft main section ($\frac{1}{2}$-inch socket wrench).

(4) Remove Harness from Magneto.

Screw driver   Wrench, socket, $\frac{3}{8}$-inch

(a) Remove two nuts, front and rear, on the top of the distributor block radio shield ($\frac{3}{8}$-inch socket wrench).

Figure 52. Ignition harness (exhaust manifold removed).

(b) Remove two screws located near the top (one on each side) of the distributor block radio shield (screw driver).

(c) Unfasten the two clips at the bottom of the distributor block radio shield and separate the two halves of the shield.

(d) Remove the distributor block assembly.

(e) Lift the entire harness assembly out over the top of the engine.

c. Maintenance and Inspection.

Brush, wire   Solvent, dry-cleaning
Carbon tetrachloride

(1) Electrical wiring, shielding, and conduits require frequent inspection and checking.

(2) Replace the ignition harness if the shielding is crushed. Clean oily or dirty spark plug shields or shielding fittings, and tighten all coupling nuts. Replace the harness if the high-tension wires have become oil-soaked.

(3) In cleaning couplings or spark plug shields, use a solution of carbon tetrachloride. If carbon tetrachloride is not available use dry-cleaning solvent. After cleaning and drying, clean the threads of each coupling with a small wire brush to remove high-resistance oxidation which sometimes forms on the inside of aluminum couplings.

d. Conduit Removal from Harness.

Wrench, leather strap

(1) Loosen connection between conduit and body of harness.

(2) Break connection at spark plug and remove conduit.

e. Ignition Harness Installation.

(1) Replacing Harness on Magneto.

Screw driver   Wrench, socket, ⅝-inch.

(a) Beginning with the plug connection end, work the harness down through the engine and pull the distributor block end through the support to the magneto.

(b) Replace the blocks in position and clamp the two halves of the distributor block radio shield in place.

(c) Replace the two screws holding the shield together.

(d) Replace the two nuts, front and rear, on the top of the distributor block radio shield, securing the harness to the shield.

(2) Secure Harness to Engine.

Wrench, open-end, ⅝-inch.   Wrench, socket, ½-inch.

(a) Place the four clips holding the harness to the crankshaft main section over the studs and turn down the nuts (½-inch socket wrench.)

(b) Insert the plug-type connections at each cylinder and tighten the retaining nut (⅝-inch open-end wrench).

(3) **Replace Air Deflector Disks.**
Screw driver

Slide the air deflector disk and grommet along each conduit and into place on the intercylinder air deflector, securing each disk with two screws.

(4) **Install Intake Pipe.**

Wrench, lug, intake pipe packing nut   Wrench, open-end, $\frac{9}{16}$ inch.

(a) Insert intake pipe into intake manifold port. Install cap screws through flange at top of pipe. Install new gasket over the screws and turn screws into cylinder finger tight.

(b) Make sure that intake pipe is centered in intake manifold. With the intake packing nut, push the packing down into the intake manifold. Turn the packing nut down about two threads, being very careful to see that it is not cross threaded on the aluminum manifold.

(c) Tighten down intake pipe flange cap screws.

(d) With intake pipe packing nut wrench, tighten nut, being careful to avoid excessive pressure which will distort the packing and cause an air leak. When packing nut is correctly tightened, the shoulder should be approximately flush with the manifold housing.

f. Conduit Installation in Harness.
Wrench, leather strap

(1) Connect new conduit at spark plug.

(2) Tighten connection which attaches conduit to body of harness.

53. BOOSTER COIL.

a. Description. A booster coil is provided for use with the Scintilla magneto to supply an intense auxiliary spark across the points of the spark plugs to facilitate engine starting. The booster coil switch is located on the instrument panel (fig. 13), and is used to connect the booster coil at starting. The right magneto is used to distribute the booster coil spark to both sets of spark plugs.

b. Equipment.
Pliers   Wrench, open-end, $\frac{3}{8}$-inch
Screw driver   Wrench, open-end, $\frac{3}{4}$-inch

c. Procedure.

(1) **Booster Coil Removal.**
Pliers   Wrench, open-end, $\frac{3}{8}$-inch
Screw driver   Wrench, open-end, $\frac{3}{4}$-inch

(a) At time of 100-hour check, or whenever booster coil replacement is found necessary, the unserviceable unit will be replaced with a serviceable one. Disassembly, servicing, and adjustment will be done only by ordnance personnel.

FIGURE 53. — *Removing booster coil retaining screws.*

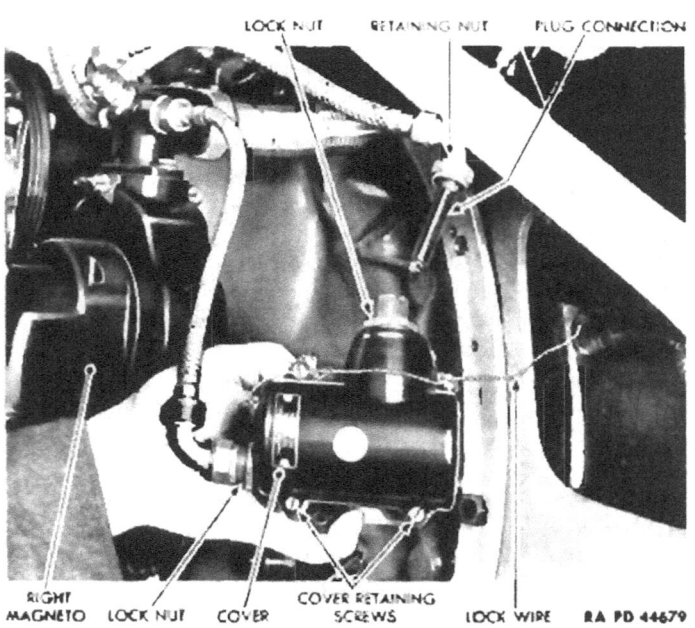

FIGURE 54. — *Booster coil and leads.*

130

Figure 55. *Booster coil, cover removed, showing low-tension lead to rear terminal box.*

(b) Remove screws and washers which hold booster coil to engine support beam (screw driver) (fig. 53).

(c) Loosen lock nut (⅝-inch open-end wrench) and remove cable with shielded connection which plugs in top of booster coil (pliers) (fig. 54).

(d) Remove locking wire from screws which hold booster coil cover to body (pliers) (fig. 54). Remove screws and lift off cover (screw driver) (figs. 54 and 55).

(e) Disconnect lead wire inside body from rear terminal box (⅝-inch open-end wrench) (fig. 55). Loosen lock nut (¾-inch open-end wrench) and remove cable with shielded connection from end of booster coil, and pull out lead wire.

(2) **Booster Coil Installation.**
    Pliers   Wrench, open-end, ⅝-inch
    Screw driver   Wrench, open-end, ¾-inch

(a) Pass lead wire from rear terminal box through cable with shielded connection. Install lead wire on lower terminal inside booster coil (⅝-inch open-end wrench) (fig. 55).

(b) Place booster coil cover on body and install screws (screw driver) (figs. 54 and 55). Install lock wire on all screws (pliers) (fig. 54). Attach shielded cable to booster coil (pliers), and tighten lock nut (¾-inch open-end wrench).

(c) Install cable with shielded connection which plugs in top of booster coil (pliers). Tighten lock nut (⅝-inch open-end wrench).

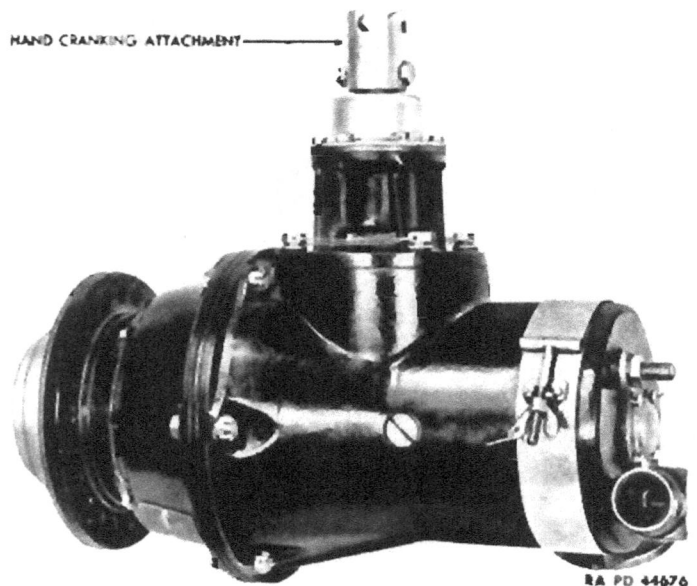

Figure 56. Starter.

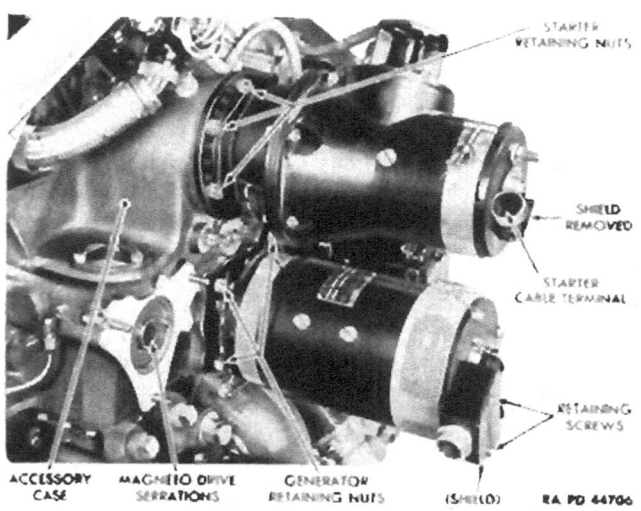

Figure 57. Left magneto removed to show retaining nuts for starter and generator.

(d) Place booster coil in position on engine support beam and install attaching screws (screw driver).

54. STARTER (figs. 56 and 57).
a. Equipment.
Pliers
Pliers, diagonal cutting
Screw driver
Wrench, open-end, 11/32-inch
Wrench, open-end, 1/2-inch
Wrench, open-end, 9/16-inch
Wrench, socket, 9/16-inch

b. Procedure.
(1) **Description.** A 24-volt electric starter is used on this engine. Provision is made at top of the starter for turning the engine over by hand cranking (fig. 56). To connect the starter to the batteries, a solenoid-operated starter switch is located on the hull inside the battery compartment (fig. 138). It is operated by a switch on the instrument panel.

(2) **Starter Removal.**
Pliers, diagonal cutting
Screw driver
Wrench, open-end, 11/32-inch
Wrench, open-end, 1/2-inch
Wrench, open-end, 9/16-inch
Wrench, socket, 9/16-inch

(a) If starter fails to operate properly, it must be replaced with a serviceable starter.

(b) See that battery switch is open (fig. 14).

(c) Remove terminal shield (screw driver) and disconnect cable from starter switch (11/32-inch open-end wrench).

(d) Remove cotter pins and eight bolts and nuts which secure generator and starter support bracket to bands on generator and starting motor (pliers and 1/2-inch open-end wrench) (fig. 21).

(e) Remove locking wire or cotter pins and nuts which hold starter to crankcase (9/16-inch open-end wrench and 9/16-inch socket wrench) (fig. 57). Remove starter and gasket from studs.

(3) **Starter Installation.**
Pliers
Screw driver
Wrench, open-end, 1/2-inch
Wrench, open-end, 9/16-inch
Wrench, socket, 9/16-inch

(a) Install new starter and gasket in place on studs and install nuts which hold motor to accessory case (9/16-inch open-end wrench and 9/16-inch socket wrench) (fig. 57).

(b) Install cotter pins on nuts or install locking wire (pliers).

(c) Place generator and starter support bracket in position (fig. 21). Install bolts and nuts (1/2-inch open-end wrench). Install cotter pins (pliers).

(d) Install starter cable (9/16-inch socket wrench) (fig. 57).

(e) Install terminal shield (screw driver) (fig. 57).

Figure 58. Generator.

(4) **Lubrication of Starter.** All starters are sufficiently lubricated at time of issue, and should not require lubrication between major engine overhaul periods.

55. GENERATOR (figs. 57 and 58).

a. Equipment.

Pliers, diagonal cutting Wrench, box, ½-inch
Screw driver Wrench, socket, ⅝-inch

b. Description. The generator is flange-mounted on the accessory case below the starter and is held in place by four studs and nuts (fig. 57). The generator regulator described in paragraph 152, controls the current output of the generator. A generator filter is provided in the generator-battery circuits to help eliminate radio interference.

c. Lubrication. Generators are properly lubricated at engine overhaul periods and should not require additional lubricant between overhaul periods.

d. Procedure.

(1) **Generator Removal.**

Pliers, diagonal cutting Wrench, box, ½-inch
Screw driver Wrench, socket, ⅝-inch

(a) Remove generator terminal shield from generator (screw driver) and disconnect generator cable from generator cable terminal (⅝-inch socket wrench).

(b) Remove lock wire and four nuts which hold generator to accessory case (diagonal cutting pliers and ½-inch box wrench) (fig. 57).

(c) Loosen generator retaining bracket nuts (fig. 21), slide generator off studs (½-inch box wrench), and remove gasket.

(2) **Generator Installation.**
    Pliers   Wrench, box, ½-inch
    Screw driver   Wrench, socket, ⅜-inch

(a) Place gasket and generator over studs, and tighten retaining bracket nuts (½-inch box wrench) (fig. 21).

(b) Install four nuts that hold generator to accessory case (½-inch box wrench) (fig. 57). Wire all nuts securely (pliers).

(c) Connect generator cable to generator cable terminal (⅜-inch socket wrench) and install terminal guard (screw driver).

56. FUEL PUMP (fig. 59).

a. Equipment.
    Wrench, open-end, ½-inch   Wrench, open-end, ⅜-inch

b. Description. The fuel pump is mounted on the engine accessory drive housing to right of generator. The pump is mounted on a base installed on the housing.

c. Maintenance. At daily inspection, check for leakage at joints and fittings. If there is doubt about pump operation, disconnect inlet line at carburetor, hold it over an open container, and turn engine over with starter. If no fuel is delivered at inlet line, pump is defective and must be replaced.

d. Procedure.

(1) **Fuel Pump Removal** (fig. 59).
    Wrench, open-end, ½-inch   Wrench, open-end, ⅜-inch

(a) Close fuel tank valves.

(b) Disconnect fuel lines at each side of pump. Rotate T-fitting on right side of pump back slightly to break connection (⅜-inch open-end wrench).

(c) Disconnect fuel lines at elbow above carburetor screen. Rotate elbow to left, slightly, to break line connection at fuel pump (⅜-inch open-end wrench). Remove four palnuts, nuts, and washers which hold pump to accessory drive housing, and move pump off studs (½-inch open-end wrench). Remove gasket.

(2) **Fuel Pump Installation** (fig. 59).
    Wrench, open-end, ½-inch   Wrench, open-end, ⅜-inch

(a) Inspect new pump to see that connections on pump come out over long sides of base. If necessary, turn pump on base by removing four screws on underside of base, and locate pump in correct position.

(b) Place gasket over the four studs on the accessory case.

Figure 59. Fuel pump removal.

(c) Install base of pump on studs; install and tighten four nuts, palnuts, and washers which hold pump to accessory drive housing (½-inch open-end wrench).

(d) Remove the shaft seal drain plug on the bottom of the pump.

(e) Connect fuel lines at each side of pump and at elbow (⅞-inch open-end wrench).

(f) Open fuel tank valves.

57. CARBURETOR (figs. 60 and 61).

a. Description. The carburetor is attached directly to the intake manifold in the bottom of the engine crankcase.

b. Maintenance. Once the carburetor is properly installed, very little attention is needed between major engine overhauls. A fuel inlet screen is located in the rear left center of the carburetor where the bypass line leaves the carburetor, and may be reached by removal of the large square-head plug at the bottom of carburetor. A small square-head plug is provided as a drain in the bottom of each float chamber. Remove strainer and drain plugs frequently to remove accumulated dirt. Shut off fuel supply before removing plugs. Inspect entire carburetor to see that all parts are tight and properly safety-wired. In case of faulty operation that cannot be cured by adjusting the slow idle, replace carburetor. No repairs are to be made by using arms.

c. Equipment.
    Pliers   Wrench, open-end, 13/16-inch
    Solvent, dry-cleaning

d. Procedure.

(1) **Cleaning of carburetor inlet screen** (fig. 62).
    Pliers   Wrench, open-end, 13/16-inch
    Solvent, dry-cleaning

(a) Close fuel tank valves.

(b) Remove locking wire from nut on bottom of screen (pliers).

(c) Remove nut and lead gasket that retain screen (13/16-inch open-end wrench) (fig. 62).

(d) Wash screen with dry-cleaning solvent.

(e) Install screen with gasket (13/16-inch open-end wrench), taking care not to turn retaining nut too tight, since excessive tension may spread or injure lead gasket. Install locking wire (pliers).

(f) Open fuel tank valves. Remedy all leaks.

(2) **Carburetor Adjustment** (fig. 61). After a new carburetor is installed, it should be adjusted to the engine. Two adjustments are provided, one for idling speed and the other for mixture quality. Throttle adjustment on the carburetor is set so that engine idles between 300 to 400 revolutions per minute (tachometer reading).

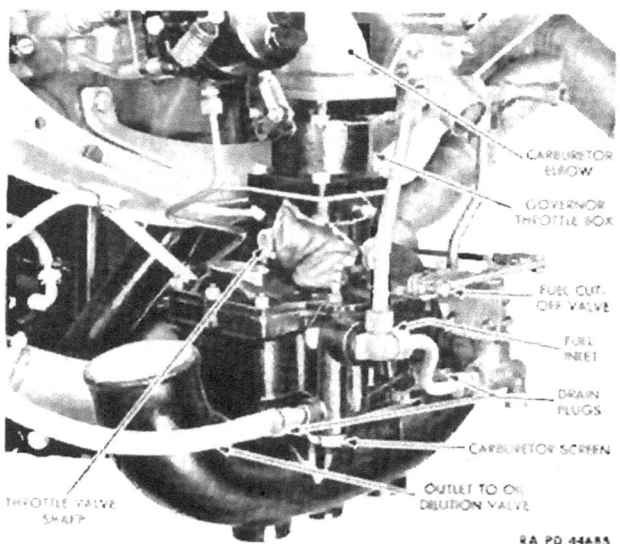

Figure 60. Left side of carburetor.

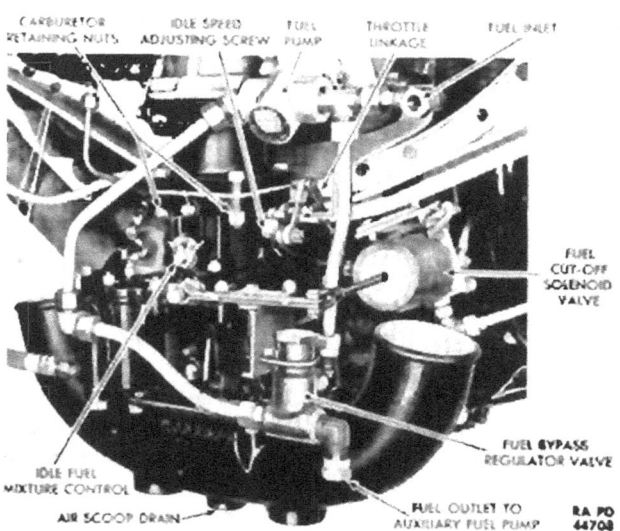

Figure 61. Right side of carburetor.

Mixture control lever is moved to right (lean) until engine begins to run unevenly. Lever is then moved one notch at a time to left (rich) until engine operates evenly.

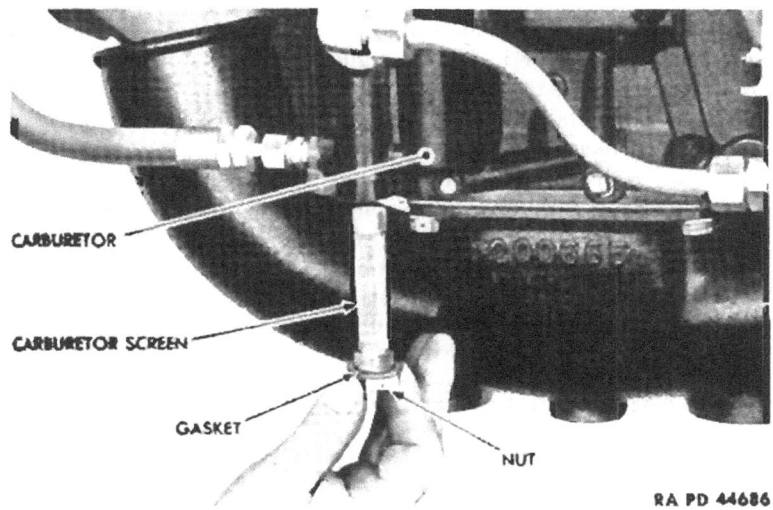

Figure 62. Carburetor screen removal.

58. CARBURETOR REMOVAL.
a. Equipment.
Cloth, wiping
Pliers
Screw driver
Wrenches, open-end, ⅜-inch (2)
Wrench, open-end, ¾-inch
Wrench, open-end, ⅝-inch
Wrench, socket, ⅝-inch

b. Procedure.
(1) Close fuel tank valves.

(2) Remove cap screws which hold plates to rear engine bulkhead (⅝-inch socket wrench) (fig. 26). Lift off plates.

(3) Remove generator as instructed in paragraph 55d.

(4) Disconnect lower clamps from bellows-type flexible couplings at lower end of lower air intake tubes, and remove couplings from carburetor air scoop (screw driver) (fig. 26). Cover scoop openings with wiping cloth.

(5) Disconnect fuel line from fuel pump to carburetor (fig. 59) at the elbow on the carburetor (¾-inch open-end wrench).

(6) If engine has oil dilution valve (see note, par. 11c), disconnect oil dilution fuel line from carburetor at solenoid valve (⅝-inch open-end wrench) (fig. 60).

(7) Disconnect fuel line from fuel pump to fuel bypass regulator valve (⅝-inch open-end wrench) (fig. 61).

139

(8) Disconnect the fuel line from auxiliary fuel pump (see note, par. 40b(11)(b)) at elbow in bypass regulator (¾-inch open-end wrench) (fig. 61).

(9) Remove cotter pin and clevis pin from fuel cut-off valve, disconnecting linkage (pliers) (fig. 60).

(10) Remove cotter pin and clevis pin from clevis on accelerator rod (right side of carburetor), disconnecting throttle linkage (pliers) (fig. 61).

(11) Remove lacing wire from retaining nuts which hold carburetor to lower flange of governor throttle box (pliers) (fig. 61). While supporting carburetor, remove nuts, then remove carburetor from studs (two ⁵⁄₁₆-inch open-end wrenches). Remove gasket.

59. CARBURETOR INSTALLATION.

a. Equipment.

Pliers
Screw driver
Wrenches, open-end, ⁵⁄₁₆-inch (2)
Wrench, open-end, ⅝-inch
Wrench, open-end, ¾-inch
Wrench, socket, ⁵⁄₁₆-inch

b. Procedure.

(1) Install gasket and carburetor on studs (lower flange of governor throttle box). Install and tighten nuts (two ⁵⁄₁₆-inch open-end wrenches) (fig. 61). Install lacing wire (pliers).

(2) Install accelerator rod clevis over throttle arm, insert clevis pin, and secure with cotter pin (pliers) (fig. 61).

(3) Install clevis on fuel cut-off solenoid switch rod over fuel cut-off arm (fig. 58). Insert clevis pin and secure with cotter pin (pliers).

(4) Connect fuel line from auxiliary fuel pump (see note, par. 40b(11)(b)) into elbow in bypass regulator (¾-inch open-end wrench) (fig. 61).

(5) Connect oil dilution line to oil dilution valve (¾-inch open-end wrench) (fig. 60). (Refer to note, par. 11c.)

(6) Connect fuel line from fuel pump to elbow on carburetor (¾-inch open-end wrench) (fig. 59).

(7) Connect fuel line from fuel pump to fuel bypass regulator (¾-inch open-end wrench) (fig. 61).

(8) Place air intake bellows connections over air horns, and tighten clamps (screw driver.)

(9) Place plate in position or near engine bulkhead, and install screw (⁵⁄₁₆-inch socket wrench).

(10) Open fuel tank valves. Inspect for and remedy all leaks.

60. CARBURETOR AIR SCOOP. When exchanging carburetors, the air scoop can be removed from the inoperative carburetor and placed on the serviceable one, unless a new air scoop is provided.

   a. Air Scoop Removal.

   Wrench, open-end, ½-inch

(1) Remove four ½-inch bolts holding air scoop to carburetor.

(2) Lift off air scoop and gasket.

   b. Air Scoop Installation.

   Wrench, open-end, ½-inch

(1) Put gasket in place on bottom of carburetor.

(2) Lay on air scoop so that it will face the fan when carburetor is in place.

(3) Insert and tighten the four ½-inch bolts that secure air scoop to carburetor.

61. AIR CLEANERS. The two air cleaners are of the heavy-duty, oil-bath type. Air enters the cleaner housing through the upper air intake tube and is drawn down along the sides of the housing to the air cleaner oil reservoir. With a swirling motion, the dust-ladened air hits the pool of oil in the reservoir (fig. 64), dropping most of dirt into the oil. Air is then drawn up through the removable filter sections (fig. 64) and the fixed filter above that, and out into the lower intake tube. Some oil is picked up by the air and carried up into the filter elements, adding to their effectiveness in removing any dirt remaining in the air. When the engine is stopped, this oil, with its dirt, drains down into the oil reservoir. All dirt is collected in the bottom of the reservoir.

62. CHANGING OIL IN AIR CLEANER.

   a. General. Change the oil in the air cleaners daily, unless the vehicle has been operating in wet weather, snow, or under unusually dust-free conditions. If the vehicle has been operated under particularly dusty conditions, it may be necessary to service air cleaners even more frequently. To change oil, it is necessary to remove only the oil reservoir of the air cleaner (fig. 63). In the event complete replacement of the air cleaner is required, disconnect the air cleaner from the air intake tubes, loosen the bracket retaining bolts (fig. 63), and lift out the air cleaner.

   b. Changing Oil in Air Cleaner.

(1) Loosen two wing nuts on air cleaner retaining bolts (fig. 63), sufficiently to allow bolts to be pulled free of latches. This releases oil reservoir of air cleaner.

(2) Remove two wing nuts that hold removable filter sections (fig. 64) in place and remove sections (fig. 65).

Figure 63. Air cleaner installation.

(3) Pour out oil and scrape dirt from reservoir.

(4) Clean removable filter section with dry-cleaning solvent.

(5) Install removable filter section in place on upper filter body.

(6) Fill oil reservoir up to oil line indicated on reservoir, using seasonal grade engine oil.

(7) Install reservoir on upper filter body, hooking latching bolts and wing nuts beneath latches, and tightening into place (fig. 63).

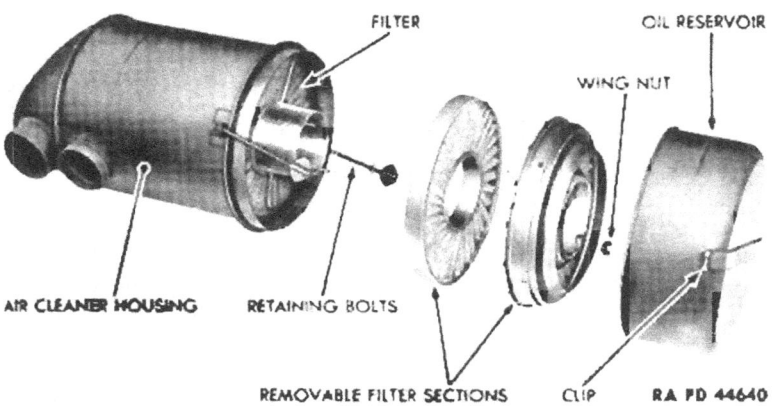

Figure 64. Component parts of air cleaner.

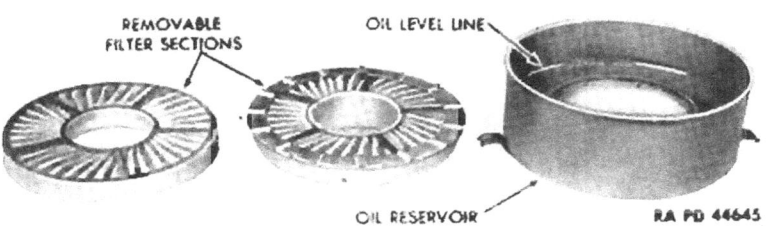

Figure 65. Air cleaner parts requiring cleaning.

63. MAINTENANCE OF AIR CLEANERS. In addition to daily service given to air cleaners (par. 62), the condition of the entire air intake system, including air cleaners, requires regular and careful inspection. It is particularly important to check all flexible connections between cleaner and carburetor air horns, since presence of even a very small hole or leak would allow a surprising amount of dirt or dust to enter the engine. Under ordinary conditions, it is sufficient to service air cleaners once a day. However, under very dusty conditions it may be necessary to change oil in air cleaners at every halt. It is seldom necessary to clean the upper filter section of the cleaner.

However, if inspection of the lower, removable filter sections indicates presence of an excessive amount of dirt, the cleaner housing and upper filter section can be removed and flushed with dry-cleaning solvent.

NOTE: Although the carburetor air cleaner will handle very heavy concentrations of dust, the cleaner will not function properly if the oil in the reservoir is allowed to become too thick with dust to wash the filter element properly. Service the cleaner frequently. To service, remove the oil reservoirs from the air cleaner housings and remove the removable filter sections (par. 62). Empty dirt and oil from the reservoir and fill with seasonal grade engine oil. The filter elements, being self-washing, should require no attention if the oil in the cups is kept reasonably clean. However, inspect the filters occasionally and wash in dry-cleaning solvent if they should appear to be clogging (fig. 49). An air cleaner can clean only the air passing through it. The air connections between the air cleaner and engine must be kept airtight. Small leaks, no larger than a pinhole, will allow a surprising amount of dust to pass through.

64. GOVERNOR REMOVAL AND INSTALLATION. To prevent damage due to overspeeding the engine, a governor is provided. Centrifugal weights within the governor vary in position with the speed of the engine, and actuate a lever and linkage which close a butterfly valve in the throttle box when the engine reaches the prescribed maximum speed. Proper adjustment of the governor is made when it is installed on the engine, and the cap over the adjusting screw wired and sealed. This adjustment should not be tampered with. If the governor operation is faulty, notify ordnance personnel. The governor must be removed if the oil pump is to be replaced.

a. Equipment.
Pliers   Wrench, open-end, ⅜-inch
Wrench, open-end, ⁷⁄₁₆-inch

b. Procedure.

(1) **Governor Removal** (fig. 67).
Pliers   Wrench, open-end, ⅜-inch
Wrench, open-end, ⁷⁄₁₆-inch

(a) Disconnect oil return line at connection on governor (⅜-inch open-end wrench) (fig. 66).

(b) Disconnect governor control linkage by removing cotter pin from lower link clevis pin, and removing clevis pin (pliers) (fig. 66).

(c) Remove four retaining palnuts, nuts, and washers that hold governor to oil pump, and pull governor off studs (⁷⁄₁₆-inch open-end wrench) (fig. 67).

(d) **Remove gasket.**

(2) **Governor Installation.**
Pliers   Wrench, open-end, ⅜-inch
Wrench, open-end, ⁷⁄₁₆-inch

(a) Place gasket in position on four studs on oil pump.

(b) Install governor on four studs on oil pump, and secure by installing and tightening four retaining palnuts, nuts, and washers (⁹⁄₁₆-inch open-end wrench) (fig. 67).

(c) Connect governor linkage by inserting clevis pin in lower link and throttle box arm (fig. 66). Secure with cotter pin (pliers).

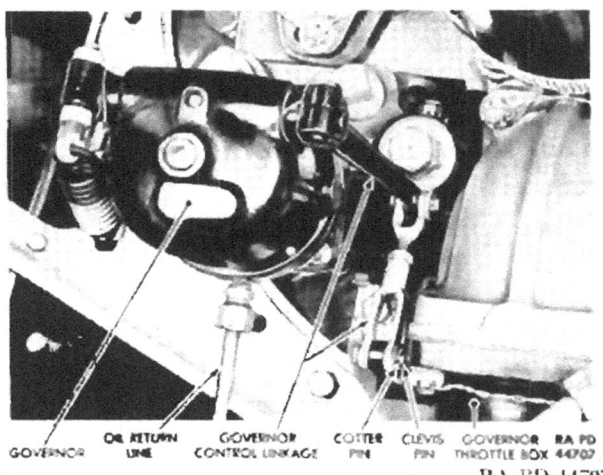

Figure 66. Governor installation.

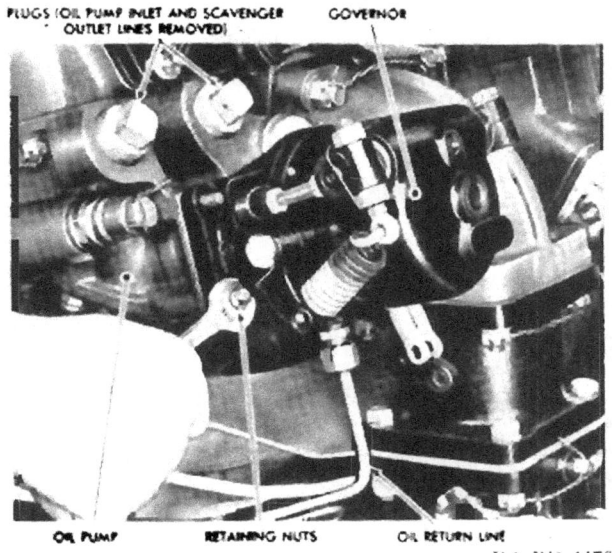

Figure 67. Governor removal.

145

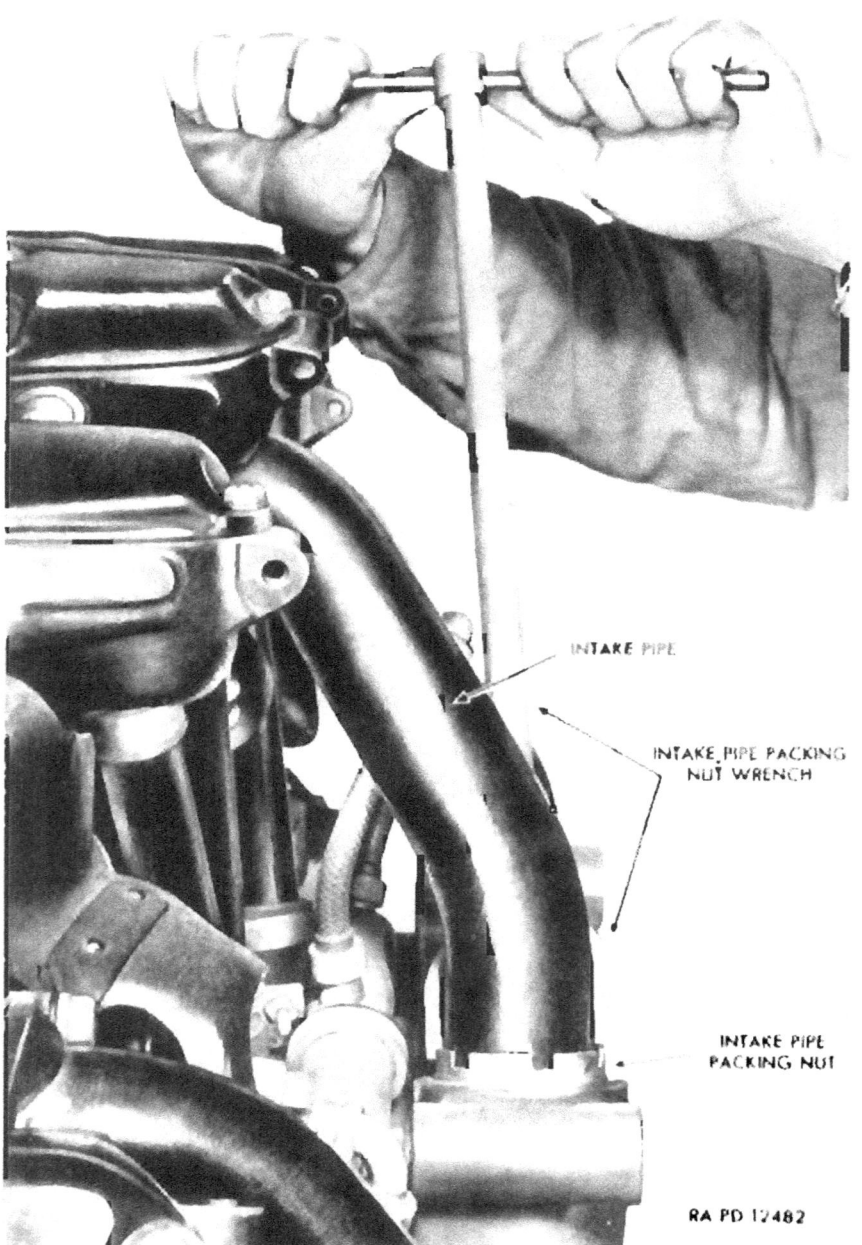

Figure 68. Use of special wrench to tighten intake pipe packing nut.

(**d**) Connect up governor oil return line (⅜-inch open-end wrench) (fig. 66).

Note: This procedure applies to installation of governor after oil pump has been installed. If new governor is to be installed, it must be installed and adjusted by ordnance maintenance personnel. If old governor does not operate satisfactorily, notify ordnance personnel.

65. ROCKER BOX COVER.
a. Equipment.
Wrench, socket, ⅞₆-inch
b. Procedure.
(1) **Rocker Box Cover Removal.** Remove four ⅞₆-inch nuts holding cover to box and lift off cover. Remove gasket from box studs.

(2) **Rocker Box Cover Installation.**
(a) Put gasket in place over four studs on rocker box.
(b) Lay on cover and secure with four ⅞₆-inch nuts. Tighten nuts with tension wrench using 7 foot-pounds for give.

66. ROCKER ARMS (fig. 69).
a. General. Rocker arms at the top of each cylinder open and close intake and exhaust valves, and are operated by means of the cam and push rod mechanism. A roller at the valve end of the rocker arm contacts the end of the valve stem. Two roller bearings are used where the rocker arm is supported on the rocker bolt. The push rod end of each rocker arm is split and tapped to provide for a valve clearance adjusting screw. A clamping screw keeps the adjusting screw from turning, once it has been properly set.

b. Inspections. Each rocker arm roller and bearing should be inspected at the 100-hour inspection period. If the roller has flat spots on it, or if bearing of a rocker arm assembly is defective, replace the rocker arm assembly with a serviceable unit.

c. Precautions. Remove, inspect, and install one rocker arm assembly and push rod at a time. The rocker arms are not all interchangeable.

d. Equipment.
Drift, ⅜-inch	Screw driver
Hammer	Wrench, box, 1¹⁄₁₆-inch
Pliers	Wrench, open-end, ⅜-inch

e. Procedure.
(1) **Rocker Arm Removal** (fig. 71).
Drift, ⅜-inch	Wrench, box, 1¹⁄₁₆-inch
Hammer	Wrench, open-end, ⅜-inch
Pliers	

(a) Remove rocker box cover and gasket (par. 65**b**(1)).

(b) Turn the engine with starting crank until rocker arm is free (valve closed).

(c) Remove cotter pin from nut on rocker arm shaft (pliers). Hold shaft and remove nut ($1\frac{1}{16}$-inch box wrench and $\frac{3}{4}$-inch open-end wrench).

(d) Drive out rocker arm shaft (hammer and $\frac{3}{4}$-inch drift). Remove spacers and lift out rocker arm assembly.

(2) **Push Rod Removal.** Perform operations in paragraph 68.

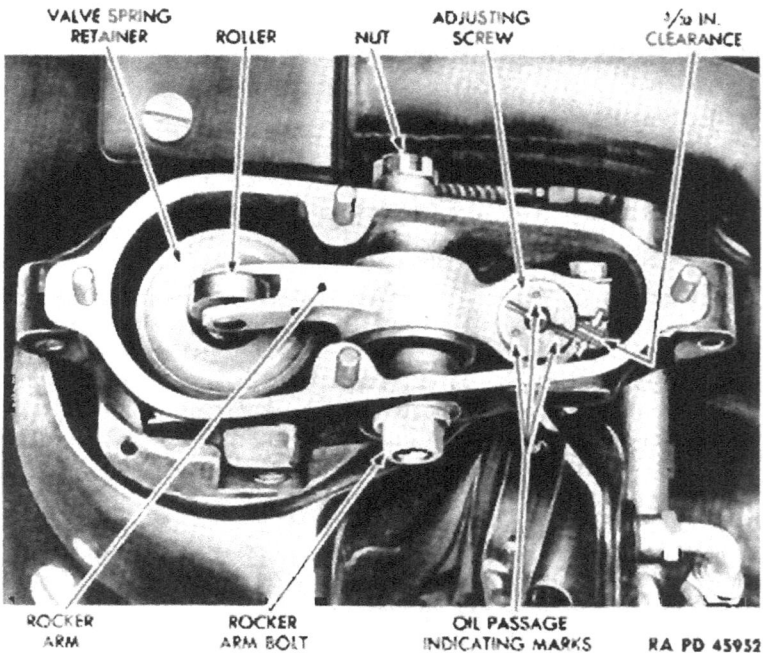

Figure 69. Interior of rocker box.

(3) **Rocker Arm Installation** (fig. 69.)

Pliers   Wrench, box, $1\frac{1}{16}$-inch
Screw-driver   Wrench, open-end, $\frac{3}{4}$-inch

(a) Install push rod (par. 68).

(b) Put rocker arm into place, with spacer on each side. Use small screw driver or rod to line up spacers with rocker arm. *Caution:* In installing new rocker arms, be sure that an exhaust rocker arm is used in an exhaust rocker box. An intake rocker arm will have the wrong pitch, and can be used only in an intake rocker box.

Figure 70. Adjusting valve clearance.

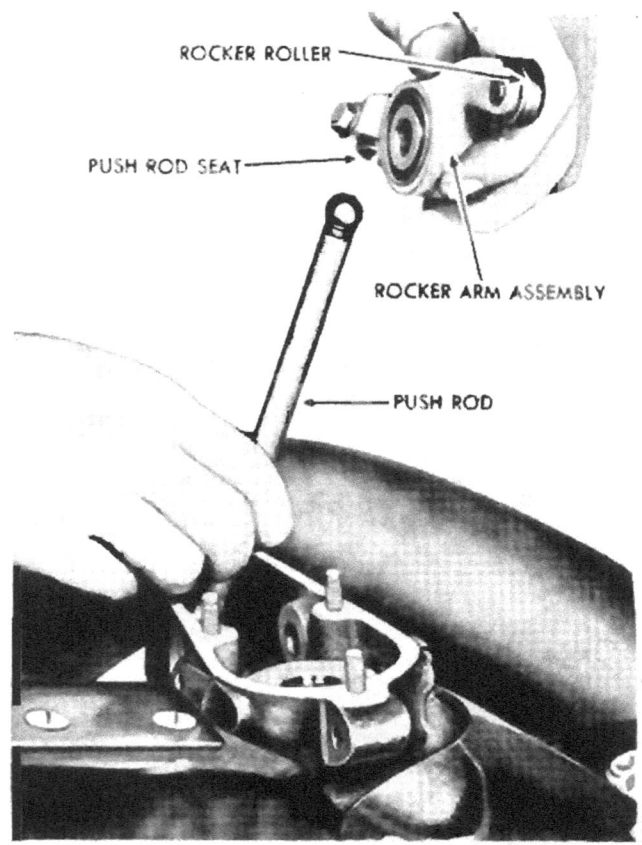

Figure 71. Rocker arm and push rod removal.

(c) Put one outside spacer on rocker arm shaft and install shaft from inner side of rocker box so that nut will be on the side toward outer wall of cylinder.

(d) Install outside spacer on other end of shaft; install and tighten nut ($^{11}/_{16}$-inch box wrench and $\frac{3}{4}$-inch open-end wrench). Secure nut with cotter pin (pliers).

(e) Adjust valve clearance (par. 69).

(f) Install rocker box covers and gaskets (par. 64b(2)).

67. ENGINE COWL.

a. Engine Cowl Removal.

Screw driver, 8-inch Wrench, socket, $^{9}/_{16}$-inch
Wrench, open-end, $\frac{1}{2}$-inch

(1) Disengage the Dzus fasteners which secure the intercylinder and cylinder-head air deflectors to the cowl. The Dzus fasteners may be disengaged by inserting a screw driver in the slotted head of the fasteners and turning in a counterclockwise direction.

(2) Remove the nuts (⅜-inch socket wrench) and bolts (½-inch open-end wrench) which secure the cowl to the bosses on the front of the rocker boxes and remove the cowl.

(3) The inspection plates of the cowl are also secured by Dzus fasteners and can be removed with a screw driver as directed above.

b. Engine Cowl Installation.

Pliers	Wrench, open-end, ⅜-inch
Screw driver	Wrench, open-end, ½-inch
Wrench, box, 11/16-inch	Wrench, socket, ⅜-inch

(1) Put the cowl in place and fasten it to the bosses on the front of the rocker boxes with bolts (½-inch open-end wrench) and nuts (⅜-inch socket wrench).

(2) Install intercylinder and cylinder-head air deflectors with Dzus fasteners in holes provided. The Dzus fasteners may be engaged by inserting a screw driver and turning in a clockwise direction.

68. PUSH RODS (fig. 71). Oil for valve mechanism lubrication is delivered through oil passages in the push rods. When a rocker arm is removed, the push rod should be lifted out and inspected for clogged oil passage. After blowing passage out with compressed air, prime passage with clean oil and put push rod back in place.

69. VALVE CLEARANCE ADJUSTMENT (figs. 69 and 70).

a. Equipment.

Gauge, feeler, 0.006-inch	Wrench, valve clearance adjusting
Screw driver	or
	Wrench, open-end, ⅜-inch

b. Procedure.

(1) Preliminary Steps.

(a) Remove rocker box covers and gaskets from all cylinders (par. 65b(1)).

(b) Remove front or rear spark plugs (par. 51).

(c) Turn engine crankshaft until piston in cylinder to be adjusted is at top dead center. If rocker arm rollers on both intake and exhaust valves are free, valve adjustment can be made. If either valve is not free when piston is at top dead center, the piston is on the exhaust stroke. Turning the crankshaft a complete revolution will bring the piston to top dead center on the compression stroke and free both rocker arm rollers.

(d) Before adjusting valve clearance, check rocker arm roller for flat spots or burs. Check rocker arm for excessive side play. Replace rocker arm assembly if necessary (par. 66).

(2) **Adjusting Valve Clearance** (fig. 70).

Gauge, feeler, 0.006-inch Wrench, valve clearance adjusting
Screw driver

(a) Using valve clearance adjusting wrench, loosen clamping screw.

(b) Insert a 0.006-inch feeler gauge between valve stem and rocker roller, and, with a screw driver, turn adjusting screw until gauge will slide in and out smoothly, causing roller to turn. *Caution:* Hold down on push rod end of rocker arm when checking clearance.

(3) **Checking Position of Adjusting Screw** (fig. 69). The adjusting screw has three radially-drilled oil holes to permit pressure lubrication of the upper eight rocker arms. To be sure that each of these holes is blanked off from the split in the rocker arm, the location of which is marked by three zeros on the head of the screw, no hole should be closer than $\frac{3}{32}$ of an inch from nearest edge of the split. If proper 0.006-inch clearance should bring one of the zeros closer than $\frac{3}{32}$ of an inch to the split in the rocker arm, move adjusting screw in shortest direction to bring zero (mark) to $\frac{3}{32}$ of an inch from the split.

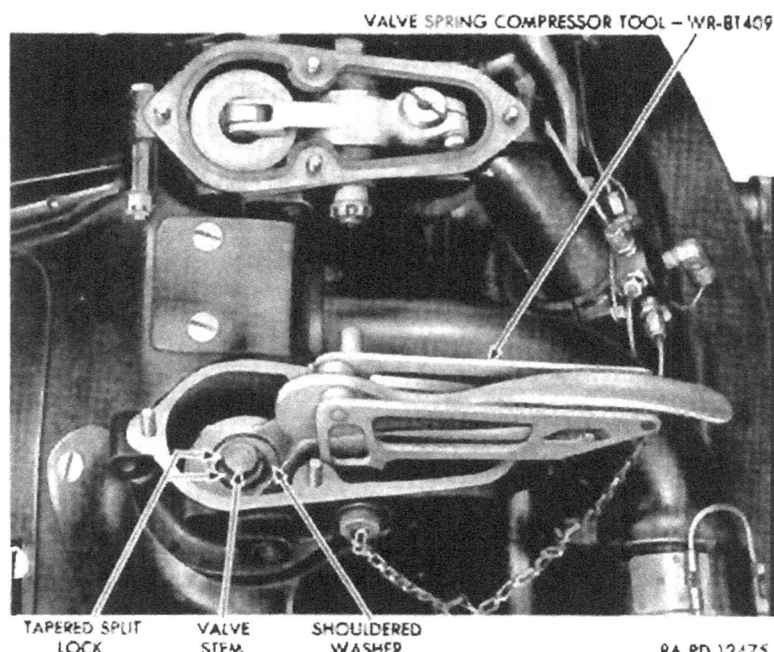

Figure 72. Compressing valve springs to remove split lock.

(4) **Tightening Clamping Screw.**

Screw driver	Wrench, valve clearance adjusting

(**a**) Tighten clamping screw with special wrench or 5/8-inch open-end wrench. Test tightness by trying to turn adjusting screw with screw driver. If clamping screw is properly tightened, it will be impossible to turn adjusting screw.

(**b**) Install rocker box covers and gaskets on all cylinders (par. 65**b**(**2**)).

70. ENGINE INSTALLATION.

a. Equipment.

Cable, length	Wrenches, open-end, 9/16-inch (2)
Extensions, long (2)	Wrench, open-end, 5/8-inch
Handle, sliding T	Wrench, open-end, 3/4-inch
Hoist	Wrench, open-end, 15/16-inch
Pliers	Wrench, open-end, 1-inch
Screw driver	Wrench, open-end, 1 1/4-inch
Sling, engine lifting	Wrench, open-end, 1 3/8-inch
Stand, engine	Wrench, socket, 7/16-inch
Wrench, open-end, 9/16-inch	Wrench, socket, 9/16-inch
Wrench, open-end, 11/32-inch	Wrench, socket, 3/4-inch
Wrench, open-end, 5/8-inch	Wrench, socket, 1 1/4-inch
Wrench, open-end, 7/16-inch	

b. Procedure. Operations involving the 155-mm gun are applicable only to the 155-mm Gun Motor Carriage M12. In all other respects, the installation of the engine in either the gun carriage or cargo carrier is identical.

(1) **Attach Lifting Sling to Engine** (figs. 36 and 37).

Sling, engine lifting	Wrench, open-end, 9/16-inch
Stand, engine	

(**a**) Remove bolts which secure bracket arms of engine nose breather (9/16-inch open-end wrench) (figs. 18 and 35).

(**b**) Lay engine lifting sling on top of engine in engine stand. The two hooked cables at the rear of sling pass down the rear of the engine, inside the engine support beam, and hook to the engine support beam.

(**c**) Wrap the hooked cable on the front of the engine sling twice around the clutch spring housing hub, then hook the cable to the eye on the sling.

(**2**) **Place Engine in Position in Vehicle** (figs. 37 and 38).

Hoist	Wrench, socket, 3/4-inch

(**a**) Place the gun in the position described in paragraph 40**b**(**21**).

(**b**) Attach hoist to lifting sling already installed on engine (fig. 35). Lift engine up over engine compartment. Rotate engine clockwise

564206°—44——11

slightly, so that the terminal shields on the starting motor and generator will clear the top of the rear engine compartment bulkhead. Have one or two men "ride" the front of the engine so that it tilts downward as it enters the engine compartment. Lower the engine slowly into the engine compartment. *Caution:* Take extreme care to see that no part of the engine catches on any projection in the engine compartment. Place steady-rest adapters in the steady-rest tube before the engine is completely lowered. After the engine has been lowered sufficiently beneath the recoil mechanism to clear, carefully turn the engine back to a straight position and lower remainder of way into engine compartment.

(3) **Connect Engine to Engine Supports.**

Extensions, long (2)
Handle, sliding T
Pliers
Wrenches, open-end, 9/16-inch (2)
Wrench, open-end, 1¼-inch
Wrench, socket, 1¼-inch

(a) Install bolts which hold steady-rest adapters to steady-rest brackets (1¼-inch open-end and 1¼-inch socket wrenches, with extension and sliding T-handle) (figs. 23 and 37). This will require two men; one to hold the bolt and one to turn the nut.

(b) Place ground strap in position on engine support beam bracket and install retaining bolt and nut (two 9/16-inch open-end wrenches) (fig. 35).

(c) Install the eight bolts and nuts (four on each side) which secure engine support beam to brackets (two 9/16-inch open-end wrenches) (fig. 35). Install cotter pins through nuts (pliers).

(d) Remove engine lifting sling.

(4) **Connect Engine to Propeller Shaft.**

Wrenches, open-end, 9/16-inch (2)

Install the eight bolts and nuts that secure the clutch companion flange to the universal joint companion flange (two 9/16-inch open-end wrenches) (fig. 33).

Note: Turn the engine with the crank, to rotate companion flange into a position from which bolts may more easily be installed.

(5) **Install Oil Tank Breather** (fig. 33).

Wrench, open-end, 15/16-inch

Screw oil tank breather into position in top of oil tank (15/16-inch open-end wrench).

(6) **Install Left Front Fire Extinguisher Horn and Bracket** (fig. 33).

Wrenches, open-end, 9/16-inch (2)   Wrench, socket, 9/16-inch
Wrench, open-end, 1-inch

(a) Place horn bracket in position on recessed compartment bulkhead and install two retaining bolts and nuts (two 9/16-inch open-end wrenches).

(b) Install horn on connector (1-inch open-end wrench). Rotate horn into position above bracket, and install two cap screws ($\frac{5}{16}$-inch socket wrench).

(7) **Install Exhaust Tubes** (fig. 32).

Wrench, open-end, $\frac{5}{16}$-inch   Wrench, socket, $\frac{5}{16}$-inch

Place exhaust tube clamps on lower end of exhaust tubes (right and left), and secure to shorter section of exhaust tube clamped to flame arresters ($\frac{5}{16}$-inch open-end and $\frac{5}{16}$-inch socket wrenches) (fig. 32).

(8) **Connect Accelerator Linkage** (fig. 32).

Pliers

(a) Place accelerator rod clevis on carburetor throttle arm (beneath engine). Install clevis pin and lock with a cotter pin (pliers).

(b) Hook throttle return spring to bracket on floor of compartment and to throttle arm (pliers).

(9) **Connect Oil Temperature Gauge Line and Tachometer Drive Cable** (fig. 32).

Pliers   Wrench, open-end, $\frac{5}{8}$-inch

(a) Place tachometer drive cable in position. Tighten cable knurled nut (pliers).

(b) Remove plug from oil pump finger strainer. Insert oil temperature gauge bulb and tighten retaining nut ($\frac{5}{8}$-inch open-end wrench).

(10) **Connect Oil Pressure Gauge Line and Primer Distributor Line** (fig. 32).

Pliers   Wrench, open-end, $\frac{5}{8}$-inch
Wrench, open-end, $\frac{5}{16}$-inch

(a) Remove cloth wired to openings (pliers).

(b) Connect oil pressure gauge line ($\frac{5}{8}$-inch open-end wrench) and primer distributor line ($\frac{5}{16}$-inch open-end wrench) to connectors under left side of engine support beam.

(11) **Connect Fuel Lines.**

Pliers   Wrench, open-end, $\frac{5}{8}$-inch

(a) Remove cloth covering ends of all exposed fuel lines (pliers).

(b) Connect fuel inlet line at fuel pump ($\frac{5}{8}$-inch open-end wrench) (fig. 27).

(c) Connect fuel bypass line at auxiliary fuel pump ($\frac{5}{8}$-inch open-end wrench) (fig. 32). (Refer to note, par. 40**b**(11)(**b**).)

(12) **Connect Oil Lines.**

Pliers   Wrench, open-end, 1-inch
Wrench, open-end, $\frac{5}{8}$-inch   Wrench, open-end, 1$\frac{5}{8}$-inch

(a) Remove cloth from end of all open oil lines (pliers).

(b) Connect oil dilution valve line at T-connection with oil inlet line ($\frac{5}{8}$-inch open-end wrench) (fig. 34).

(c) Connect front section scavenger line at Y-connection with oil outlet line (1-inch open-end wrench) (fig. 34). **Caution:** When making connections at Y-fittings, take care not to injure one connection while connecting other line.

(d) Connect main oil pump inlet connection and main scavenger outlet connection (1⅝-inch open-end wrench) (figs. 27 and 34). Avoid tightening the wrong part of the coupling. (Tighten the large lower nut.) Attempting to turn the two upper hexes may shear the hose.

(13) Connect Electrical Connections on Engine (fig. 27).

    Pliers   Wrench, open-end, 11/32-inch

    Screw driver   Wrench, open-end, 5/16-inch

(a) Connect starter cable at starting motor (11/32-inch open-end wrench). Connect conduit knurled nut (pliers). Install shield (screw driver).

(b) Connect generator cable (5/16-inch open-end wrench). Connect conduit knurled nut (pliers). Install shield (screw driver).

(14) Connect Rear Terminal Box Electrical Connections (figs. 30 and 31).

    Pliers   Wrench, open-end, 5/16-inch

    Screw driver   Wrench, open-end, ⅜-inch

(a) Insert oil dilution valve lead (see note, par. 11c) and fuel cut-off solenoid leads through opening in bottom of rear terminal box, and install wires on their proper terminals (5/16-inch open-end wrench). Leads should have been tagged at removal to assure correct installation. In addition, the wires are colored the same as wires connecting to corresponding terminals within the terminal box.

(b) Connect knurled nut which secures oil dilution valve and fuel cut-off solenoid conduit leads (pliers).

(c) Insert both magneto ground leads and the booster primary lead through conduit opening in upper side of terminal box, and install leads on the proper terminals (⅜-inch open-end wrench). The wires are colored the same as lead-in wires already connected to terminals.

(d) Connect knurled nut which secures magneto grounds and booster primary conduit (pliers).

(e) Place rear terminal box cover in position, and install the four retaining screws and lock washers (screw driver).

(15) Install Lower Intake Tubes (fig. 27).

    Screw driver.

Place lower intake tubes (right and left) in position. Install hose connections which clamp tubes to air cleaners.

(16) Install Upper Half of Engine Compartment Shroud (figs. 28 and 29).

    Wrench, open-end, 5/16-inch

(a) Place upper half of engine compartment shroud in position.

(b) Install bracket on right lower side of shroud ($\frac{7}{16}$-inch open-end wrench) (through which passes right rear fire extinguishing line).

(17) Install Upper Air Intake Tubes (figs. 28 and 29).
    Screw driver

(a) Slide upper air intake tubes (right and left) forward into upper half of engine compartment shroud. Place hose connections on tubes and on air cleaners. Tighten clamp screws (screw driver).

(b) Prime the oil pump and start the engine. After the engine has been running a short time, and all tubes have been inspected for leaks, the rear plates can be installed.

(18) Install Engine Compartment Rear Plates (fig. 26).
    Wrench, socket, $\frac{9}{16}$-inch

Place engine compartment rear plates in position, and install retaining cap screws ($\frac{9}{16}$-inch socket wrench).

(19) Install Engine Compartment Top Plate (figs. 24 and 25).

Cable, length	Screw driver
Hoist	Wrench, open-end, $\frac{9}{16}$-inch
Pliers	Wrench, socket, $\frac{9}{16}$-inch

(a) Place a length of cable completely around the center of the engine compartment top plates. Hook a hoist to the cable. Lift the plate and place in position.

(b) Install cap screws which secure upper half of engine compartment shroud to top plate ($\frac{9}{16}$-inch socket wrench).

(c) Install remaining cap screws which secure top plate ($\frac{9}{16}$-inch socket wrench).

(d) Thread the two outside fire extinguisher handle cables through their mounting on the top plate (fig. 24) and into the recessed compartment behind the driver. Working within the compartment, thread the cables through the tubes located between the top plate and the control heads on the fire extinguishers. Thread the cables into the control heads, around the circular position within the head, then secure cables to small block inside control heads (screw driver).

(e) Install covers on control heads (screw driver). Install lock wire on retaining screws (pliers).

(f) Connect tube through which cables run at top plate and at control head ($\frac{9}{16}$-inch open-end wrench).

(20) Install Engine Compartment Top Guards (fig. 24).
    Wrench, socket, $\frac{9}{16}$-inch

Place engine compartment top guards (front and rear) in position and install retaining cap screws ($\frac{9}{16}$-inch socket wrench).

(21) Concluding Steps (fig. 14).

(a) Open the two fuel shut-off valves.

(b) Return the 155-mm gun to its traveling position and tighten the stop bolts.

SECTION V
FUEL SYSTEM

71. DESCRIPTION (fig. 73). Fuel is supplied to the carburetor at a pressure of approximately 3½ pounds per square inch. From the fuel tanks, fuel goes to a collector or header line. From there it is drawn through the fuel filter, and up into the engine fuel pump located just above and to the right of the carburetor. From the fuel pump it is forced to the carburetor. A bypass line at the carburetor fuel inlet permits fuel in excess of carburetor requirements to return through a bypass pressure regulator valve which regulates the pressure of the fuel fed to the carburetor. A fuel line connects the header to the primer pump, located on the instrument panel. Upon operation of the pump plunger, the fuel drawn in is forced back to the priming distributor and priming lines into the top five cylinders. An oil dilution valve is fed by a fuel line connecting with the fuel system at the carburetor inlet (see note, par. 11c). When the valve is opened, fuel is allowed to enter the oil inlet line to thin the oil in the engine for cold weather starting.

Note: On vehicles of later manufacture, a new type fuel filter is installed in the gas tank.

72. INSPECTION.

a. All connections and fuel lines, fuel pump, and carburetor will be inspected periodically for signs of leaks and evidence of damage or wear.

b. Care must be taken to maintain all flexible lines in position so that they are not subjected to twisting or to abrasion through rubbing against some other line or part.

c. Carburetor must be checked for signs of leakage and carburetor elbow examined for cracks.

d. All air intake connections must be periodically checked for tightness and for presence of even the smallest holes.

73. TROUBLE SHOOTING. See paragraph 39.

74. FUEL TANKS (figs. 77 and 78).

a. Description. Two fuel tanks, with a total capacity of 200 gallons, are provided. Tanks are placed horizontally, one on each side of the engine compartment in the sponsons.

b. Drain Fuel Tanks.

Screw driver, offset, ¾-inch   Wrench, Allen, ⁵⁄₁₆-inch bit

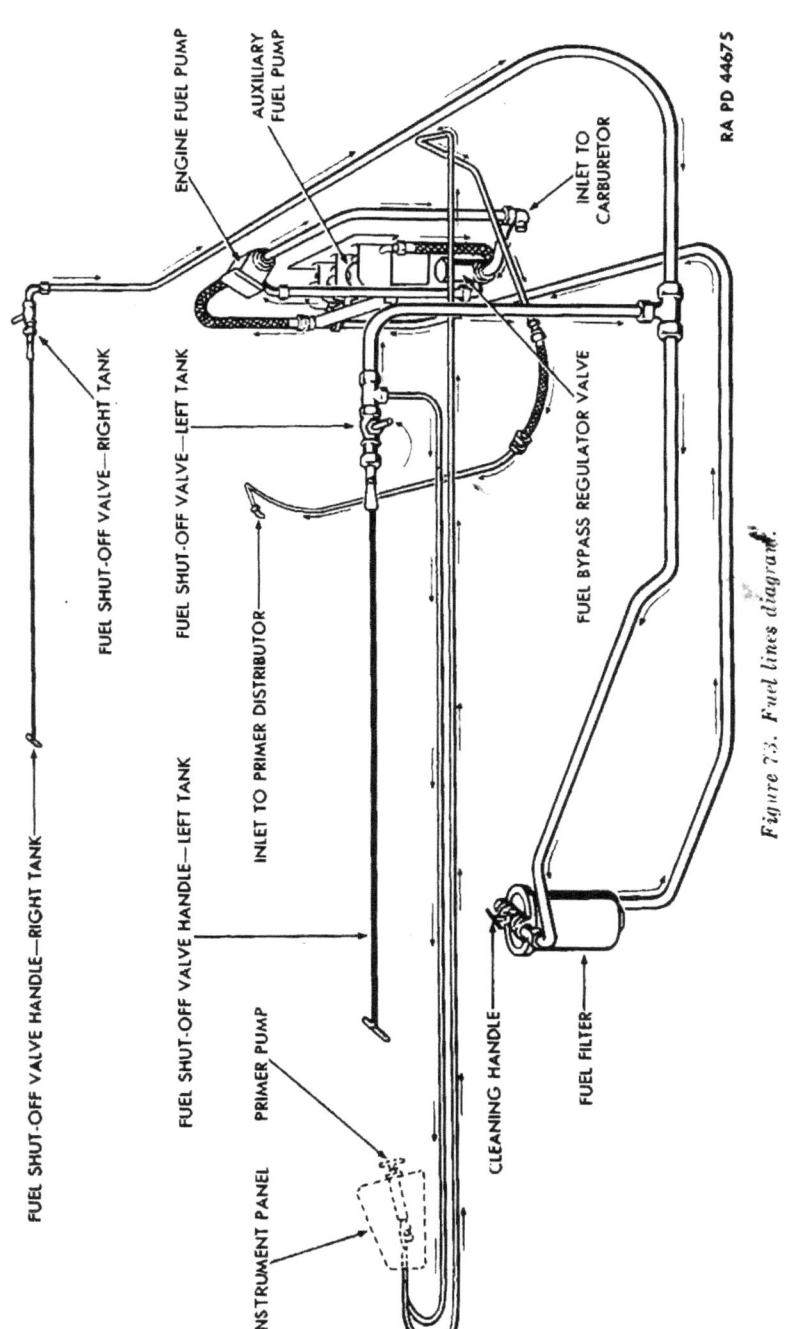

Figure 73. Fuel lines diagram.

(1) Remove three screws holding the circular plate in place on the bottom of the sponson, allowing plate to fall free (screw driver).

(2) Remove plug ($\frac{5}{16}$-inch Allen wrench) and drain fuel.

75. FUEL TANK REMOVAL. In the procedure in **b** below, the right tank in the 155-mm Gun Motor Carriage M12 is removed. Some differences exist between the removal of the left and right fuel tanks on the Gun Motor Carriage M12 and the Cargo Carrier M30. These differences are noted wherever they occur in the text.

a. Equipment.

Cable, length of	Wrench, open-end, $\frac{1}{2}$-inch
Cable, sling	Wrench, open-end, $\frac{9}{16}$-inch
Hoist	Wrench, open-end, $\frac{3}{4}$-inch
Pliers	Wrenches, socket, $\frac{9}{16}$-inch (2)
Screw driver	Wrench, socket, $\frac{3}{4}$-inch
Wrench, adjustable	

b. Procedure.

(1) Open Battery Switch.

(2) Remove Tools and Equipment. Remove all pioneer tools and equipment stowed on the fuel tank top plate or stowed in such a position that they will interfere with the removal of the fuel tank top plate.

(3) Disconnect Spade Winch Cable (155-mm Gun Motor Carriage).

Wrench, open-end, $\frac{9}{16}$-inch

(a) Release ratchet arm and brake on spade winch, and lower spade.

(b) Turn out winch manually until end of cable is visible. Remove two nuts which fasten U-bolt through winch drum, securing end of cable ($\frac{9}{16}$-inch open-end wrench) (fig. 74). Remove and lay cable aside.

(4) Drain Fuel Tanks (par. 74b).

(5) Remove Shut-off Valve and Nipple From Tank (fig. 75).

Pliers	Wrench, open-end, $\frac{1}{2}$-inch
Screw driver	Wrench, open-end, $\frac{9}{16}$-inch
Wrench, adjustable	Wrench, open-end, $\frac{3}{4}$-inch

(a) Remove cap screws which secure top rear engine compartment guard ($\frac{9}{16}$-inch open-end wrench). Lift off guard.

(b) Remove cotter pin that holds tank shut-off valve handle rod to valve stem universal joint (pliers). Lift rod up out of joint. Unscrew and remove the valve stem (half of universal joint) which remains in the valve (screw driver).

(c) Disconnect fuel line from valve ($\frac{3}{4}$-inch open-end wrench) (fig. 75).

Figure 74. Spade winch installation (155-mm gun motor carriage)

(d) Disconnect primer fuel line from valve (⅞-inch open-end wrench) (fig. 75).

NOTE: This operation applicable only when removing left tank.

(e) Turn valve, nipple and T-fitting out of fuel tank (adjustable wrench).

(6) **Remove Fuel Tank Top Plate** (figs. 76 and 77).

Cable, length of Wrench, socket, ⁹⁄₁₆-inch
Hoist Wrench, socket, ¾-inch

(a) Remove battery compartment top plate (par. 148).

Note: This operation necessary only for removal of right fuel tank.

161

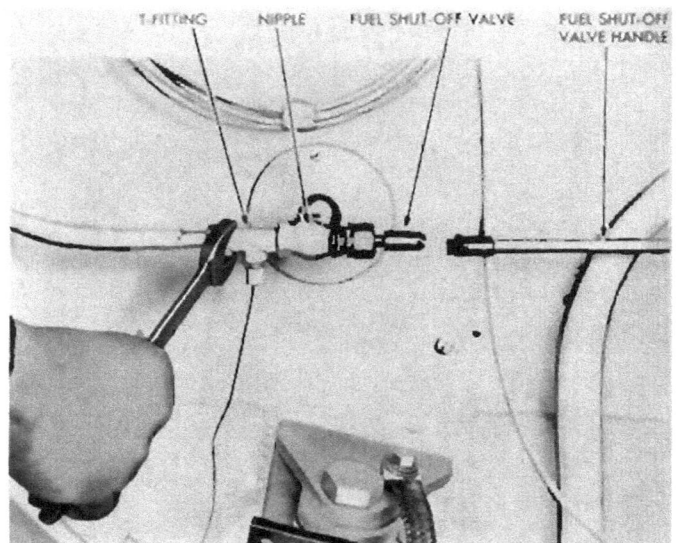

Figure 75. Disconnecting fuel shut-off valve.

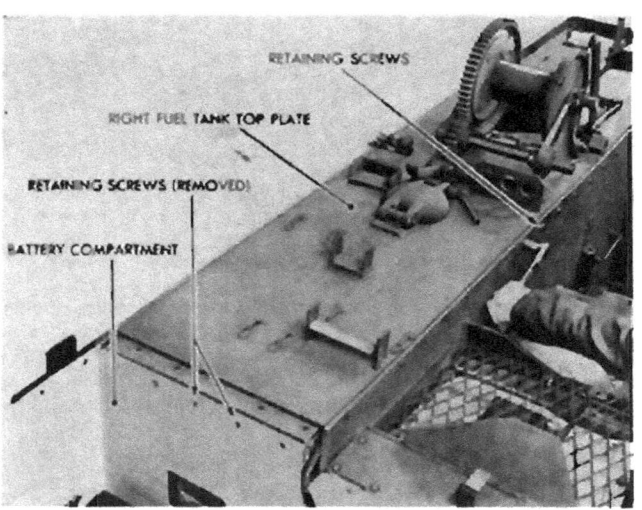

Figure 76. Right fuel tank top plate removal.

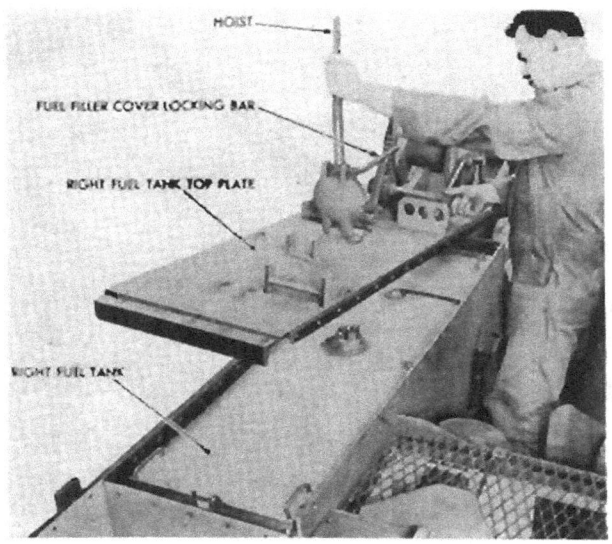

Figure 77. Lifting right fuel tank top plate.

(b) Remove cap screws which secure forward end of fuel tank top plate (inside battery compartment) (⁹⁄₁₆-inch socket wrench).

(c) Remove cap screws which secure inner side of fuel tank top plate (⁹⁄₁₆-inch socket wrench).

(d) Remove cap screws which secure outer and rear sides of top plate (½-inch socket wrench).

(e) Attach a length of cable to the fuel filler cover locking bar, in place in the cover (fig. 77). Attach a hoist to the cable. Remove the fuel tank top plate.

(7) **Disconnect and Remove Fuel Tank** (figs. 78 and 79).

Cable, sling
Hoist
Screw driver
Wrench, open-end, ⅝-inch
Wrenches, socket, two, ⁹⁄₁₆-inch

(a) Remove the retaining bolt which secures the fuel tank hanger (one at each end of tank) (two ⁹⁄₁₆-inch socket wrenches). Bend the hangers up away from tank (fig. 78).

(b) Disconnect fuel gauge electrical lead to tank (⅝-inch open-end wrench).

(c) Pry the three cable brackets up from the side of the tank and lift the electrical cable up and outside the compartment (screw driver).

(d) Hook a cable sling, equipped with hooks, to the four handles provided on the fuel tank. Attach a hoist to the sling. Carefully

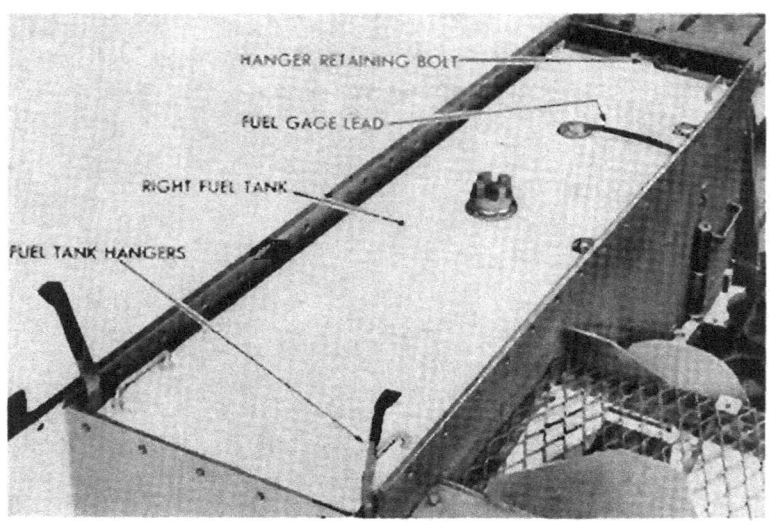

Figure 78. Fuel tank installed.

lift the fuel tank out of the compartment. Take care not to break off the handles (fig. 79).

(8) **Remove Fuel Tank Spacers** (fig. 80).

    Screw driver

  (a) Unless it is necessary to replace the wooden spacers because of damage or wear, they need not be removed.

  (b) Lift out the two wooden spacers in the bottom of the fuel tank compartment.

  (c) Remove wooden screws which secure two wooden spacers to inner wall of fuel tank compartment (screw driver). Lift out spacers.

76. FUEL TANK INSTALLATION. In the following procedure the right tank in the 155-mm Gun Motor Carriage M12 is installed. Some differences exist in the installation of the left and right fuel tanks on the Gun Motor Carriage M12 and the Cargo Carrier M30. These differences are noted wherever they appear in the text.

  a. Equipment.

Cable, length of	Wrench, open-end, ⅝-inch
Cable, sling	Wrench, open-end, ⁹⁄₁₆-inch
Hoist	Wrench, open-end, ¾-inch
Pliers	Wrenches, socket, two, ⁹⁄₁₆-inch

  b. Procedure.

(1) **Install Fuel Tank Spacers** (fig. 80).

Screw driver

(a) Place two wooden spacers (long strips) along inner side of compartment and install retaining wooden screws.

(b) Place smallest wedge-shaped spacer at bottom front of compartment; tapered end toward gasoline drain hole.

(c) Place large wedge-spacer in bottom of compartment, tapered end slanting toward gasoline drain hole. Thick end of spacer must be about 16 inches from rear wall of compartment.

(2) **Install and Connect Fuel Tank in Compartment** (fig. 79).

Cable, sling
Hoist
Screw driver
Wrench, open-end, ⅝-inch
Wrench, socket, ⁹⁄₁₆-inch (2)

(a) Hook a cable sling equipped with hooks, to the four handles provided on the fuel tank. Attach hoist to sling. Lift tank and place it carefully in position in compartment. Remove sling.

(b) Slip fuel gauge cable under the three cable brackets along inner

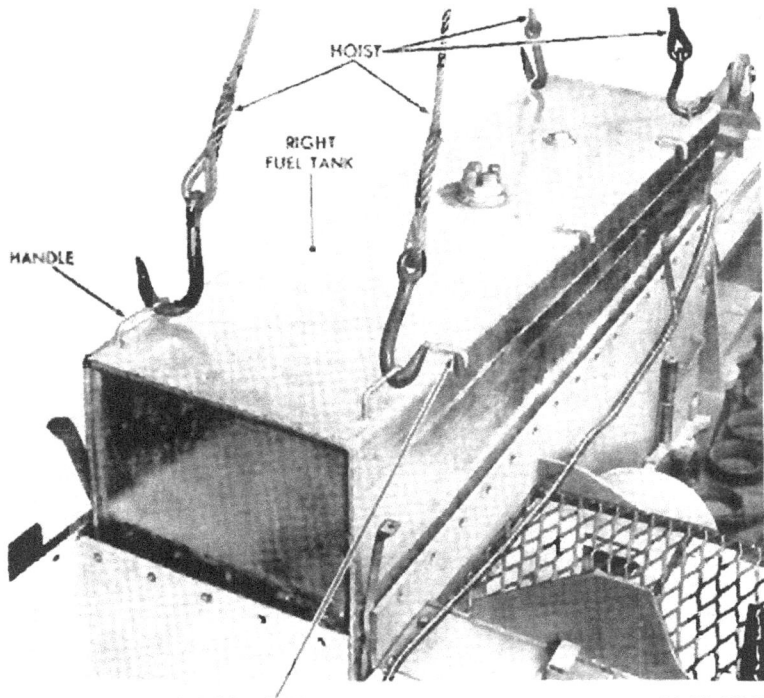

Figure 79. Lifting out right fuel tank.

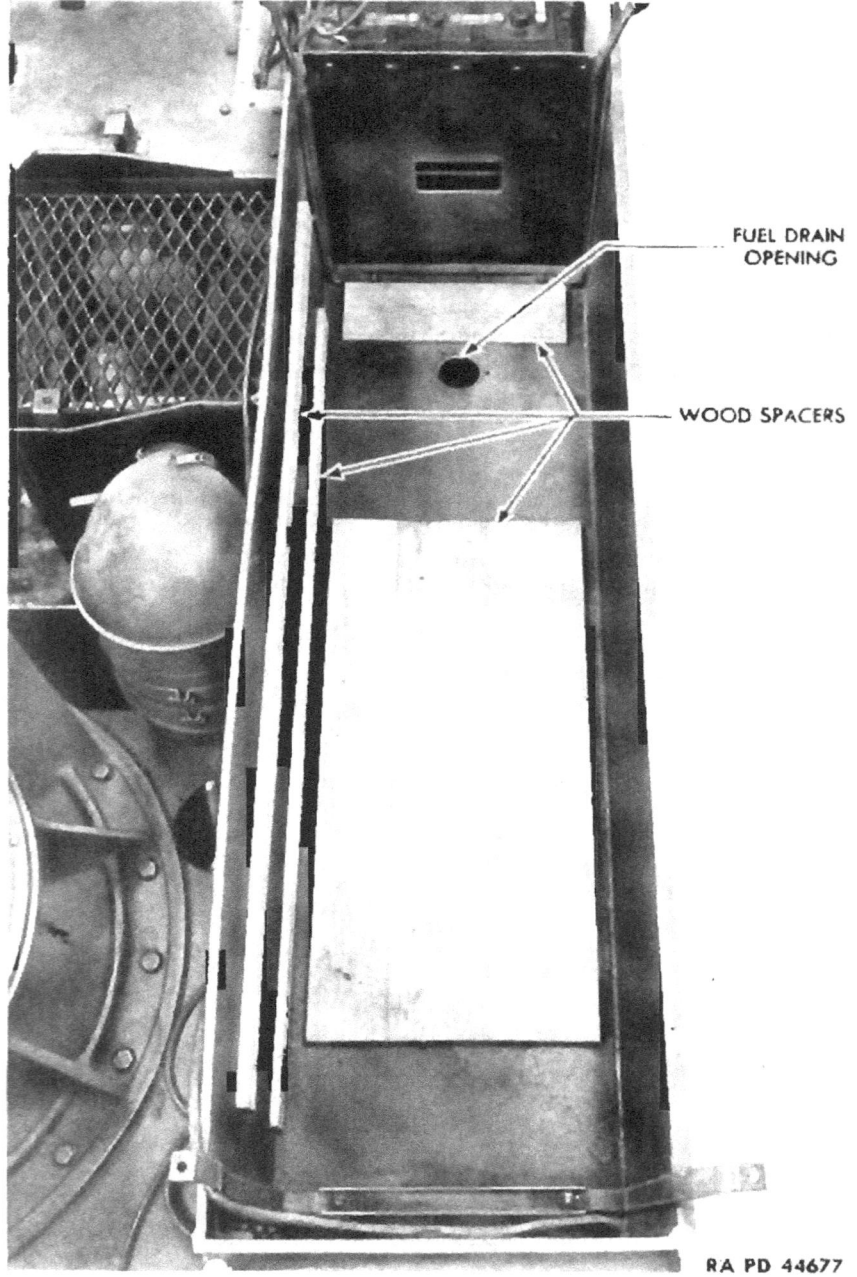

Figure 80. Position of wooden spacers in fuel tank compartment.

edge of fuel tank. Wedge brackets down over cable (screw driver).

(c) Connect fuel gauge electrical lead to tank (¾-inch open-end wrench).

(d) Bend fuel tank hangers down into position. Install retaining bolts (two ⁹⁄₁₆-inch socket wrenches).

(3) **Install Fuel Tank Top Plate** (figs. 76 and 77).

| Cable, length of | Wrench, socket, ⁹⁄₁₆-inch |
| Hoist | Wrench, socket, ¾-inch |

(a) Attach a length of cable to fuel filler cover locking bar, in place in cover. Attach a hoist to cable. Lift top plate up and into position. Remove cable.

(b) Install cap screws which secure outer and rear sides of top plate (¾-inch socket wrench).

(c) Install cap screws which secure forward and inner sides of top plate (⁹⁄₁₆-inch socket wrench).

(d) Install battery compartment top plate (par. 148).

Note: This operation necessary only for installation of right fuel tank.

(4) **Install Shut-off Valve and Nipple** (fig. 75).

Pliers	Wrench, open-end, ½-inch
Screw driver	Wrench, open-end, ⁹⁄₁₆-inch
Wrench, adjustable	Wrench, open-end, ¾-inch

(a) Turn valve, nipple, and T-fitting into position in fuel tank (adjustable wrench).

(b) Connect primer fuel line to valve (⅜-inch open-end wrench).

Note: This operation applicable only when installing the left fuel tank.

(c) Connect fuel line to valve (¾-inch open-end wrench).

(d) Screw valve stems (half of universal joint) into valve (screw driver). Drop fuel tank shut-off valve handle rod into place in universal joint. Install connecting cotter pin (pliers).

(e) Place top rear engine compartment guard in position, and install retaining cap screws (⁹⁄₁₆-inch open-end wrench).

(5) **Connect Spade Winch Cable (155-mm Gun Motor Carriage)** (fig. 74).

Wrench, open-end, ⁹⁄₁₆-inch

(a) Place cable in position around winch. Slip U-bolt over cable, through winch drum, and install retaining nuts (⁹⁄₁₆-inch open-end wrench).

(b) Place winch crank in position, wind cable, and raise spade.

(6) **Install Tools and Equipment.** Stow tools and equipment which were removed to facilitate the removal of the fuel tank.

(7) **Install Drain Plugs and Covers.**

(8) **Fill Fuel Tanks.** Do not attempt to fill the tanks beyond the level to which they fill rapidly. This is done to allow for fuel expansion.

77. PRIMER PUMP (figs. 13, 81 and 82).

a. Description and Operation.
A primer pump, located on the instrument panel (fig. 13), provides a means of injecting a spray of fuel into the engine intake manifold to facilitate starting. The pulling stroke of the primer draws a charge of gasoline into the primer cylinder. During the return or charging stroke, the charge is delivered to a small distribution housing from which five lines run to the top five intake manifold pipes (fig. 82). To thoroughly atomize the gasoline, pull the primer plunger out slowly and push it in quickly.

b. Maintenance
(fig. 81). If more than a few strokes are required to prime the engine, the leather packing on the instrument panel end of the plunger should be checked for leakage. To stop leakage, compress packing by half turns of the packing nut located behind the priming plunger button on instrument panel until the leakage stops. If the primer pump no longer delivers gasoline to the engine, as evidenced by lack of resistance to pump-handle operation, replace primer pump. To test the discharge valve of the primer pump for leaks, proceed as follows: Start the engine and idle it at 800 revolutions per minute. Disconnect the outlet line of the primer pump and close the end of the line with a finger. If the idle speed differs from what it was with the line attached, the discharge valve of the primer pump leaks. Replace the defective primer pump with a serviceable unit.

78. PRIMER PUMP REMOVAL (figs. 13 and 81).

a. Equipment.

Wrench, open-end, ¾-inch   Wrench, open-end, 1-inch
Wrench, open-end, ⅝-inch

b. Procedure.
(1) Close fuel tank shut-off valves and open battery switch (par. 40b (1)).
(2) Remove bottom plate from instrument panel by removing six nuts from studs on bottom of instrument panel (¾-inch open-end wrench).
(3) Loosen screws that hold panel to mounting bracket (¾-inch open-end wrench).
(4) Tilt instrument panel forward and remove inlet and outlet lines from back end of primer pump (¾-inch open-end wrench). Mark or tag lines for correct installation.

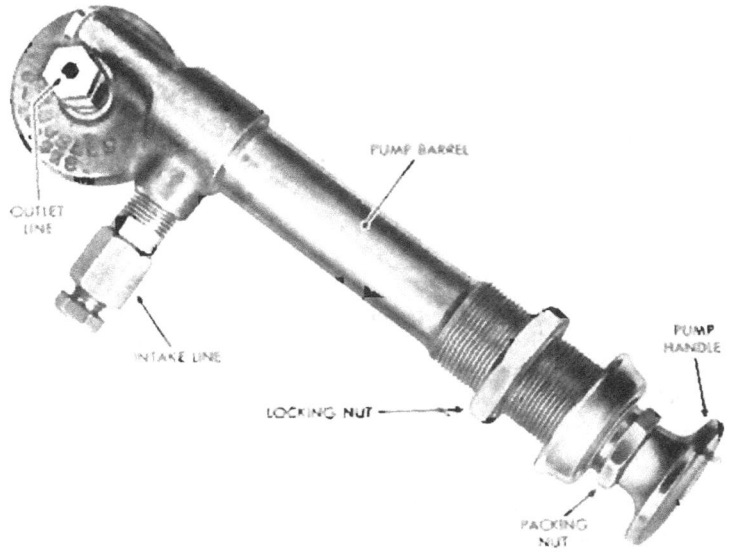

Figure 81. Primer pump.

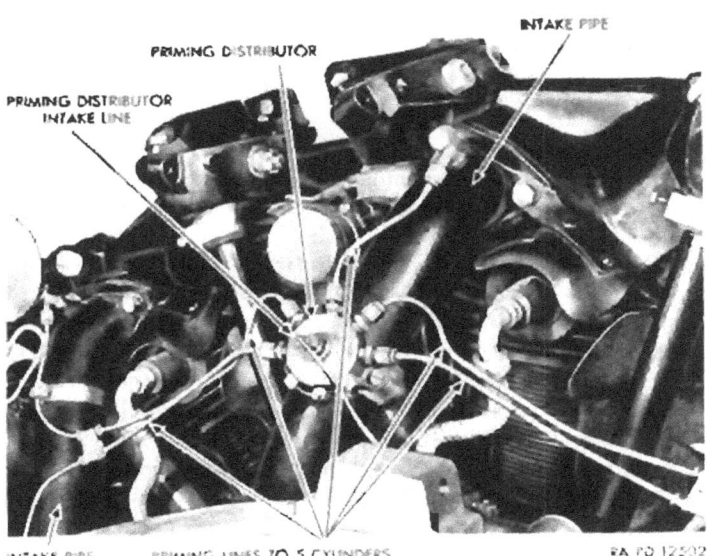

Figure 82. Priming distributor and lines.

(5) At rear of panel, loosen locking nut that holds pump to panel (1-inch open-end wrench).
(6) Remove nut on front end of pump barrel, and pull out plunger (⅜-inch open-end wrench). Pull pump out of panel from rear.

79. PRIMER PUMP INSTALLATION (figs. 13 and 81).
a. Equipment.
    Wrench, open-end, ⅜-inch   Wrench, open-end, 1-inch
    Wrench, open-end, ⅝-inch.

b. Procedure.
(1) Remove plunger and special nut from new or replacement primer pump, and install threaded end of barrel from rear of panel.
(2) Install plunger and special nut (⅜-inch open-end wrench) on panel end of pump and tighten the locking nut from back (1-inch open-end wrench).
(3) Connect inlet and outlet lines at side and end of pump (⅝-inch or open-end wrench).
(4) Push instrument panel in correct position, and install bottom plate on studs at bottom of instrument panel. Install and tighten six nuts that hold bottom plate (⅝-inch open-end wrench).
(5) Open fuel tank valves.

80. PRIMER LINE REMOVAL AND INSTALLATION (fig. 82).
a. Equipment.
    Screw driver   Wrench, open-end, ⅞₆-inch
    Wrench, open-end, ⅝-inch

b. Procedure.
(1) Broken or damaged priming lines can be replaced without special tools. Do not dent new lines, or constrict them by making sharp bends.
(2) Replace intake line to distributor by disconnecting it at the center fitting on the distributor (⅞₆-inch open-end wrench) and at the flexible coupling underneath the engine (⅝-inch open-end wrench). Also loosen two clips that hold line to engine support beam (screw driver).
(3) If replacing lines to any but No. 1 (top) cylinder, keep the two adjoining lines separated by means of the spacing clip. Shape lines so that they will not rub against projections on engine, particularly such parts as spark plug elbows and ignition harness.
(4) In replacing inlet line to distributor, secure line with three clips on engine support beam, to protect it against vibration.

81. DESCRIPTION OF AUXILIARY FUEL PUMP.
    a. Three fuel pumps, mounted together in parallel, form the auxil-

iary fuel pump. The pump is mounted on the lower-right-hand side of the rear engine bulkhead (fig. 66) in the engine compartment.

Note: The auxiliary pump on later type vehicles has been replaced by AC type fuel pump.

b. All fuel flows from the fuel strainer, past the inoperative auxiliary fuel pump, and then to the fuel pump mounted on the engine. In the event of a vapor lock in the engine fuel pump, stopping pump operation, the auxiliary fuel pump may be switched on by a toggle switch mounted on the instrument panel (fig. 81). This starts the auxiliary fuel pump which then furnishes the proper amount of fuel to the carburetor. ***Caution:*** Never operate the auxiliary fuel pump when the engine fuel pump is functioning properly. An excessive amount of fuel will be forced through the bypass regulator valve, flooding the carburetor.

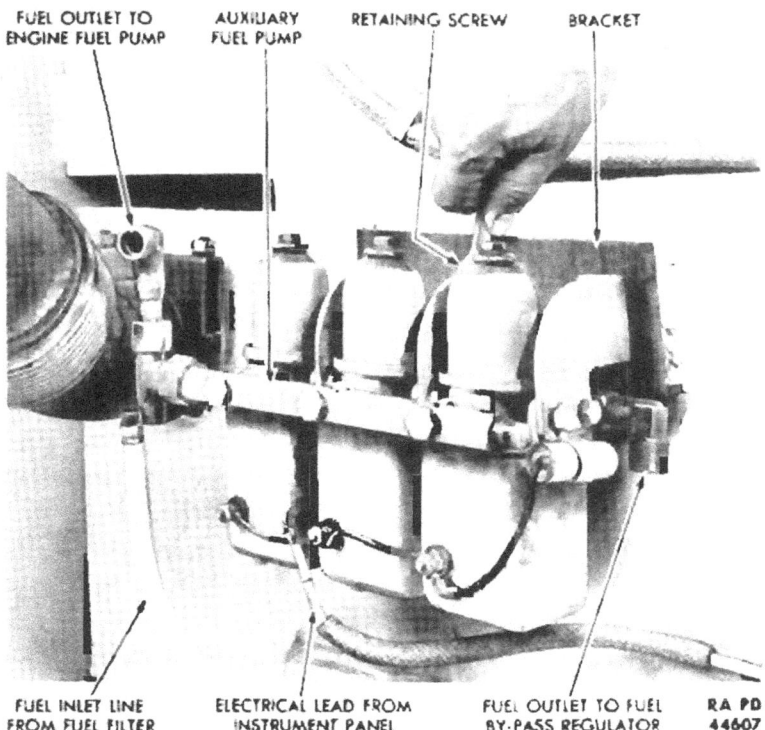

Figure 83. Auxiliary fuel pump removal.

82. AUXILIARY FUEL PUMP REMOVAL FROM VEHICLES OF EARLY MANUFACTURE (fig. 83).

a. Equipment.
Wrench, open-end, ¾-inch   Wrench, open-end, ¹³⁄₁₆-inch
Wrench, open-end, ⁷⁄₁₆-inch   Wrench, open-end, ⅜-inch

b. Procedure.

(1) Preliminary Steps.
(a) Close fuel shut-off valves (par. 40**b**(1)).
(b) Open battery switch.
(c) Remove engine compartment rear plates (par. 40**b**(4)).

(2) Disconnect Fuel Lines.

Wrench, open-end, ¹³⁄₁₆-inch   Wrench, open-end, ¾-inch

(a) Disconnect flexible fuel line from auxiliary fuel pump to fuel bypass regulator valve, at the auxiliary fuel pump end (¾-inch open-end wrench).

(b) Disconnect flexible fuel line from auxiliary fuel pump to fuel pump, at auxiliary fuel pump end (¹³⁄₁₆-inch open-end wrench).

(c) Disconnect metal tubing from fuel filter to auxiliary fuel pump, at auxiliary fuel pump end (¹³⁄₁₆-inch open-end wrench).

(3) Disconnect Electrical Lead.

Wrench, open-end, ⅜-inch.

Disconnect electrical lead at center terminal on front of auxiliary fuel pump (⅜-inch open-end wrench).

(4) Remove Auxiliary Fuel Pump.

Wrench, open-end, ⁷⁄₁₆-inch.

Remove six cap screws which secure auxiliary fuel pump to bracket welded on bulkhead (⁷⁄₁₆-inch open-end wrench) (fig. 83). Lift off pump.

83. AUXILIARY FUEL PUMP INSTALLATION ON VEHICLES OF EARLY MANUFACTURE (fig. 83).

(a) Equipment.
Wrench, open-end, ¾-inch   Wrench, open-end, ¹³⁄₁₆-inch.
Wrench, open-end, ⁷⁄₁₆-inch   Wrench, open-end, ⅜-inch

b. Procedure.

(1) Install Auxiliary Fuel Pump.

Wrench, open-end, ⁷⁄₁₆-inch

Place auxiliary fuel pump in position on bracket on bulkhead. Install six cap screws which hold pump in position (⁷⁄₁₆-inch open-end wrench).

(2) **Connect Electrical Leads.**

Wrench, open-end, ½-inch

Connect electrical lead to center terminal on front of auxiliary fuel pump (½-inch open-end wrench).

(3) **Connect Fuel Lines.**

Wrench, open-end, 13/16-inch   Wrench, open-end, ½-inch

(a) Connect metal tubing from fuel filter to right lower side of auxiliary fuel pump (13/16-inch open-end wrench).

(b) Connect flexible fuel line from engine fuel pump to upper right fitting on auxiliary fuel pump (½-inch open-end wrench).

(c) Connect flexible fuel line from fuel bypass regulator valve (½-inch open-end wrench).

(d) Turn on fuel shut-off valves. Remedy all fuel leaks.

(4) **Install Engine Compartment Rear Bulkhead Plates.**

Install engine compartment rear bulkhead plates (par. 70b(8)).

84. FUEL CUT-OFF SOLENOID (fig. 21).

a. **Description.** The fuel cut-off solenoid (fig. 21) operates a valve located on the right side of the carburetor. This valve is used to stop the engine by preventing passage of fuel from carburetor to engine. The solenoid is connected to this valve by means of an adjustable linkage and operated by a toggle switch mounted on the instrument panel.

b. **Maintenance.** The fuel cut-off solenoid requires little attention other than to be sure that the solenoid is in working order and that the linkage is adjusted to the length necessary to assure full closure of the valve and free return of the solenoid to the open position when the switch is turned off. The linkage is adjusted by removing the clevis pin and turning the clevis in or out to secure the correct position. If the solenoid is inoperative, it must be immediately replaced.

c. **Solenoid Shut-off Valve Removal.**

Pliers   Wrench, open-end, ½-inch.

(1) Remove the cotter pin and clevis pin from the linkage at the rod end (pliers).

(2) Remove the cotter pin and clevis pin holding the linkage to the bracket (pliers).

(3) Slide the linkage off the valve assembly.

(4) Remove the ½-inch nut holding the valve assembly in place on the fuel pump and slide out valve assembly.

d. **Solenoid Shut-off Valve Installation.**

Pliers   Wrench, open-end, ½-inch

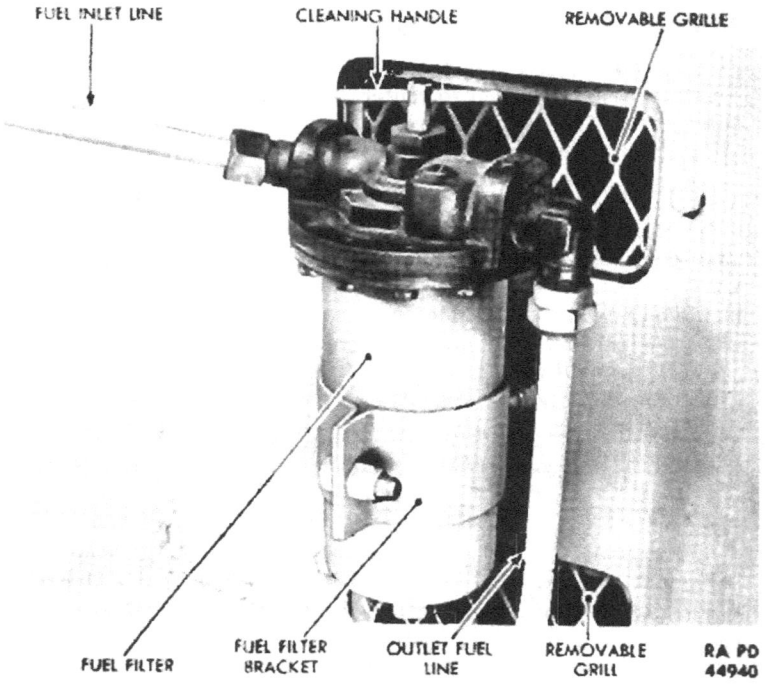

Figure 84. Fuel filter installed.

(1) Slide the valve assembly into the fuel pump, and tighten the ⅞-inch nut to hold it there.

(2) Slide the linkage into place on the valve assembly, and replace the clevis pin and cotter pin in the bracket.

(3) Replace the clevis pin and cotter pin in the linkage at the rod end.

85. GASOLINE FUEL FILTER REMOVAL, CLEANING, AND INSTALLATION ON VEHICLES OF EARLY MANUFACTURE (figs. 33 and 84).

a. Description. Vehicle fuel is cleaned by means of a disk-type fuel filter, located in the fuel system between the fuel tanks and the fuel pumps (fig. 73). The fuel filter is mounted in the left front corner of the engine compartment on the front engine compartment bulkhead (fig. 33). Fuel enters the filter body and is forced up through the filter element or cartridge. The cartridge consists of a series of fine, closely spaced, metal, slotted disks. The accumulated dirt is removed from the edges of the disks when the cartridge is turned by means of an external cleaning handle (fig. 84). This may be reached by removing the grille opposite the handle, from within the driving compartment. A complete turn in either direction is called for in the daily inspection. At the 50-hour inspection, drain, and flush the filter. At the 100-hour inspection, remove the filter, disassemble, and clean.

b. Equipment.
    Screw driver   Wrench, open-end, 5/16-inch
    Solvent, dry-cleaning   Wrench, open-end, 1/2-inch

c. Procedure.

(1) **Fuel Filter Removal** (fig. 84).
    Wrench, open-end, 5/16-inch.   Wrench, open-end, 1/2-inch.

  (a) Close fuel shut-off valves (par. 40**b**(1)).

  (b) Disconnect fuel inlet and outlet lines (1/2-inch open-end wrench).

  (c) Loosen bracket clamp bolt (5/16-inch open-end wrench), and lift out fuel filter.

(2) **Cleaning Fuel Filter.**
    Screwdriver   Wrench, open-end, 1/2-inch
    Solvent, dry-cleaning

  (a) In appearance, operation, and service requirements, the fuel filter is identical with the manually-operated oil filter (fig. 84).

  (b) Remove screws which secure filter cover to body (screw driver). Lift off filter cover and gasket.

  (c) Remove drain plug and washer from bottom of filter body (1/2-inch open-end wrench).

  (d) Thoroughly clean filter element, or cartridge, in dry-cleaning solvent. Flush out the filter body. *Caution:* Do not disassemble the disk assembly.

(3) **Assembling Fuel Filter.**
    Screwdriver   Wrench, open-end, 1/2-inch

  (a) Install drain plug and washer in bottom of filter body) (1/2-inch open-end wrench).

  (b) Place new gasket on filter cover. Assemble filter cover and filter body, and install retaining screws (screw driver).

(4) **Fuel Filter Installation** (fig. 84).
    Wrench, open-end, 5/16-inch   Wrench, open-end, 1/2-inch

  (a) Place fuel filter in position in bracket. Tighten bracket clamp screw (5/16-inch open-end wrench).

  (b) Connect fuel inlet and outlet lines (1/2-inch open-end wrench).

  (c) Open fuel tank valves and remedy all leaks.

86. FUEL LINES AND VALVES.

  a. Fuel Lines. All fuel lines are loom-covered where necessary for protection against rubbing. Frequent inspections of all lines and connections are necessary to insure against fuel leaks.

(1) **Maintenance.** At 100-hour check, carefully inspect fuel lines. Blow them out with compressed air if necessary.

(2) **Replacement.** When a defective fuel line is to be replaced, it is of extreme importance that the new line be made to conform exactly to the shape of original line.

b. Fuel Valves (fig. 75). A fuel shut-off valve is provided for each tank. the valve handles are located on the left and right sides of the driving compartment (figs. 14 and 146). At 100-hour check, remove and clean the fuel valves.

87. FUEL BYPASS REGULATOR VALVE REMOVAL (figs. 21 and 59).

a. General. If fuel bypass regulator valve fails to give proper fuel pressure, causing either too lean or too rich a fuel mixture, it must be replaced. No attempt should be made by the using arms to regulate or adjust the valve.

b. Equipment.
  Wrench, open-end, 5/16-inch   Wrench, open-end, 5/8-inch

c. Procedure.

(1) Close fuel tank shut-off valves and open battery switch (par. 40b (1)).

(2) Disconnect valve inlet line (line from carburetor inlet) at valve end (5/8-inch open-end wrench).

(3) Disconnect valve outlet line (line to fuel pump) at valve elbow (5/8-inch open-end wrench).

(4) Disconnect second valve inlet line (line from auxiliary fuel pump) at valve end (5/8-inch open-end wrench).

(5) Remove retaining nuts and U-bolt that holds valve to bracket, and remove valve (5/16-inch open-end wrench).

88. FUEL BYPASS REGULATOR VALVE INSTALLATION.
a. Equipment.
  Wrench, open-end, 5/16-inch   Wrench, open-end, 13/16-inch
  Wrench, open-end, 5/8-inch

b. Procedure.

(1) Insert new valve in U-bolt; install, and tighten nuts (5/16-inch open-end wrench).

(2) Connect valve outlet line to carburetor (5/8-inch open-end wrench) at elbow, and valve inlet line from auxiliary fuel pump (see note, par. 81a) (5/8-inch open-end wrench).

(3) Connect valve outlet line to engine fuel pump (¾-inch open-end wrench).

(4) Open fuel tank valves. Remedy all leaks.

89. TESTING FUEL BYPASS REGULATOR VALVE. If engine still shows evidence of too lean a mixture (lack of power, popping, uneven running, etc.) or too rich a mixture (black smoke from exhaust), the action of the fuel bypass regulator valve should be tested with a pressure gauge, and adjusted by ordnance maintenance personnel to the recommended pressure of 4½ pounds per square inch with full fuel tanks at 1,000 revolutions per minute before trying to locate any other cause of trouble.

90. GRADES OF ENGINE GASOLINE. Commercial motor gasoline, having an octane rating of 80, is preferred and should be used.

SECTION VI
LUBRICATION SYSTEM

91. GENERAL DESCRIPTION (fig. 85). The engine lubrication system depends upon the forced circulation of oil from a remote oil tank or reservoir, through the engine to the oil sump, out to the oil filter, through the engine oil cooler and back to the expansion hopper and oil reservoir. The amount of pressure built up in the system is determined by an oil pressure regulator valve located in the engine oil pump. Oil is picked up from the rear end of the oil sump by means of a main scavenging pump. Because the engine tilts forward, making the front end of the sump somewhat lower than the rear end, a second, or nose scavenging pump is used to pick up the oil from this end of the sump. The engine is vented by a single breather pipe with a "hat-type" filter. This breather is located just back of the No. 1 cylinder. The engine oil tank is vented to the atmosphere.

92. INSPECTIONS. The entire engine lubrication system must be given regular periodic inspections to detect any leaks or damage to lines before they can cause loss of engine oil pressure. It is particularly important to check the proper seating of the oil temperature gauge bulb, since air leaks may develop at this point. Regular inspections of the lubrication system are included in the daily, 50- and 100-hour periodic inspections (pars. 17 and 34).

93. TROUBLE SHOOTING.

   a. Testing Oil Pump Operation. If no oil pressure is shown on the instrument panel gauge, stop the engine immediately. The action

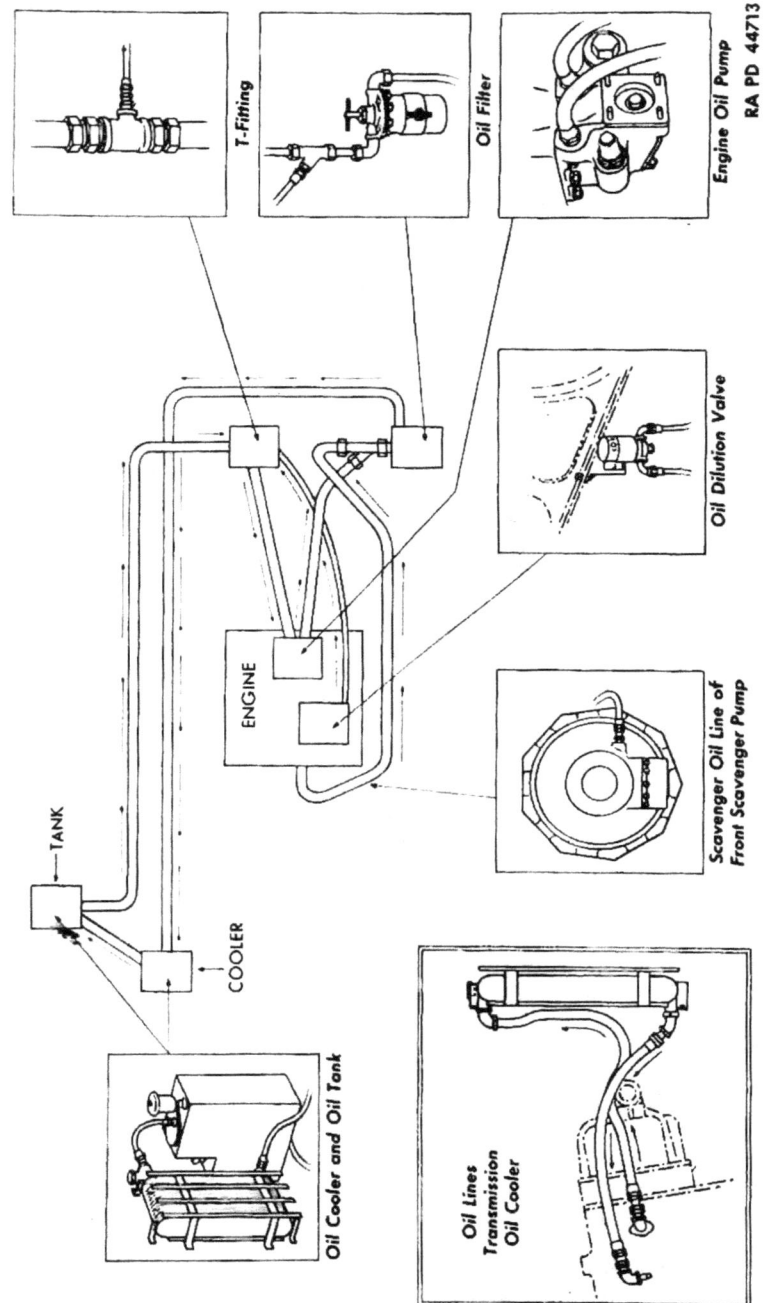

Figure 85. Lubrication system.

of the oil pump can be tested by disconnecting the oil inlet line from the top of the oil filter and priming the pump by pouring oil into the line. If the pump is operating properly it will suck oil while the engine is being turned over with the starting motor. If no oil is drawn in, remove the sump screen, clean, and replace. If pump still refuses to suck in oil, the pump must be replaced.

b. Checking the High-Pressure Relief Valve (fig. 40). If the oil pump is in working order, and oil pressure is low, check the high-pressure relief valve. To check the valve, stop the engine, remove it from the oil pump body, and inspect it for dirt particles in the valve seat. This valve must be kept absolutely clean, since even a minute dirt particle may hold the valve open, allowing oil from the supply line to be pumped back into the return line instead of into the engine.

c. Special Precautions. Too much emphasis cannot be placed upon the importance of tightening all oil line connections and installing the oil temperature gauge bulb securely, so that any possible air leak into the lubrication system is eliminated. Extreme care should be taken to prevent dirt, sand, or other foreign material from entering the lubrication system, particularly when filling the oil supply tank. The oil pressure gauge on the instrument panel should be carefully watched at all times; and, if the oil pressure is lost at any time, the engine should be stopped immediately and investigation of the failure made.

d. Reference to Engine Trouble Shooting. Because the lubrication system is so vital to engine operation, much of the trouble shooting procedure in the system has been included under engine trouble shooting (par. 39).

94. OIL FILTER REMOVAL, CLEANING AND INSTALLATION (disk type, manual) (figs. 32, 34, and 87).

a. Description. Lubricating oil in the vehicle is cleaned by means of a disk-type oil filter, located in the oil return line between the scavenging pumps and the oil cooler. The oil filter is in the left rear, bottom of the engine compartment, just inside and to the lower left of the bottom engine plate (fig. 32). The oil enters the filter body and is forced up through the filter element or cartridge. The cartridge (fig. 87) consists of a series of fine, closely spaced, metal, slotted disks. The accumulated dirt is removed from the edges of the disks when the cartridge is turned by means of an external filter handle. A complete turn in either direction is called for in the daily inspection. At the 50-hour inspection, drain, and flush the filter. At the 100-hour inspection, remove the filter, disassemble, and clean (fig. 87).

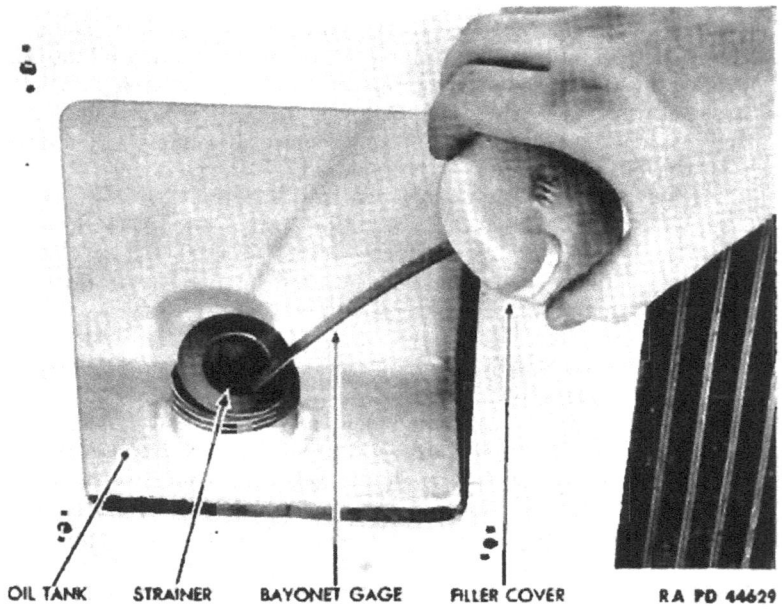

OIL TANK   STRAINER   BAYONET GAGE   FILLER COVER   RA PD 44629

Figure 86. Checking oil supply line with bayonet gauge.

b. Equipment.

Hammer
Pliers, diagonal cutting
Screw driver
Solvent, dry-cleaning

Wrench, open-end, 7/16-inch.
Wrench, open-end, 9/16-inch.
Wrench, open-end, 1 1/8-inch.
Wrench, open-end, 1 1/4-inch.

c. Procedure.

(1) **Removing Oil Filter.**

Wrench, open-end, 9/16-inch   Wrench, open-end, 1 1/8-inch.

(a) Disconnect inlet and outlet couplings on filter (1 1/8-inch open-end wrench).

(b) Loosen bracket clamp screw which holds filter and remove filter (9/16-inch open-end wrench) (fig. 34).

(2) **Cleaning Oil Filter** (fig. 87).

Pliers, cutting diagonal   Wrench, open-end, 7/16-inch.
Solvent, dry-cleaning   Wrench, open-end, 1 1/4-inch.

(a) Remove cap screws and washers from filter cover (7/16-inch open-end wrench).

(b) Remove filter cover and gasket from filter body.

(c) Remove locking wire on drain plug at bottom of filter body (pliers).

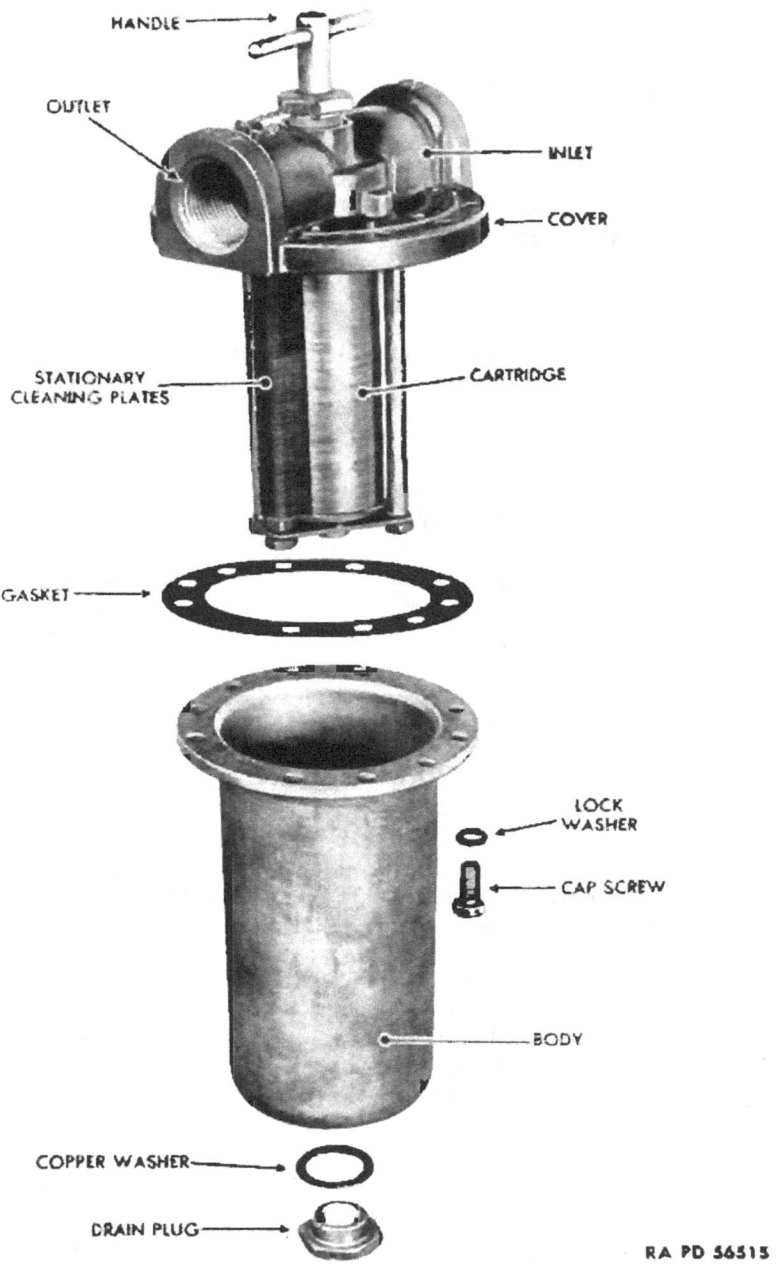

Figure 87. Parts of disk-type filter.

(d) Remove drain plug and copper washer from filter body (1⅛-inch open-end wrench).

(e) Thoroughly clean filter element, or cartridge, in dry-cleaning solvent, or other cleaning fluid. Flush out the filter body. *Caution:* Do not disassemble the disk assembly.

(3) **Assembling Oil Filter** (fig. 87).

    Hammer   Wrench, open-end, ⅝-inch
    Pliers   Wrench, open-end, 1⅛-inch
    Screw driver

(a) Install copper washer and drain plug into bottom of filter body and secure with locking wire (1⅛-inch open-end wrench and pliers).

(b) Place new gasket on filter cover, assemble body to cover, and install and tighten cap screws and locking lug washers (⅝-inch open-end wrench).

(c) Bend up locking lugs on washers (screw driver and hammer).

(4) **Installing Oil Filter** (fig. 32).

    Wrench, open-end, ⅝-inch   Wrench, open-end, 1⅛-inch

(a) Install filter in place and tighten bracket clamp screw (⅝-inch open-end wrench).

(b) Connect inlet and outlet couplings to filter and tighten (1⅛-inch open-end wrench).

Note: Connect the oil line from the scavenger pumps to the side of the cover marked in. If the connections are reversed, the filter will not function.

95. AUTOMATIC OIL FILTER.

a. Description. Some vehicles are equipped with a disk-type oil filter whose cleaning plates are turned automatically when the engine is operating. The plates are turned by means of an oil motor operated by engine oil pressure which is built into the head of the filter. Oil is forced through the filter and then out through the motor, causing the plates to turn.

b. Maintenance. Drain and clean the sump of automatic filters in the same way and at the same intervals as those manually operated. Check operation of filter at the end of operation by removing cap at top of filter and turning plates by hand. If stiff action indicates clogging, replace the filter with a serviceable unit.

96. OIL COOLERS (fig. 88).

a. Description. Individual oil coolers are provided to cool engine oil and transmission oil. Oil from engine enters at the bottom of the cooler and is forced upward through finned passages and returned to the oil tank. A thermostatic control allows cold or thick oil to pass directly to the oil tank without passing through the oil cooler passages, if undue pressure is developed in the cooler. Both oil coolers

Figure 88. Installation of oil tank and oil cooler.

are located in the engine compartment, and are mounted on the front engine bulkhead. The engine oil cooler is at the right, directly beside the engine oil tank. The transmission oil cooler is at the left, below the auxiliary generator set.

b. Equipment.
Air, compressed   Wrench, open-end, 1 13/16-inch
Solvent, dry-cleaning

c. Procedure. Oil coolers must be removed and thoroughly cleaned at every 100-hour inspection. Clean out interior oil passages with steam or flush with dry-cleaning solvent. Blow out fins and air passages with compressed air. Remove the bypass valve by turning out plug (1 13/16-inch open-end wrench) and lifting out spring and valve assembly. Clean all parts and check for free action before assembling. Replace if operation is in any way faulty.

97. OIL COOLER REMOVAL. Both the engine and transmission oil coolers are removed in the same manner, except that engine oil cooler inlet elbow is located on the engine compartment side of the bulkhead. They can be removed with the engine in the vehicle, but to do so will prove very difficult. They are more easily removed after engine removal.

a. Equipment.
Wrench, open-end, 7/16-inch   Wrench, open-end, 1 3/8-inch

b. Procedure.
(1) Remove engine compartment top guard (par. 40b(2)).
(2) Disconnect cooler inlet line at top, and oil return line at base of cooler (1 3/8-inch open-end wrench) (fig. 88).

Note: In removing elbow on cooler inlet line (engine compartment side of bulkhead), be sure to hold fitting sweated to cooler, to avoid damage to the fitting.

(3) Remove four cap screws that hold oil cooler guard to bulkhead, and remove guard (7/16-inch open-end wrench) (fig. 89).
(4) Remove cap screws which secure oil cooler to front engine bulkhead (7/16-inch open-end wrench) (fig. 89).
(5) Lift out cooler (fig. 90).

98. OIL COOLER INSTALLATION.

a. Equipment.
Wrench, open-end, 7/16-inch.   Wrench, open-end, 1 3/8-inch

b. Procedure.
(1) Slide oil cooler into place, first inspecting felt buffer strip on side guards, and replacing if needed (fig. 90).
(2) Install and tighten cap screws which secure cooler in place on front engine bulkhead (7/16-inch open-end wrench) (fig. 89).
(3) Connect cooler inlet and oil return lines (1 3/8-inch open-end wrench) (fig. 88).

(4) Install and tighten oil cooler guard in position on bulkhead (⁹⁄₁₆-inch open-end wrench) (figs. 88 and 89).
(5) Install engine compartment top front guard (par. 70**b**(**20**)).

99. ENGINE OIL TANK REMOVAL.
The oil tank can be replaced only after the engine and engine oil cooler have been removed. It is regularly removed, inspected, flushed with dry-cleaning solvent, and reinstalled at the 100-hour inspection.

a. Equipment.
Pan, oil drain
Screw driver
Wrench, open-end, 1⅝-inch
Wrench, open-end, 1¾-inch
Wrench, socket, ⁹⁄₁₆-inch
Wrenches, socket, ¾-inch (2)
Wrench, socket head set screw, ⁵⁄₁₆-inch

b. Procedure.
(1) Drain oil tank by removing drain plug plate and gasket from floor of hull (screw driver) and remove drain plug (⁵⁄₁₆-inch socket head set screw wrench). Catch oil (about 8 gallons) in an open container (oil drain pan).
(2) Remove engine (par. 40).
(3) Disconnect oil tank inlet line at cooler elbow (1⅝-inch open-end wrench) (fig. 88). *Caution:* If connection is tight, protect the elbow by bracing it with a 1¾-inch open-end wrench.
(4) Remove engine oil cooler (par. 97).
(5) Disconnect outlet line from tank to engine by disconnecting flexible coupling in floor of engine compartment (1⅝-inch open-end wrench). *Caution:* Do not attempt to disconnect the outlet line at the bottom of the tank before the tank is removed.
(6) Remove cap screws which secure steady-rest tube support to bracket welded to hull (¾-inch socket wrench).
(7) Remove bolts which secure engine steady-rest tube support to floor (two ¾-inch socket wrenches).

Note: This will require two men, one beneath hull to hold nut while man in compartment turns out bolt. Lift off support.

(8) Remove two cap screws at bottom front of oil tank (inside driving compartment) and two cap screws at bottom rear of oil tank (⁹⁄₁₆-inch socket wrench). Slide tank sideways and out.

Note: To avoid spilling oil while removing oil tank, replace drain plug during removal.

100. OIL TANK INSTALLATION (fig. 89).

a. Equipment.
Oil, engine, seasonal grade
Screw driver
Wrench, open-end 1⅝-inch
Wrench, socket, ⁹⁄₁₆-inch
Wrenches, socket, ¾-inch (2)
Wrench, socket head set screw, ⁵⁄₁₆-inch.

Figure 89. Oil cooler guard removal.

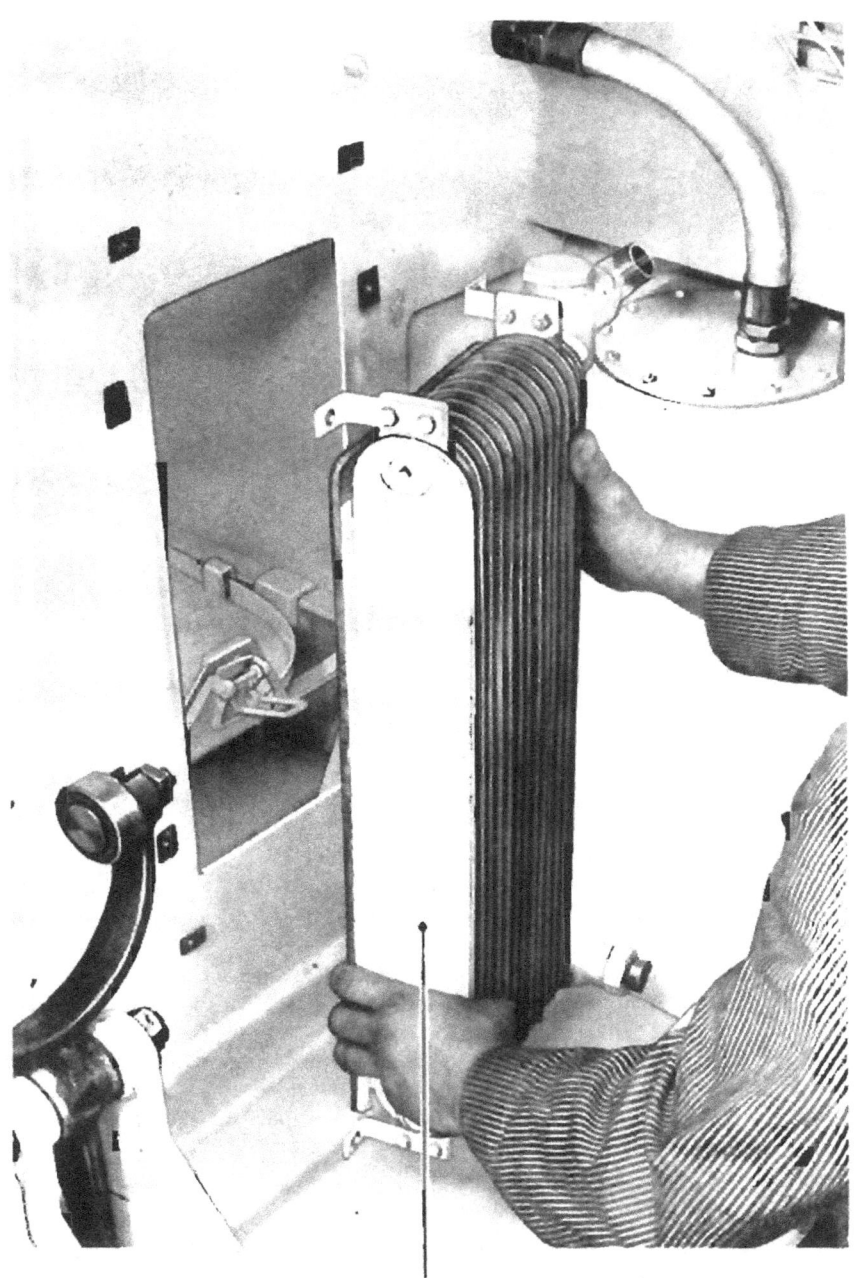

Figure 90. Oil cooler removal.

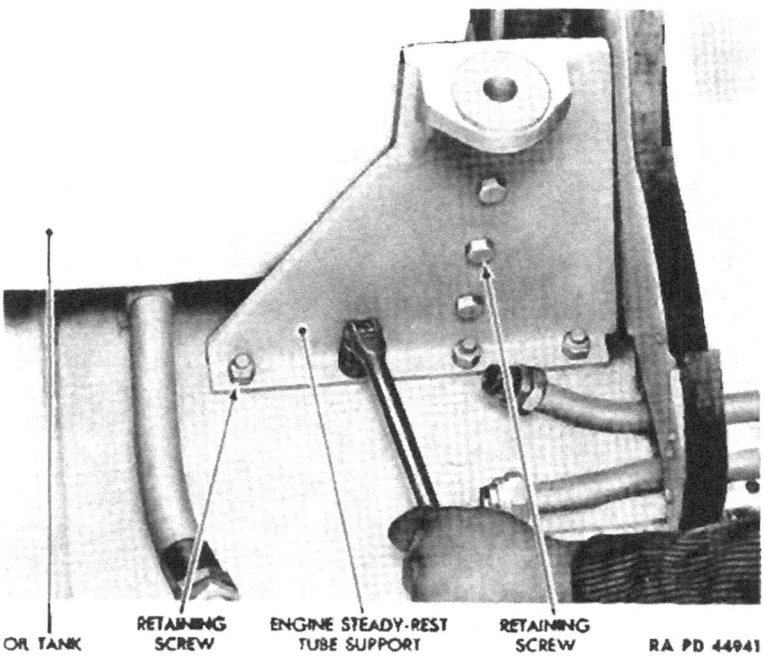

Figure 91. Engine steady-rest tube support removal.

b. Procedure.

(1) With drain plug in place, install reservoir in position in engine compartment, with flexible outlet line projecting out in engine compartment.

(2) Install two cap screws which secure bottom front of oil tank (inside driving compartment) and two cap screws which secure bottom rear of oil tank (%-inch socket wrench).

(3) Place engine steady-rest tube support in position. Install cap screws which hold support to bracket welded to hull (%-inch socket wrench). Install bolts which secure support to floor of hull (two %-inch socket wrenches).

Note: The latter operation will require two men, one beneath hull to hold nut, while man above installs bolt.

(4) Connect oil outlet and inlet lines oil tank (1%-inch open-end wrench) (fig. 88).

(5) Tighten drain plug (%-inch socket head set screw wrench) and install a new gasket and drain plug plate to under side of hull.

(6) Install engine oil cooler (par. 98).

(7) Install engine (par. 70).

(8) Fill tank with seasonal grade engine oil, of amount and grade specified in Lubrication Guide (sec. IV, ch. 1).

SECTION VII
COOLING SYSTEM

101. DESCRIPTION (fig. 92). The engine is cooled by an air blast produced by a fan mounted on the engine flywheel. The fan draws air through a screen in the engine compartment front top plate, and forces it between and around the finned cylinders of the engine. The warm air passes out through a screen in the engine compartment rear top plate. Air ducts are formed on the engine by baffles bolted around and between each cylinder and cylinder head. A shroud forms a further duct for the inlet of air through the guard. A shroud

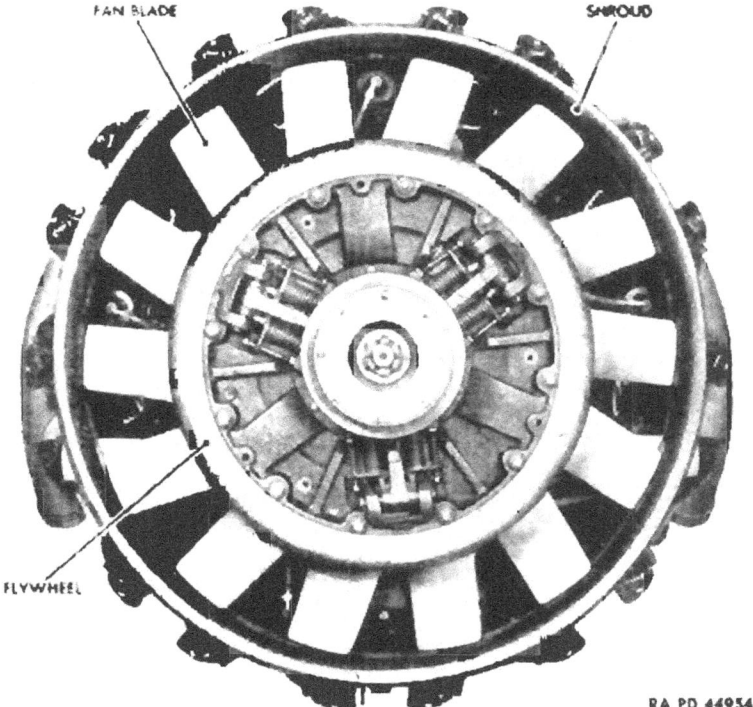

Figure 92. Front view of engine with clutch and flywheel installed.

in the engine compartment forces all air drawn in by the fan to pass directly around the engine cylinders.

102. INSPECTION AND MAINTENANCE. At the 100-hour inspection, the cooling fins on the engine cylinders and the surfaces of the shrouding and baffles must be cleaned thoroughly of all accumulated oil and dirt. All bolts on the fan ring, baffles, and fan cowling must be tightened and the fan cowling cleaned. Dirt must be cleaned from the cylinder heads, especially the lower two, at the 50-hour check. Inspect stowage on top of the vehicle to see that it does not interfere with the free entrance of air into the fan compartment. Refer to paragraph 107 for the removal of the fan and flywheel assembly.

SECTION VIII
CLUTCH

103. DESCRIPTION (fig. 92).

a. The clutch serves to engage the engine to, and disengage the engine from, the propeller shaft and the power train, which consist of transmission, differential, final drives, and sprockets. The clutch is built into the engine flywheel and is located in the front of the engine.

b. Some vehicles may have a ventilated clutch with a large inclosed ball throwout bearing installed on the clutch sleeve and some vehicles will have two small ball release bearings attached to the clutch yoke (fig. 94). The ventilated clutch is readily distinguished by the three openings in the face of the flywheel ring exposing the pressure plate driving lugs (fig. 101). Both clutches are completely interchangeable and the same instructions for disassembly, adjustment, and reassembly apply. However, to interchange a clutch equipped with large release bearing with one equipped with the two small release bearings, it is also necessary to change the clutch yoke.

104. OPERATION. The clutch allows the engine to pick up the load gradually after shifting gears. When the clutch is engaged, the pressure springs exert their full pressure against the clutch driving and driven plates, enabling engine power to go through driven plates, spindle, and companion flange to the propeller shaft. When the clutch pedal is depressed, the pressure plate is pulled back from the driven plates, compressing the clutch springs, allowing the clutch driven plates to stop rotating.

105. TROUBLE SHOOTING.

Possible cause — *Possible remedy*

a. Slipping.

Possible cause	Possible remedy
Weak spring action.	Notify ordnance personnel.
Facing torn loose from plate.	Replace plate (par. 107).
Sticking pressure plate.	Clean outer edge of pressure plate assembly.
Excessive oil soaking of plates.	Replace plates (par. 107). Notify ordnance personnel if the grease seals have been leaking.

b. Grabbing.

Possible cause	Possible remedy
Oil on lining.	Replace plates (par. 107).
Worn splines on spindle.	Replace spindle (par. 107).

c. Vibration.

Possible cause	Possible remedy
Release yoke loose on stud.	Replace yoke (par. 110).

d. Dragging (usually evidenced by hard shifting).

Possible cause	Possible remedy
Dirt in clutch.	Work clutch pedal rapidly up and down with transmission in neutral and engine idling. Blow out with compressed air. In extreme cases, disassemble clutch.
Improper adjustment of intermediate plate separating pins.	Adjust according to paragraph 109**b**(6).
Improper linkage adjustment.	Adjust according to paragraph 106**h**(6).
Bent clutch yoke preventing full sleeve travel.	Replace yoke (par. 110).
Broken intermediate plate.	Replace intermediate plate.
Binding linkage.	Clean and free linkage.
Bent release lever.	Replace lever.

106. CLUTCH CONTROLS AND CLUTCH CARE AND ADJUSTMENT.

a. Equipment.

Gun, pressure lubricating
Pliers
Wrench, open-end, 5/16-inch
Wrench, open-end, 1-inch
Wrench, socket, 5/16-inch

b. Procedure.

(1) Lubrication of Pilot Bearing and Outer Hub Bearing. Clutch pilot bearing and outer hub bearing are lubricated as prescribed in the Lubrication Guide (fig. 16).

(2) **Cleaning.** Cleaning will usually be required from accumulation of dust and dirt which is usually manifested by gradual reduction in full pedal travel and incomplete release (par. 105d). With engine at idling speed and transmission in neutral, work clutch pedal up and down rapidly, being careful not to force the pedal past the blocked position on the down stroke, with more pressure than is normally required to release the clutch, as clutch yoke and release levers may be sprung by excessive pedal pressure. Repeat this procedure until full pedal travel is obtained. Blow out clutch with compressed air when available. Clutch is disassembled for cleaning whenever required by some unsatisfactory operating condition which cannot be remedied in the vehicle. Clutch may also be disassembled and cleaned at any time that the engine is removed, if conditions warrant, particularly after operation in severe dust conditions.

(3) **Lubricating Clutch Release Bearing and Checking Clutch Bearing Clearance** (fig. 94).

Gun, pressure lubricating   Wrench, socket, $\frac{9}{16}$-inch

(a) Remove top engine compartment screen guard cap screws and lift off screen guard ($\frac{9}{16}$-inch socket wrench).

(b) Grease clutch bearings with pressure lubricating gun. Hand-oil outer surfaces of bearings with oil.

Note: These lubrication instructions apply only to the small type release bearings mounted on the clutch yoke (fig. 92). The large release bearing mounted on the clutch sleeve is prelubricated at the factory and should require no further lubrication during its normal life. If, for any reason, lubrication of this type bearing is necessary, it should be accomplished by ordnance personnel, as the special grease required is not available to the using arm.

(c) Check clearance between front flange and release bearing. Clearance should be $\frac{1}{4}$ inch (fig. 94).

(d) If bearing is too close to flange, adjust free play in clutch pedal linkage, as outlined in next step.

(e) Check clutch release. There should be at least $\frac{1}{2}$-inch travel of the clutch throw-out sleeve and propeller shaft should turn freely by hand (transmission in neutral) when the clutch pedal is depressed.

(4) **Lubricating Clutch Release Bearing and Checking Clutch Bearing Clearance** (fig. 94).

Gun, pressure lubricating   Wrench, socket, $\frac{9}{16}$-inch

(a) Remove top engine compartment screen guard cap screws and lift off screen guard ($\frac{9}{16}$-inch socket wrench).

(b) Grease clutch bearings with pressure lubricating gun. Hand-oil outer surfaces of bearings with oil.

(c) Check clearance between front flange and release bearing. Clearance should be $\frac{1}{4}$ inch (fig. 94).

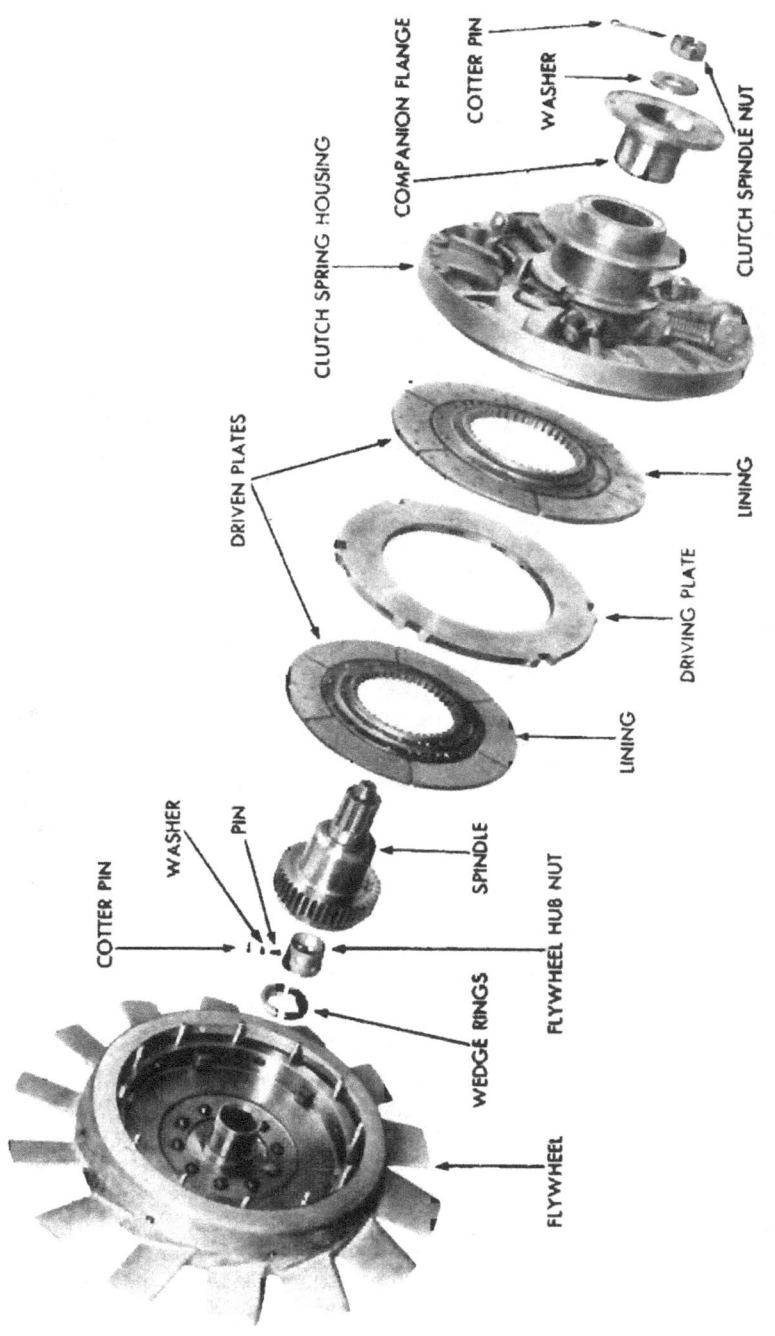

Figure 93. Flywheel and clutch assembly.

Figure 94. Clutch release bearings.

(d) If bearing is too close to flange, adjust free play in clutch pedal linkage, as outlined in next step.

(5) Adjusting Clutch Pedal Free Play (fig. 95).

Pliers
Wrench, socket, 9/16-inch.
Wrench, open-end, 1-inch

(a) Remove inspection plate from side of propeller shaft guard (9/16-inch socket wrench) (fig. 107).

(b) Remove cotter pin from clutch throw-out arm clevis pin (pliers) (fig. 93).

(c) Loosen locking nut on clevis and remove clevis pin (1-inch open-end wrench).

(d) Turn clevis to provide necessary ⅛-inch clearance at release bearing (pliers). Tighten locking nut (1-inch open-end wrench).

(e) Place clevis in position and install clevis pin and cotter pin (pliers).

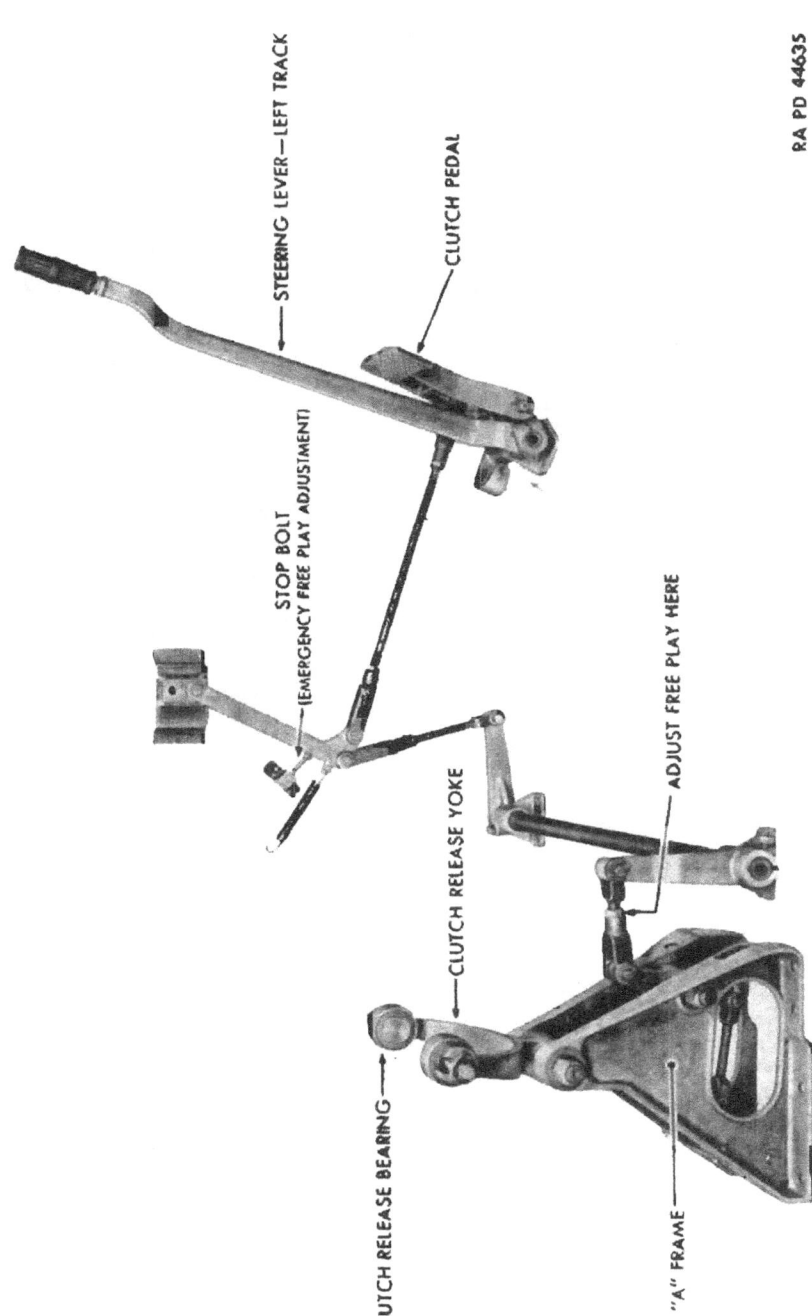

Figure 95. Clutch linkage.

Figure 96. Removing clutch companion flange.

(f) Install inspection plate on side bottom of propeller shaft guard (⁹⁄₁₆-inch socket wrench).

(6) **Emergency Adjustment of Clutch Free Play** (fig. 95).

Wrench, open-end, ⁹⁄₁₆-inch.

(a) Because of time or other limitations, it may be impossible to adjust the linkage as outlined in (5) above. In that case, an emergency adjustment of clutch pedal free play may be made. This is done by adjusting the stop bolt located on the hull next to the driver (⁹⁄₁₆-inch open-end wrench) (fig. 93). Turning the bolt down moves the clutch bearing toward the front flange; raising the bolt has the reverse effect. In making the adjustment, first loosen the clevis locking nut. Then adjust the bolt and tighten clevis locking nut *Caution:* Make proper adjustment on the clutch pedal clevis at the earliest opportunity and reset the emergency stop bolt to its original position.

(b) If it is necessary to replace any of the clutch control rods, remove cotter pins and clevis pins, and remove control rod. After installing new rod, check adjustment as outlined above.

107. CLUTCH AND FLYWHEEL REMOVAL. Engine must be removed to permit removal of clutch and flywheel.

a. Equipment.
Pliers
Wrench, socket, ¾-inch
Wrench, socket, heavy-duty, 1¹³⁄₁₆-inch, with sliding T-handle.

b. Procedure.

(1) Remove Clutch Companion Flange.
Pliers
Wrench, socket, heavy-duty, 1¹³⁄₁₆-inch, with sliding T-handle.

(a) Remove cotter pin from the clutch spindle nut (fig. 96) (pliers).

(b) Remove nut and then remove flat washer (1¹³⁄₁₆-inch heavy-duty socket wrench with sliding T-handle).

(c) Pull off companion flange.

(2) Remove Flywheel Ring.
Wrench, socket, ¾-inch.

(a) Remove twelve self-locking nuts that hold the flywheel ring to flywheel (¾-inch socket wrench) (fig. 97).

(b) If flywheel ring and flywheel are not already punch marked, mark these parts before removal to assure proper alignment in reassembly.

(c) Lift off flywheel ring and pressure plate assembly (fig. 98). If difficulty is encountered removing flywheel ring, insert three jack

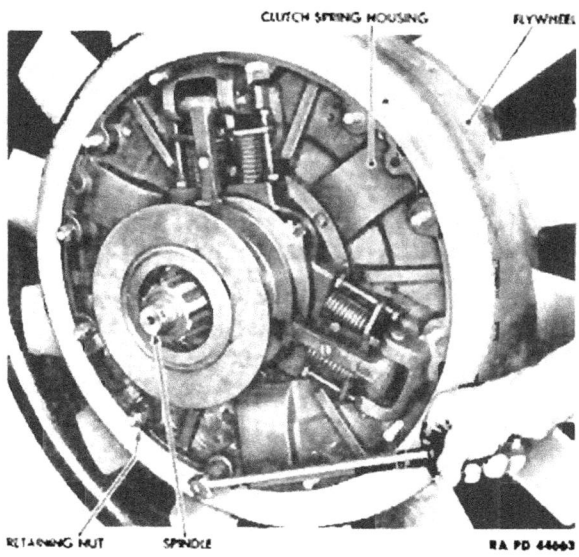

Figure 97. Clutch spring housing nuts removal.

Figure 98. Clutch spring housing removal.

screws in the three tapped holes provided near the outer circumference of the ring, tightening the jack screws alternately until ring is forced away from the flywheel.

(3) **Remove Clutch Plates.** Remove clutch driving and driven plates from flywheel.

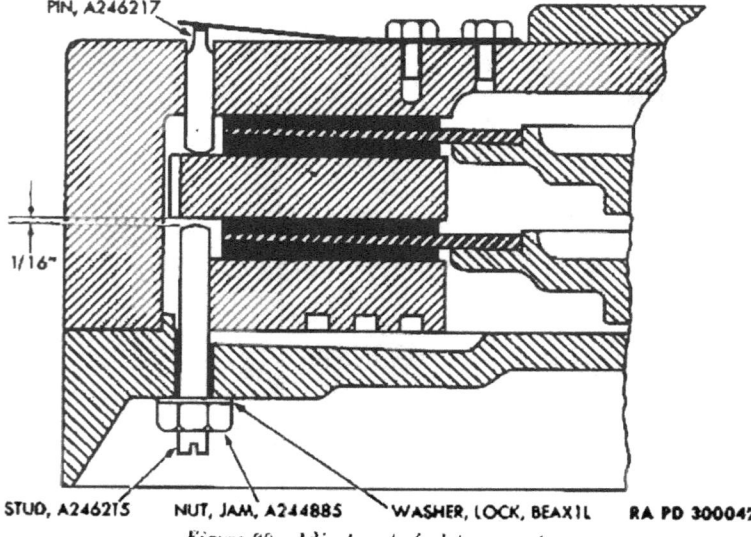
Figure 99. Adjustment of plate separators.

198

SPINDLE　　　　FLYWHEEL HUB　　RA PD 44665
Figure 100. Clutch spindle removal.

(4) **Remove Clutch Spindle.** Lift clutch spindle from hub of flywheel (fig. 100).

(5) **Remove Flywheel.**

Handle, sliding T　　　Pliers

(a) Do not remove the flywheel except in a case of emergency. If the facing of the flywheel is scored or the fan damaged, notify ordnance personnel.

(b) Remove cotter pin which secures flywheel hub nut locking pin (pliers). Lift off pin and washer.

(c) Place a sliding T-handle through flywheel hub nut, unscrew hub nut, and remove nut and front cone.

(d) Slide flywheel off crankshaft taking care not to damage crankshaft splines.

(e) If rear brass cone is damaged or scored, remove by inserting a screw driver in slot to spread cone and slide it off the shaft.

Figure 101. Ventilated clutch and fan assembled.

RA PD 44668

Figure 102. Flywheel hub nut removal.

RA PD 44671

Figure 103. Flywheel removal.

108. CLUTCH AND FLYWHEEL INSPECTION.

a. Clean clutch plates and examine them for glazing, warpage, cracks, scoring, and worn or damaged linings. Replace any defective plates.

b. Return plates with faulty linings to ordnance personnel for installation of new linings.

c. Inspect spindle inner and outer bearings for wear.

d. Inspect bearing races for pitting or wear. Bearings should not be removed unless it is necessary to replace them, in which case they should be removed by ordnance personnel. Inspect the oil seals. If the seals have been damaged in the disassembly procedure, or have been leaking grease, notify ordnance personnel.

e. Clean flywheel and inspect for score marks, cracks, or other damage.

f. Check splines of spindle for burs. If the splines of the spindle are excessively worn, replace the spindle with a serviceable unit.

109. CLUTCH AND FLYWHEEL INSTALLATION.

a. Equipment.

Drift, brass
Hammer
Handle, sliding T
Pliers
Wrench, socket, ¾-inch
Wrench, socket, heavy-duty, 1 13/16-inch.

b. Procedure.

(1) Install Flywheel.

Handle, sliding T   Pliers

(a) Install flywheel rear brass cone on crankshaft, and slide flywheel carefully on crankshaft (fig. 103).

(b) Place flywheel hub nut with front cone on crankshaft (fig. 102). Tighten flywheel hub nut to 550 to 650 foot-pound torque (approximately 150 pounds at the end of a 4-foot bar).

(c) Install flywheel hub nut locking pin, washer, and cotter pin (pliers) (fig. 101).

(2) Install Clutch Spindle.

(a) Clean roller bearing in spindle and pack with ball and roller bearing grease. Place film of grease on leather seal.

(b) Slide the clutch spindle carefully on the hub of the flywheel (fig. 100).

(3) Install Clutch Plates.

(a) Install first driven disk with flange of hub toward the outside.

(b) Install intermediate driving plate with arrow visible and pointing counterclockwise (fig. 104).

Note: Only the cored driving plate used on early production clutches has the indicating arrow. Solid driving plate used on late production clutches has no arrow and can be installed either way.

(c) The intermediate plate must drive through all six flywheel lugs to assure even load distribution. Turn engine on stand so that crankshaft is vertical, if possible. Check contact between driven face of flywheel lugs; if plate does not drive on all six lugs simultaneously, file driven side of slots as required. Special pains must be taken when filing slots to keep all surfaces square and parallel.

(d) Check side clearance on backlash side between slot in intermediate plate and flywheel lugs. Slot clearance should be 0.012 to 0.018 inch. File slots as required until proper clearance exists.

(e) Install second driven disk with flange of hub toward outside. **Caution:** Do not allow spindle to be moved forward after plates are installed as inner driven disk will fall down behind spindle which will necessitate starting over again.

Figure 104. Arrow marks on drive plate

(4) Install Flywheel Ring Assembly.

Drift, brass   Wrench, socket, ¾-inch
Hammer

(a) Clean ball bearing in flywheel ring hub and pack with ball and roller bearing grease.

(b) Slide flywheel ring assembly into place on spindle (fig. 98). Punch marks on housing must be lined up.

(c) Install and tighten the 12 self-locking nuts that hold ring to flywheel to 75 to 85 foot-pounds torque (fig. 97) (¾-inch socket wrench).

(5) Install Companion Flange (fig 96).

Pliers   Wrench, socket, heavy-duty 1¹³⁄₁₆-
                           inch, with sliding T-handle

Slide companion flange into place in clutch housing hub. Do not damage or turn back tips of grease seal in hub. If it is difficult to enter companion flange through seal, use shim stock to protect seal. Install and tighten flat washer and hub nut, securing nut with a new cotter pin (pliers and 1¹³⁄₁₆-inch heavy-duty socket wrench with sliding T-handle).

(6) Adjust Intermediate Plate Separating Stud and Pin.

(a) Loosen jam nuts on the three plate separating studs on the front face of the flywheel ring.

(b) Tighten the three plate separating studs to contact intermediate driving plate.

(c) Back off studs one turn which will give ¹⁄₁₆-inch clearance between tip of stud and intermediate driving plate as shown on figure 99.

(d) Tighten jam nuts.

110. CLUTCH RELEASE BEARINGS REMOVAL, INSPECTION, AND INSTALLATION.

a. Equipment.

Grease, OD, No. 1   Wrench, open-end, ¾-inch
Screw driver   Wrench, open-end, 1¼-inch.
Solvent, dry-cleaning

b. Procedure.

(1) The following procedure applies only to the type of clutch release bearings that are mounted on the clutch yoke (fig. 94); these bearings can be removed either with the engine in or out of the vehicle. The large release bearing mounted on the clutch throw-out sleeve cannot be disassembled from its housing; do not remove this bearing from the throw-out sleeve for inspection or cleaning, but only to replace a defective bearing.

(2) **Removal of Clutch Release Bearings.**
  (a) Remove lubrication fitting from stud (⅜-inch open-end wrench).
  (b) Remove set screw (screw driver).
  (c) Loosen nut on outer end of stud that holds right bearing to yoke and slide stud, bearing, lock washer, and nut off yoke (1¼-inch open-end wrench).
  (d) Repeat operations on left bearing.
  (e) Remove nut and lock washer from each stud and tap off bearing.

(3) **Inspection of Clutch Release Bearing.**

  Solvent, dry-cleaning

  (a) Clean bearing with dry-cleaning solvent.
  (b) Examine outer surfaces of bearings for burs, wear, or scoring.
  (c) Rotate bearings carefully to check presence of flat spots.
  (d) Replace damaged or worn bearings.

(4) **Installation of Clutch Release Bearing.**

  Grease, OD, No. 1   Wrench, open-end, ⅜-inch
  Screw driver   Wrench, open-end, 1¼-inch

  (a) Pack bearings with OD, No. 1 grease, and oil outer race.
  (b) Install bearing on stud, being sure it is properly seated.
  (c) Slide stud down into place in slot in yoke, with bearing inside of yoke.
  (d) Install and tighten lock washer and nut on outer end of stud (1¼-inch open-end wrench). Line up set screw hole in stud and nut.
  (e) Install set screw (screw driver).
  (f) Install lubrication fitting on outer end of stud (⅜-inch open-end wrench).

SECTION IX
PROPELLER SHAFT

111. DESCRIPTION (figs. 105 and 106). The propeller shaft transmits power from the clutch to the input shaft of the transmission. It is equipped with two universal joints, one at each end, which permit operation at an angle between the transmission and the clutch. A slip joint at the forward end permits lengthening and shortening of the shaft. This joint also facilitates removing or replacing the propeller shaft in the vehicle, and in the installation of the engine.

112. INSPECTIONS (figs. 105 and 106). The following inspections are made of the propeller shaft:

  a. The universal joints are inspected during 50- and 100-hour inspections to determine whether bearings and cups are worn. Symptoms of wear on these parts are excess vibration and noise while shaft

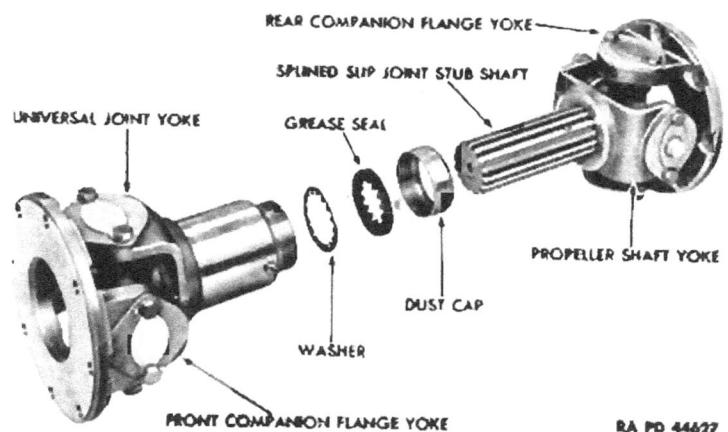

Figure 105. Propeller shaft with slip joint removed.

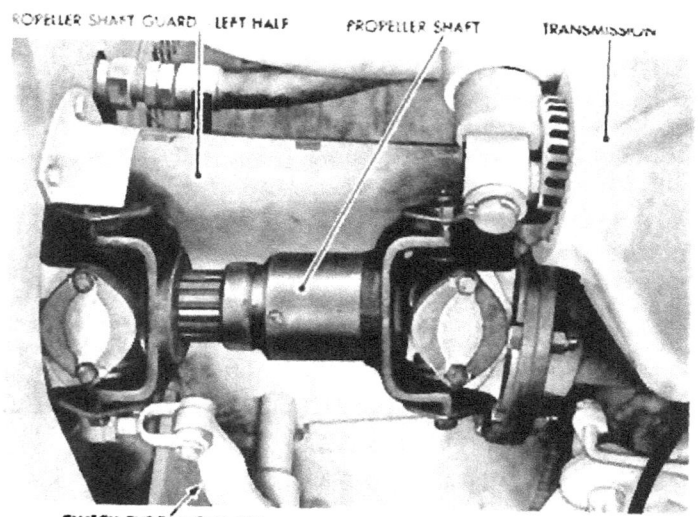

Figure 106. Propeller shaft.

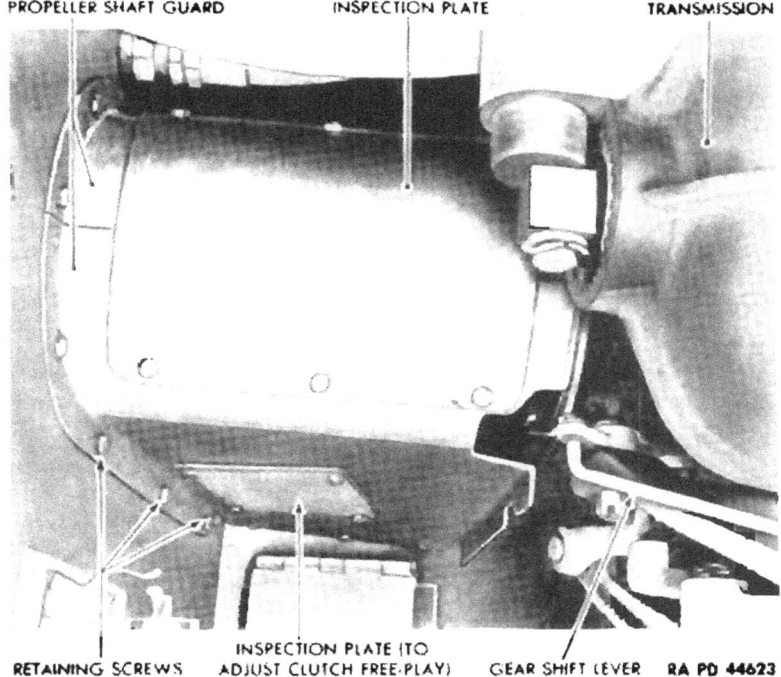

Figure 107. Propeller shaft guard.

is in motion. Worn bearings will necessitate replacement of the propeller shaft assembly.

b. Inspection must be made for leaking grease seals, indicated by streaks of lubricant around the seal. Leaky seals must be replaced. This will require separation of the slip joint (par. 114).

c. With engine stopped, test the universal joints for wear, and the bolts on the companion flanges for tightness. Tighten, if necessary.

113. LUBRICATION. The universal joints and the slip joint are lubricated as prescribed in the Lubrication Guide (sec. IV, ch. 1). Note that universal joints are to be repacked very 1,000 miles.

114. PROPELLER SHAFT REMOVAL (figs. 105, 106, and 107).

a. **Equipment.**

Wrench, socket, ⅞-inch.   Wrench, socket, ⅝-inch.

b. Procedure.
(1) Remove Propeller Shaft Guard.

Wrench, socket, 7/16-inch.

(a) Remove cap screws which secure inspection plate to top of propeller shaft guard (7/16-inch socket wrench). Lift off inspection plate (fig. 107).

(b) Remove cap screws which secure right half of propeller shaft guard to front engine compartment bulkhead and to flange on floor 7/16-inch socket wrench) (fig. 107).

(2) Remove Propeller Shaft.

Wrench, socket, 9/16-inch.

Remove bolts and nuts which secure propeller shaft companion flange yokes to companion flanges on transmission and on engine (9/16-inch socket wrench). Lift out propeller shaft.

(3) Disconnect Propeller Shaft. Unscrew dust cap, and remove grease seal and washer. Disconnect slip joint by pulling stub shaft out of hub. *Caution:* Before disconnecting at slip joint, it is important to note whether there are alignment arrows on the front and rear sections of the slip joint. If there are none, these sections must be marked so that they can be lined up properly when assembled.

115. PROPELLER SHAFT INSTALLATION (figs. 105, 106 and 107).

a. Equipment.

Wrench, socket, 7/16-inch.   Wrench, socket, 9/16-inch.

b. Procedure.

(1) Assemble Propeller Shaft. Slide stub shaft into slip joint hub, first making sure dust cap, washer, and grease seal are in place. Screw dust cap into place. *Caution:* When assembling slip joint, always make sure alignment marks on propeller shaft and on slip joint hub are exactly in line. If on disassembly of propeller shaft there were no arrows on front and rear sections of slip joint, and no marks were made as suggested in paragraph 114**b(3)**, the propeller shaft can be aligned properly on assembly by joining front and rear sections of slip joint in such a position that yokes on front and rear universal joints lie in same plane as in figure 106. It is extremely important that slip joint sections be aligned properly, as a vibration of propeller shaft will result if universal joints are out of alignment.

(2) Install Propeller Shaft.

Wrench, socket, 9/16-inch.

Place propeller shaft in position. Install bolts and nuts which secure propeller shaft's companion flange yokes to companion flanges on transmission and on engine ($\%_{16}$-inch socket wrench).

(3) Install Propeller Shaft Guard.

Wrench, socket, $\%_{16}$-inch.

(a) Place right half of propeller shaft guard in position and install retaining cap screws ($\%_{16}$-inch socket wrench).

(b) Place inspection plate in position on top of propeller shaft guard and install retaining cap screws ($\%_{16}$-inch socket wrench).

SECTION X
POWER TRAIN

116. DESCRIPTION (fig. 108).

a. General. Power is transmitted from the propeller shaft to the tracks by means of the power train, which consists of the transmission, controlled differential, final drives, and sprockets.

b. Transmission. The transmission has five forward speeds and one reverse speed. Second, third, fourth, and fifth gears are synchronized to permit easier shifting, and first and reverse gears are constant mesh. A parking brake, which acts on the output shaft, is built into the transmission, and is operated by a lever to the right of the driver. This brake is used only after the vehicle has been brought to a stop.

c. The controlled differential (fig. 109), which incorporates the device for steering, contains two "halves," each of which is separately controlled by a brake drum and brake shoe. When one brake is applied, one-half of the differential slows down while the other half speeds up. By this means the vehicle is steered. It is impossible to lock one final drive sprocket as long as the other sprocket is in motion.

117. TRANSMISSION AND DIFFERENTIAL LUBRICATION.

The transmission, together with the controlled differential, contains 64 quarts of oil, circulated through the mechanism and back to the transmission oil cooler by a small pump built into the transmission case and driven by the rear end of the countershaft. Oil is added, and the level is checked, by removing a bayonet-type oil gauge built into the oil filler cap. The level of oil must be checked at the end of each day's run, and additions made to keep the level up to the FULL mark. Oil as specified on the Lubrication Guide (figs. 16 and 18) is to be used. The transmission and controlled differential oil will be drained at intervals, as indicated in Lubrication Guide, by the removal

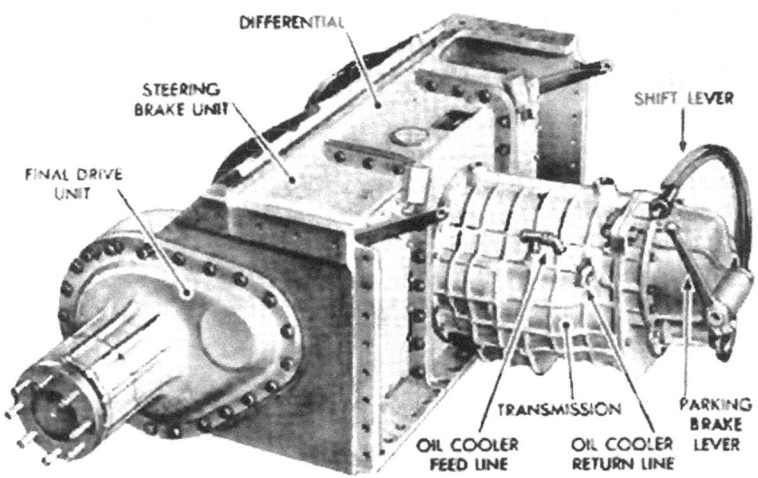

Figure 108. Power train.

of a plug in the bottom of the controlled differential case. This plug is reached from underneath the vehicle. The transmission oil cooler is of the same size and type as the engine oil cooler, and is located on the left side of the bulkhead. It is equipped with a spring-loaded bypass valve to permit cold, thick oil to circulate without passing through the cooler.

118. STEERING BRAKE ADJUSTMENT (figs. 109 and 110).

Rule, 6-inch
Straightedge
Wrench, adjustable
Wrench, deep socket, 1¼-inch

a. Remove brake adjusting hole plugs from each inspection plate. Turn the adjusting nut until it is flush with the end of the stud (¼-inch deep socket wrench).

b. Check position of the brake shafts. With brake bands properly installed, the mark on the shaft ends will be vertical.

c. Measure distance from the top of the left brake assembly housing to the top of the left brake arm (6-inch rule and straightedge). This measurement is taken with a straightedge placed along the top of the housing even with the cover plate, and a rule measuring the distance vertically from the straightedge to the brake arm. The distance should be as near to 2¾ inches as the position of the serrations on the shaft will allow (fig. 114). Any great deviation from this measurement will decrease steering leverage. If arm position is incorrect, loosen clamping bolts on the arm (adjustable wrench), slide it off, and install it on

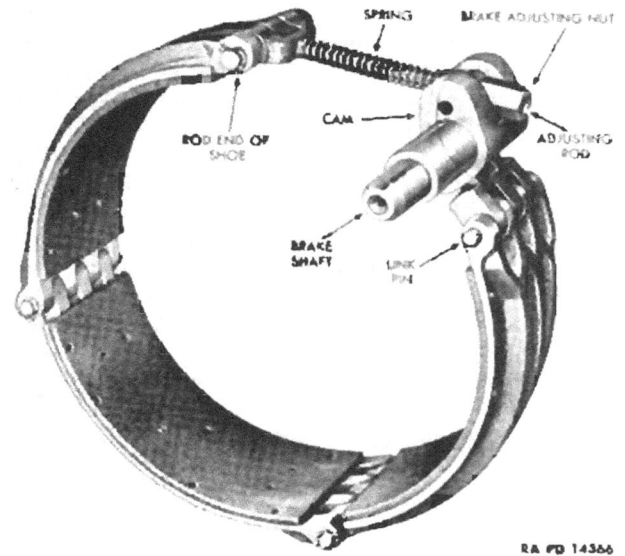

Figure 109. Brake band.

shaft on the serration bringing the arm to the point closest to required measurement.

NOTE: If new bands are being installed the steering lever linkage will have to be disconnected at brake arm. If merely an adjustment of brakes is being made, it will be necessary to disconnect linkage at brake arms before making any change in position of arm on shaft.

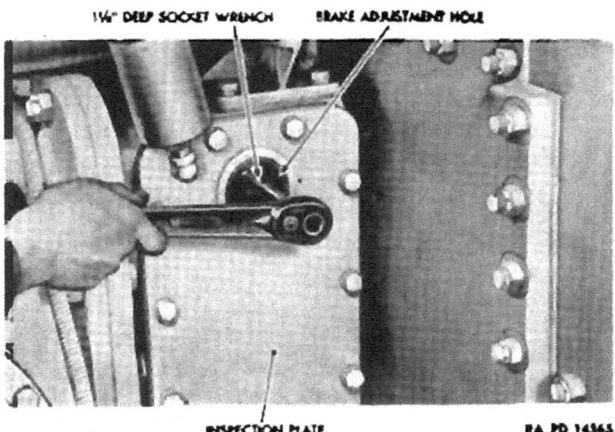

Figure 110. Steering brake adjustment.

d. If right brake arm is installed with the linkage toward the rear, the arm position should be the same as the left brake arm. If the linkage is toward the front, check position as follows:

(1) Measure distance from top of right brake assembly housing to outer end of right brake arm (just above the pin) using a straightedge and 6-inch rule (or flexible steel rule). The distance should be as near to 3½ inches as the position of the serrations will allow (fig. 115).

(2) If necessary to change position of arm, disconnect the linkage to the cross shaft, relocate arm, and connect linkage.

(3) Back off brake adjusting nut about four full turns, loosening the bands. *Caution:* Care should be taken in removing socket wrench to make sure that the socket does not come loose from the handle and drop down into housing.

(4) Set steering lever stops so that levers are even with each other and parallel with linkage rods when against stops.

(5) Connect linkage, changing positions of clevises, if necessary, to allow steering levers to rest against stops when arms are in correct position.

(6) Tighten each brake adjusting nut so band will take hold of drum after 5¼ inches of travel, measured at the top of steering lever. With

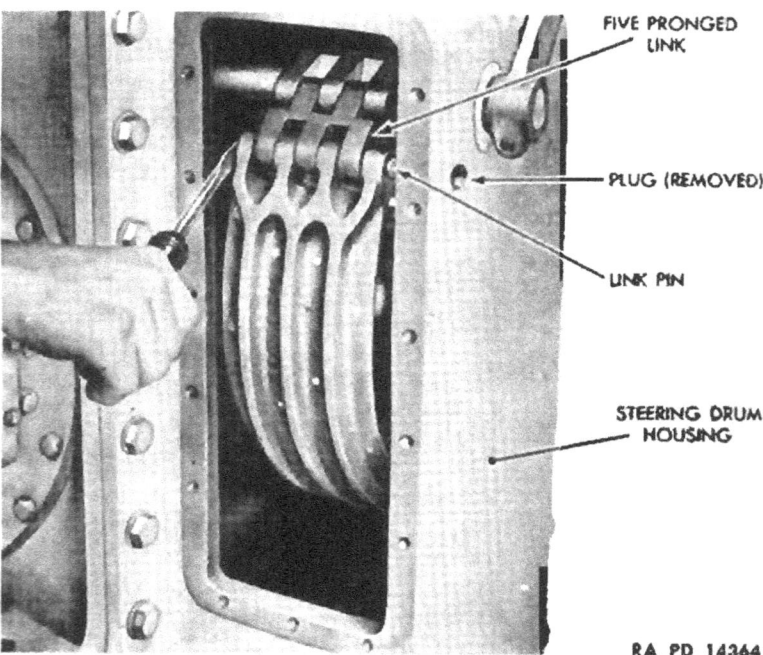

Figure 111. Disconnecting brake band from link.

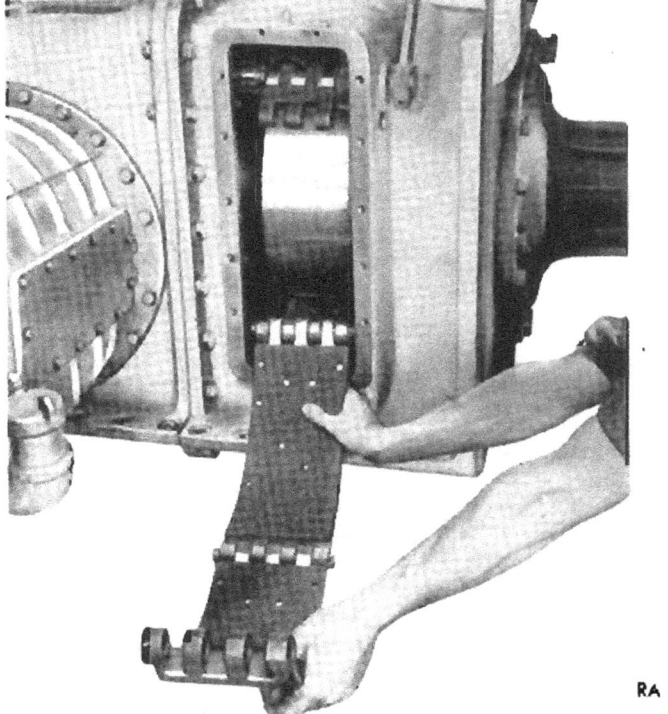

Figure 112. Pulling brake band out from below drum.

less than this amount of free travel, bands will ride on drums, causing heating and wear. More than this amount of free travel reduces total amount of braking available.

NOTE: The brake adjusting nut has a cylindrical surface on the pressure end instead of the usual flat face. It is, therefore, imperative that this cylindrical surface be seated firmly against the cross pin when above adjustment is completed.

119. BRAKE BAND REMOVAL (figs. 109, 111 and 112). When either steering lever is pulled back and track on that side fails to stop, adjustment of brakes is necessary.

a. Equipment.

Bar, pinch
Pliers
Screw driver
Wrench, deep socket, 1⅛-inch
Wrench, open-end, ⅞-inch
Wrench, open-end, 1 3/16-inch
Wrench, socket, ¾-inch

b. Procedure.
(1) Open Differential Housing.

Wrench, open-end, 1¾₆-inch   Wrench, socket, ¾-inch

(a) Drain oil from differential (1¾₆-inch open-end wrench).

(b) Remove bolts from brake inspection plate and remove plate (¾ inch socket wrench).

(2) Break Steering Linkage.

Pliers

Disconnect steering linkage at brake arms (steering levers) by removing the cotter pins and clevis pins (pliers) (fig. 113).

(3) Disconnecting Brake Bands.

Pliers   Wrench, open-end, ¾-inch
Screw driver   Wrench, open-end, 1¾₆-inch
Wrench, deep socket, 1¼-inch

(a) Turn down centering bolt beneath band to prevent brake band hinges from catching on bolt (¾-inch open-end wrench).

(b) Remove cotter pin from differential end of link pin that holds brake band to bottom end of five pronged links (pliers) (fig. 111).

(c) Remove plug in side of steering drum housing (fig. 111), and remove link pin through plug opening (1¾₆-inch open-end wrench).

(d) Remove brake adjusting nut, holding adjusting rod in place with the hand (1¼-inch deep socket wrench).

(e) Shove rod forward, securing spring and washer on a screw driver or drift so that they do not drop into housing. Both ends of brake band are now free.

(4) Remove Brake Bands.

Bar, pinch

(a) Using a pinch bar to aid in working the brake bands out, remove bands by pulling out from around bottom of drum. Care must be taken to work hinge sections over centering bolt (fig. 112).

(b) After the brake bands have been removed, examine them to determine whether they are worn through to rivet heads. Also, examine the drums to determine whether they have become scored. If bands are worn excessively, new bands must be installed. Scored drums must be reported to the next higher authority.

120. BRAKE BAND INSTALLATION (figs. 110, 111, and 112.)

a. Equipment.

Pliers   Wrench, open-end, 1¾₆-inch.
Wrench, deep socket, 1¼-inch.   Wrench, sock, ¾-inch.
Wrench, open-end, ¾-inch.

b. Procedure.
(1) Install Brake Bands.
    Pliers   Wrench, deep socket, 1¼-inch.

(a) Push rod end of brake bands up behind brake drum until it can be reached from top. A wire fastened to the end of band and pulled up tight around the drum will help hold bands away from centering bolt.

(b) Insert washer and spring over rod, and insert rod through cross arm in upper cam section, holding it in place with adjusting nut (1¼-inch deep socket wrench).

(c) Bring other ends of bands up around front of drum to link.

(d) Insert link pin through plug opening in housing and, lining up holes in link and band, insert link pin to connect band to link.

(e) Install and secure cotter pin (pliers).

(2) Close Housing.
    Wrench, open-end, ¾-inch   Wrench, socket, ¾-inch.
    Wrench, open-end, 1⅝-inch.

(a) Replace plug in side of housing (1⅝-inch open-end wrench).

(b) If centering bolt was turned down, turn it up until it is tight, and back it off one full turn (¾-inch open-end wrench). This will give the proper clearance at the bottom of the bands.

(c) Replace inspection plate, installing a new gasket (¾-inch socket wrench).

(d) Install drain plug in housing (1⅝-inch open-end wrench).

(3) Connect Steering Linkage.
    Pliers

Connect steering linkage at hole arms (steering levers) by installing clevis pins and cotter pins (pliers).

(4) Adjust the brake bands (par. 118).

(5) Fill Differential with Oil. Fill differential with proper amount and grade of oil specified in the Lubrication Guide.

121. PARKING BRAKE ADJUSTMENT (fig. 113). Parking brake should be adjusted so that brake band is firmly seated after lever is past center, otherwise it will not stay on in locked position.

a. Equipment.
    Wrench, open-end, ¾-inch.

b. Procedure.
(1) Loosen clamp screw which secures parking lever to shaft on transmission (¾-inch open-end wrench).

(2) Pull lever to the left (off shaft) until spacer and adjusting block are free.

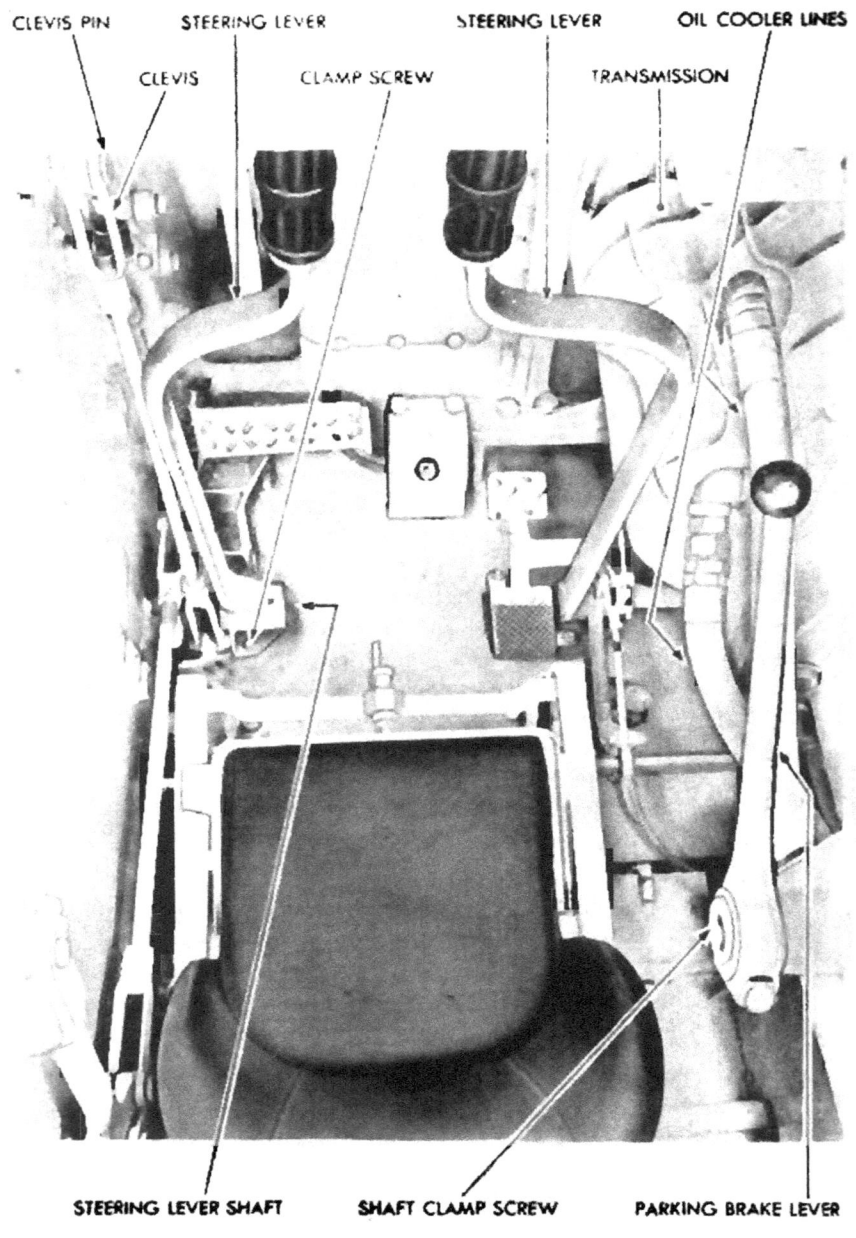

Figure 113. Steering lever linkage and installation of parking brake.

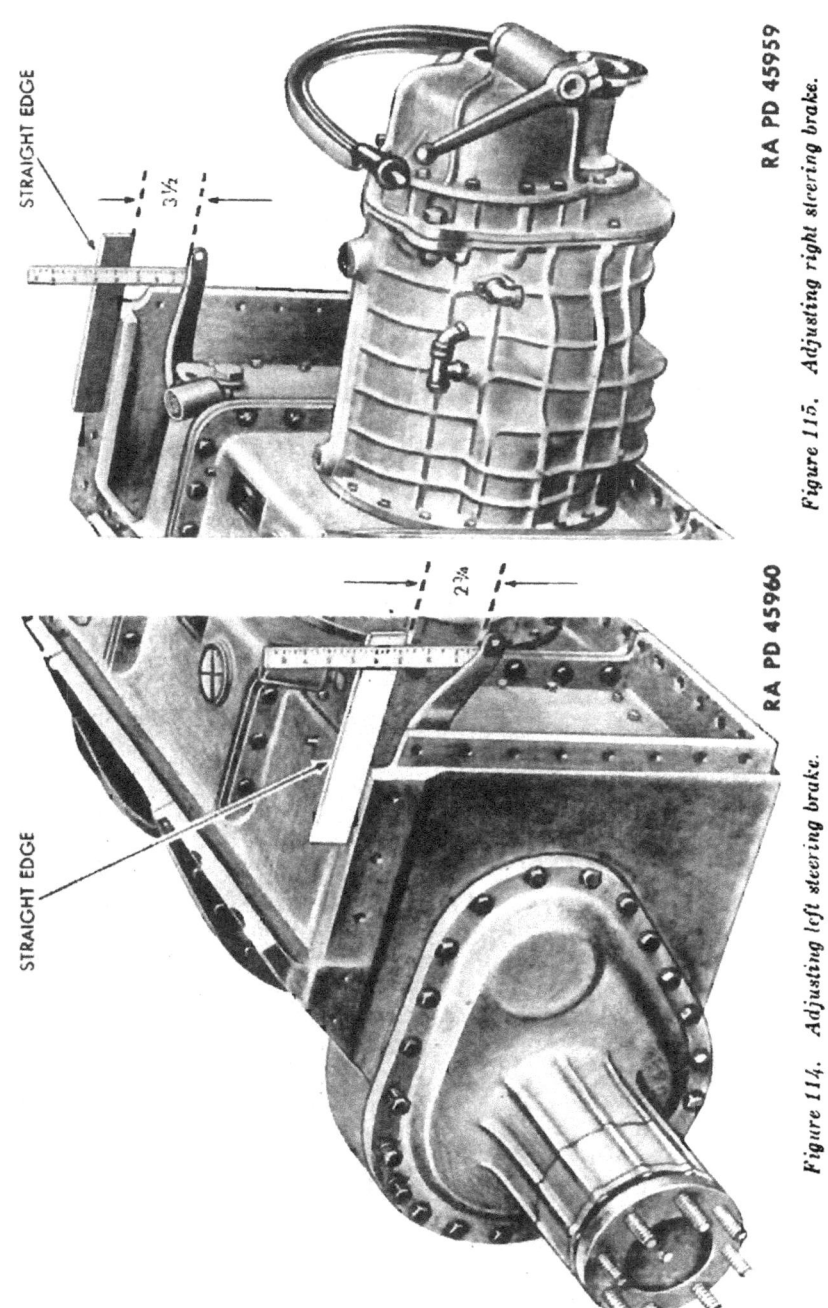

Figure 114. Adjusting left steering brake.

Figure 115. Adjusting right steering brake.

(3) Turn adjusting nut by hand counterclockwise to cause brake to take hold sooner; clockwise to allow lever to go past center before brake takes hold.
(4) Install spacer and slide lever into place on shaft.
(5) Tighten clamp screw (¾-inch open-end wrench).

122. STEERING LEVER REMOVAL (fig. 113).
a. Equipment.
Hammer   Wrench, open-end, ¾-inch.
b. Procedure.
Loosen steering lever clamp screw (¾-inch open-end wrench). Tap lever off shaft (hammer).

123. STEERING LEVER INSTALLATION (fig. 113).
a. Equipment.
Hammer   Screw driver
Pliers   Wrench, open-end, ¾-inch.
b. Procedure.
(1) Set lever, or levers, in place and connect clevis on adjustable rod to steering lever by installing clevis pin and securing with cotter pin.
(2) Line up levers and tap shaft through into place. Secure with cotter pin.
(3) If necessary, adjust stops on new levers so they will be parallel with adjustable rods, and against stops when brake shaft arms are in proper position (par. 118).
(4) Tighten steering lever clamp screw (¾-inch open-end wrench).

124. FINAL DRIVE.
a. Description (fig. 116).
The final drives transmit power from the controlled differential to the hubs of the driving sprockets. Each has a set of pinion gears, through which the power is transmitted from the final drive shaft to the final drive gears. From the final drive gears, power is then transmitted to the drive sprockets. Each set of final drive gears is mounted on a cover bolted to a final drive housing. A driving sprocket is bolted to a hub on each final drive cover.

b. Lubrication.
The final drives are lubricated by separate quantities of 36 quarts of oil for each final drive housing. The oil level is checked through plug holes on the sides of the final drive housing. The oil level must be up to ½ inch below the bottom of the plug hole. The level of the oil should be checked and oil of the specified grade as shown in Lubrication Guide (fig. 16) added to maintain the proper level. Oil may be added at the filler plugs outside the vehicle in the final drive housing. Drain the oil in the final drives at intervals as indicated in the Lubrication Guide. Drain plugs are located in the bottom of each final drive housing. Thirty-six quarts

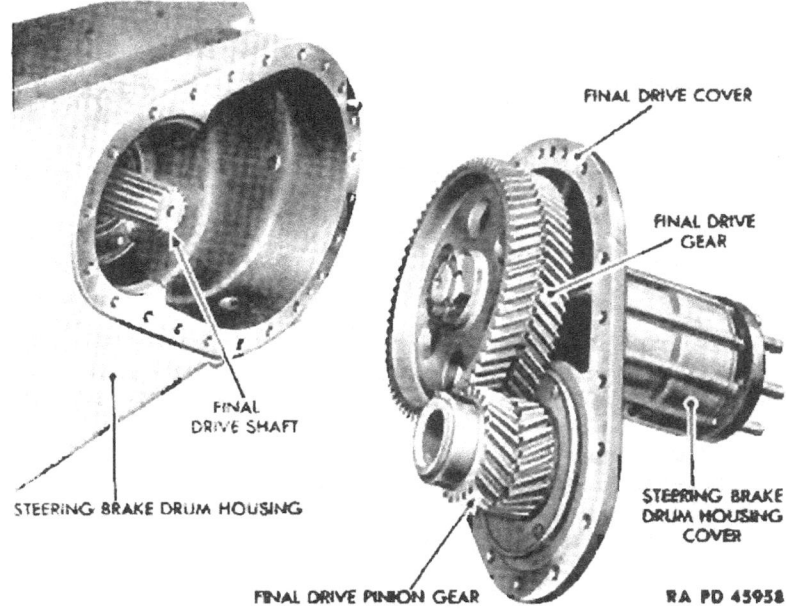

Figure 116. Final drive removed from power train.

of oil of the grade shown in the Lubrication Guide will be required for refilling each final drive.

Note: Always check oil in either final drives or transmission after the tank has stood for several hours. Never check the oil immediately after running the tank as it may be hot and foaming and will not give an accurate reading. When checking oil, be sure the vehicle is on level ground.

c. Equipment.

Hoist   Wrenches, socket, two, 1⅛-inch

d. Procedure.

(1) **Final Drive Unit Removal** (fig 116).

Hoist   Wrenches, socket, two, 1⅛-inch

(a) Drain oil from final drive.

(b) Remove track (par. 140).

(c) Remove sprockets (par. 125).

(d) Remove 22 cap screws that secure final drive unit to final drive housing (two 1⅛-inch socket wrenches).

(e) Withdraw final drive unit, using a hoist to support weight of unit.

(2) **Final Drive Unit Installation** (fig. 116).

    Hoist   Wrenches, socket, two, 1½-inch

(a) Install final drive unit in place, supporting weight with hoist.

(b) Install 22 cap screws that secure final drive unit to final drive housing (two 1½-inch socket wrenches). Remove hoist.

(c) Refill final drive with quantity and grade of oil specified in Lubrication Guide (fig. 16).

(d) Install sprockets (par. 125).

(e) Install tracks (par. 141).

125. DRIVE SPROCKETS.

a. Description. Each track drive sprocket assembly is made up of two toothed plates bolted to a sprocket hub, which in turn is bolted to the flanged end of the final drive sprocket shaft. The right and left drive sprockets are interchangeable.

b. Maintenance. Inspect the sprockets frequently to insure that tooth plate bolts and hub bolts are sufficiently tight. When appreciable wear is noted on the sprocket teeth, interchange the right and left sprockets to bring the track in contact with the unused area of the teeth.

Note: If the drive sprockets bolts cannot be kept tight, replace them with self-locking bolts, A246629 which do not require lock washers. Tightening these bolts to 100 to 110 foot-pounds torque will secure the sprocket to the hub effectively.

c. Equipment.

    Wrench, socket, 1½-inch

d. Procedure.

(1) **Drive Sprocket Removal.**

    Wrench, socket, 1½-inch

(a) Slacken and disconnect the track below the sprocket (par. 140).

(b) Remove eight hub retaining nuts securing hub to drive shaft flange (1½-inch socket wrench) (fig. 116).

(c) Remove sprocket assembly.

(2) **Drive Sprocket Installation.**

    Wrench, socket, 1½-inch

(a) Install sprocket assembly in place on drive shaft flange.

(b) Secure hub to flange by installing the eight hub retaining nuts (1½-inch socket wrench).

(c) Connect and adjust the track (pars. 139 and 141).

RA PD 14360

Figure 117. Power train on jacks prior to removal.

126. POWER TRAIN REMOVAL.

a. Equipment.

- Bar, track, 27-inch
- Blocking
- Cables (2)
- Chains (2)
- Cloth, wiping
- Extension, 12-inch
- Hammer
- Handle, ratchet, heavy duty.
- Jacks, 10-ton (3)
- Pliers
- Pliers, 8-inch
- Punch, 1/4-inch
- Screw driver
- Wrenches, open-end, 9/16-inch (2)
- Wrench, open-end, 3/4-inch
- Wrench, open-end, 1 1/8-inch
- Wrenches, open-end, 1 5/8-inch (2)
- Wrench, socket, 7/16-inch
- Wrench, socket, 1/2-inch
- Wrench, socket, 1 1/8-inch

b. Procedure.

(1) Disconnect Universal Joint.

- Wrenches, open-end, 9/16-inch (2).
- Wrench, socket, 7/16-inch

(**a**) Remove propeller shaft guard (7/16-inch socket wrench) (fig. 107).

(**b**) Disconnect front universal joint by removing eight bolts and self-locking nuts from the companion flange (9/16-inch open-end wrench) (fig. 106).

(2) **Disconnect Steering Arms** (fig. 111).

    Pliers, 8-inch   Wrench, open-end, ¾-inch
    Screw driver

Remove cotter pins from clevis pins at top end of right and left steering linkage rods, and pull out clevis pins.

(3) **Remove Hand Throttle** (fig. 12).

    Wrench, socket, ⁵⁄₁₆-inch

Remove retaining screws which hold hand throttle bracket to housing, and remove throttle.

(4) **Disconnect Speedometer Cable.**

    Pliers

Disconnect speedometer cable from right rear of transmission housing by loosening knurled nut. Tie cable up out of way.

(5) **Remove Stop Light Switches** (fig. 10).

    Wrench, open-end, ¾-inch

Remove both stop light switch brackets from steering brake arms by removing cap screws. Unhook adjustable road from each switch, and remove switches.

(6) **Disconnect Transmission Oil Cooler Lines** (fig. 88).

    Cloth, wiping   Wrenches, open-end, 1¼-inch (2)

(a) Drain transmission.

(b) Do not attempt to disconnect the lines at transmission first. This will injure the flexible lines, since fittings are sweated onto hose. Disconnect both oil lines at floor (two 1¼-inch open-end wrenches). Then remove the flexible lines by turning them out of the fittings on transmission housing. Cover fittings with wiping cloth.

(7) **Break Tracks.**

    Bar, track, 27-inch

Break tracks (par. 140). Do not remove tracks completely from vehicle. Work top section of track up and off sprocket by turning sprocket back with a bar, and pulling track to rear.

(8) **Remove Front Fenders.**

    Extension, 12-inch   Wrench, socket, ½-inch
    Handle, ratchet, heavy-duty Wrench, socket, 1⅛-inch
    Screw driver

(a) Remove bolts that hold each front fender to hull.

(b) Remove two cap screws and lock washers that hold each fender side plate to final drive housing, and remove fenders.

(9) **Disconnect Power Train Housing.**

    Hammer   Punch, ¼-inch
    Handle, ratchet, heavy-duty Wrench, open-end, 1⅛-inch
    Jacks, 10-ton (3)   Wrench, socket, 1⅛-inch

(a) Install three 10-ton hydraulic jacks under the power train (fig. 117). Place one jack under rear center of controlled differential housing, one jack under right final drive housing, and third jack under left final drive housing. Pull up jacks until they are snug.

(b) Remove 20 bolts and lock washers from the bottom edge of housing. Then remove all bolts from each side of housing and those across the top (56 bolts in all) 1¼-inch open-end wrench and 1¼-inch socket wrench). This will require two men; one to hold the nut and one to turn the bolt. If difficulty is experienced in removing bolts after nuts have been removed, it may be necessary to drive them out with a punch and hammer.

(10) **Remove Power Train from Tank** (fig. 118).

Cables (2)	Chains (2)

Lower each jack slightly in rotation until housing breaks clear of hull. Be sure that jacks are on a solid level foundation, and against flat sections of housing. As a safety precaution, anchor the housing by chains installed in clevises. With towing cable attached to towing clevises on either side of rear of vehicle, use truck or wrecker unit to pull vehicle slowly away from power train.

(11) **Blocking up Power Train.**

Blocking

For the security of power train, and to release the jacks, install blocking under final drive housings and controlled differential, and release jacks. Blocking under final drives should be about 4 feet long to avoid danger of the housing tipping, and should be placed inside jacks.

127. POWER TRAIN INSTALLATION.

a. Equipment.

Blocking	Wrenches, open-end, 7/16-inch (2)
Extension, 12-inch	Wrench, open-end, ⅝-inch
Hammer	Wrench, open-end, 1¼-inch
Handle, ratchet, heavy-duty	Wrenches, open-end, 1⅝-inch (2)
Jacks, 10-ton (3)	Wrench, socket, 7/16-inch
Pliers, 8-inch	Wrench, socket, ½-inch
Punch, ¼-inch	Wrench, socket, 9/16-inch
Screwdriver	Wrench, socket, 1¼-inch

b. Procedure.

(1) **Line up Vehicle and Power Train.**

Blocking	Jacks, 10-ton (3)

Center power train on blocks to allow sufficient room for installation of jacks (fig. 117). It must also be accurately lined up with vehicle. Install jacks under differential and power train housings (par. 126b(9)(a)), and raise unit high enough for bottom of unit to slide inside vehicle. With blocking still in place, carefully push vehicle up to the power train housing.

(2) **Install Housing in Vehicle.**

  Hammer   Punch, ¼-inch

Line up holes in housing and vehicle chassis. This may require moving vehicle slightly forward or backward, or raising or lowering one or more of the jacks. When holes are correctly lined up, install and line up shims, and drive them into place (punch and hammer). Lower jacks gradually, in rotation, until top of housing seats on hull.

(3) **Attach Housing to Vehicle.**

  Handle, ratchet, heavy- Wrench, open-end, 1⅛-inch
    duty   Wrench, socket, 1⅛-inch
  Pliers

Install and tighten top and side bolts and lock washers (56 bolts), then remove blocking and install 20 bottom bolts and lock washers (1⅛-inch socket wrench with ratchet handle and 1⅛-inch open-end wrench). This will require two men; one within the driving compartment to hold the nut, and the other man to turn the bolt. Wire two cap screws together (pliers).

DRIVE SPROCKET HUB RETAINING NUT RA PD 45951

Figure 118. Power train separated from vehicle.

(4) **Connect Propeller Shaft** (fig. 106).

Wrenches, open-end, ⁹⁄₁₆- Wrench, socket, ⁹⁄₁₆-inch
inch (2)

(a) Line up holes in slip joint companion flange and transmission flange, and install connecting bolts and lock washers (two ⁹⁄₁₆-inch open-end wrenches).

(b) Install propeller shaft guard (par. 115b(3)).

(5) **Connect Steering Levers** (fig. 113).

Pliers

Install clevis pins and cotter pins that connect linkage of right and left steering lever shafts to brake drum arms (pliers).

(6) **Connect Transmission Oil Cooler Lines** (fig. 88).

Wrenches, open-end, 1⅜-inch (2).

(a) Install and tighten flexible inlet and outlet oil cooler lines to fittings on transmission housing. The long line (inlet line) is connected to the T-fitting toward front of transmission, and the short line (outlet line) to elbow (two 1⅜-inch open-end wrenches).

(b) Connect other ends of flexible lines to oil cooler inlet and outlet lines on floor (two 1⅜-inch open-end wrenches). Make connections carefully to be sure there will be no oil or air leaks.

(7) **Install Speedometer Cable.**

Pliers.

Connect speedometer cable at right side of transmission (pliers).

(8) **Install Stop Light Switches** (fig. 10).

Wrench, open-end; ¾-inch

Connect adjustable rods from stop light switches on left brake shaft arm to left end of cross shaft; install and tighten brackets on housing (¾-inch open-end wrench).

(9) **Install Hand Throttle** (fig. 12).

Wrench, socket, ⁹⁄₁₆-inch

Install throttle and bracket cap screws which hold hand throttle bracket to housing (⁹⁄₁₆-inch socket wrench).

(10) **Install Front Fenders.**

Extension, 12-inch   Wrench, socket, ½-inch
Handle, ratchet, heavy-duty   Wrench, socket, 1⅛-inch
Screw driver

Set fenders in place. Install and tighten four fillister head bolts which hold each fender to hull at top. Install and tighten two cap screws and lock washers which hold side plate to the final drive housing.

(11) **Connect tracks** (par. 141).

SECTION XI
SUSPENSION AND TRACKS

128. DESCRIPTION AND OPERATION. Six two-wheeled, rubber-tired bogies or suspensions, bolted to the hull, support the vehicle on springs. The tracks are driven by sprockets of the final drives. Two adjustable idlers at the rear end of the hull are provided to adjust the tension of the tracks. The weight of the upper portion of the track is carried by steel track-supporting rollers mounted to the rear of the suspension brackets.

129. BOGIES.

a. Description and Operation. The bogies are the supporting and conveying units, and are sometimes called trucks or suspensions. Movement is transferred from wheels to arms and levers, and is absorbed by springs. Wear between wheel arms and spring levers is taken by upper and lower rubbing plates which are removable and can be replaced (fig. 119). To the top rear of each bogie assembly is a single steel roller to support and carry the upper, returning portion of the track.

b. Lubrication (fig. 120). Lubrication of wheels and track supporting rollers is through lubrication fittings. Relief valves are provided to prevent injury to oil seals. See Lubrication Guide for lubrication instructions (sec. IV, ch. 1).

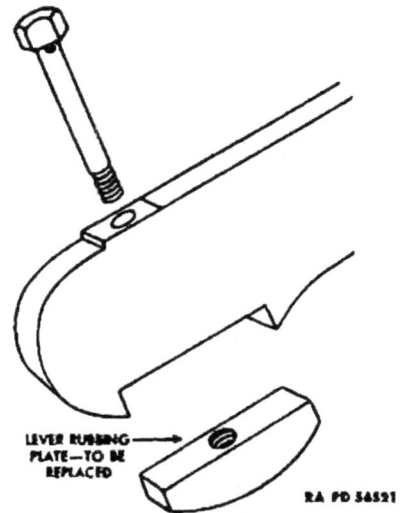

Figure 119. Bogie wheel lever and lever rubbing plate.

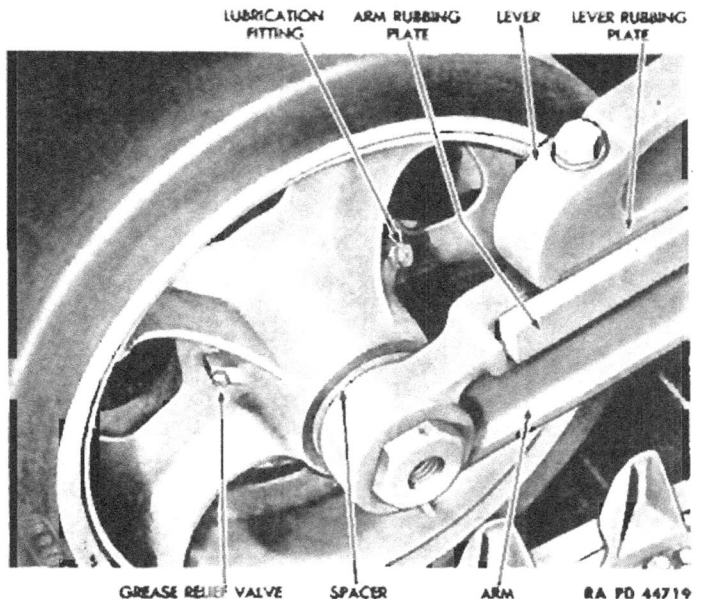

Figure 120. Bogie wheel lubrication fittings.

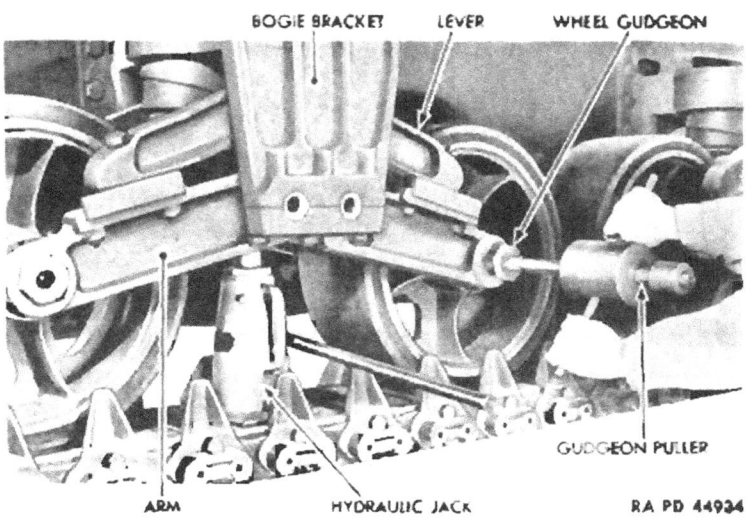

Figure 121. Pulling wheel gudgeon to remove bogie wheel.

130. BOGIE WHEEL REMOVAL (fig. 121).

a. Equipment.
Extractor, cotter pin
Jack, hydraulic, 12-ton
Lift, bogie
Puller, gudgeon (slide hammer)
Wrench, gudgeon, 2⅝-inch

b. Procedure.

(1) Bogie wheels are removed either for replacement, to repair a tire, to inspect or replace bearings, to replace grease seals or spacers, or to make other necessary repairs.

(2) Use bogie lift to raise wheel arm, or place 12-ton hydraulic jack on track directly under center of suspension assembly and jack up bottom spring seat until all suspension weight is removed from wheel arm (fig. 121).

(3) Remove cotter pin from nut on inner end of wheel gudgeon and remove gudgeon nut (cotter pin extractor and 2⅝-inch gudgeon wrench).

(4) Screw slide hammer gudgeon puller into threaded hole in outer end of wheel gudgeon and pull gudgeon (fig. 121). Lift out wheel from between arms.

131. BOGIE WHEEL GREASE SEALS AND BEARING REMOVAL.

a. Equipment.
Drift, brass
Hammer
Puller, oil seal

b. Procedure.

(1) Remove two outer spacers.

(2) Support wheel on blocks and, with a brass drift through the upper side of the wheel, drive bearing and oil seals out of the lower side of wheel (hammer). It will be necessary to move inner spacer to one side in order to seat the drift on the outer race of the bearing. Keep drift moving around entire circumference of bearing outer race, in order to drive it out evenly and with no damage to the bearing. The spacer will drop out when bearing is removed.

(3) Turn wheel over and drive out other bearing (hammer and brass drift) and oil seals (oil seal puller).

(4) Clean and inspect both bearings for wear.

> NOTE: Whenever bogie wheels are removed, retainer (seal) B132704AB and spacer (adapter) B153965 will be installed when available. Each bogie wheel requires four retainers and two spacers. End play of bogie wheels equipped with tapered roller bearings should be 0.003 inch to 0.021 inch. If end play exceeds 0.021 inch or in case of bearing failure, replace tapered roller bearings with ball bearings CABX3AL when available.

132. BOGIE WHEEL GREASE SEALS AND BEARING INSTALLATION.

a. Equipment.
Drift, brass
Guide, expander
Hammer

b. Procedure.
(1) Replace bearing by starting it in by hand and tapping it lightly into place with a brass drift and hammer. Work drift around circumference of outer race to drive it in evenly. Be sure bearing is seated against shoulder to allow proper room for grease seals.

(2) Install expander guide on spacer, and slide two new grease seals into place on spacer, being sure that seals are installed with lips toward shoulder of spacer. Start lower seal into wheel, and tap seals and spacer into place (hammer).

(3) Turn wheel over, insert center spacer, and install bearing, oil seals, and outer spacer.

133. BOGIE ARM AND LEVER RUBBING PLATE REMOVAL.

a. Equipment.
Screw driver, ⅜-inch bit   Wrench, open-end, 1-inch

b. Procedure.
(1) When bogie arms and levers have been disassembled (fig. 123) inspect arm and lever rubbing plates. If plates are scored or pitted, they must be replaced.

(2) Remove the 1-inch cap screw and lock washer securing the lever plate within the lever.

(3) Remove the two countersunk screws securing the armplate to the arm.

134. BOGIE ARM AND LEVER RUBBING PLATE INSTALLATION.

a. Equipment.
Screw driver, ⅜-inch bit   Wrench, open-end, 1-inch

b. Procedure.
(1) Place armplate in position, and install two countersunk screws.

(2) Hold the lever plate in place, and install the 1-inch cap screw and lock washer.

135. BOGIE WHEEL INSTALLATION (fig. 121).

a. Equipment.
Bar, alignment
Hammer, soft
Pliers
Wrench, gudgeon, 2⅜-inch

b. Procedure.

(1) Lift wheel into place between two arms.

(2) Raise arms in line with wheel, and start wheel gudgeon through outer arm and into the outer spacer, grease seals, and outer bearing of wheel.

(3) Line up center spacer by means of an alignment bar inserted from inner side of wheel, and tap gudgeon into spacer (soft hammer).

(4) Drive gudgeon through arm and spacer far enough to allow key slot in gudgeon to be lined up with slot in wheel arm.

(5) Line up slots, install key, and drive gudgeon in the rest of the way (soft hammer).

(6) Install and tighten gudgeon nut on inner end of gudgeon, and secure with a cotter pin (2¾-inch gudgeon wrench and pliers).

(7) Remove jack.

136. VOLUTE SPRING REMOVAL (figs. 122 and 123).

a. Equipment.

Wood block	Puller, gudgeon
Drift, brass	Wrench, open-end, 1¼₆-inch
Hammer	Wrench, open-end, 1⅜-inch
Jack, hydraulic, 12-ton	

b. Procedure.

(1) Install Jack under Springs.

Wood block   Jack, hydraulic, 12-ton

Place jack on inside of track so that head of jack is in center of spring seat plate. In placing jack, arrange blocking so that jack plunger is nearly at upper end of travel when thrust of springs has been taken up. This procedure will permit greater lowering travel later. Raise jack until thrust of springs has been taken up.

(2) Remove Gudgeon Bolts.

Drift, brass	Wrench, open-end, 1¼₆-inch
Hammer	Wrench, open-end, 1⅜-inch

(a) Loosen and remove nuts on locking bolts that secure bogie bracket and gudgeon (1¼₆-inch and 1⅜-inch open-end wrenches) (fig. 121).

(b) Drive out bolts (hammer and brass drift).

(3) Remove Arm Gudgeon.

Puller, gudgeon

Screw a gudgeon puller into threaded hole in outside end of a bogie arm gudgeon pin, and pull pin outward (fig. 122). Remove other bogie arm gudgeon.

Note: When arm gudgeons are removed, the bogie arms and spacer plates will drop.

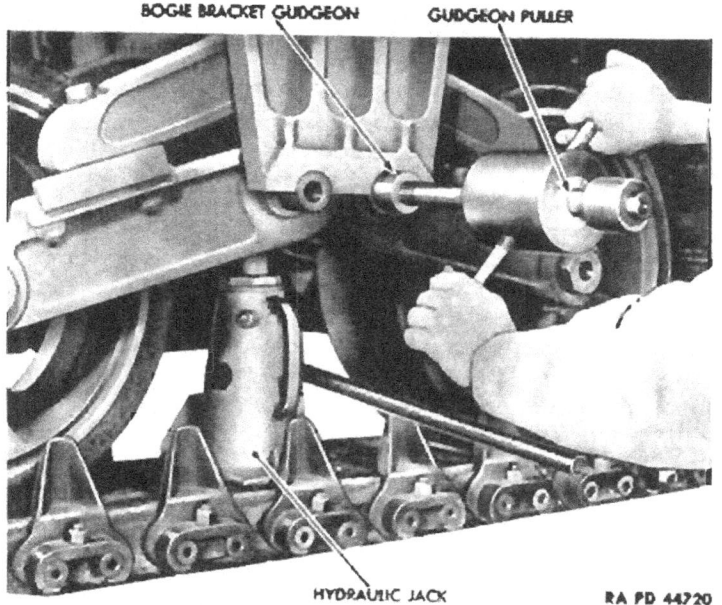

Figure 122. Pulling arm gudgeon preparatory to removing volute springs.

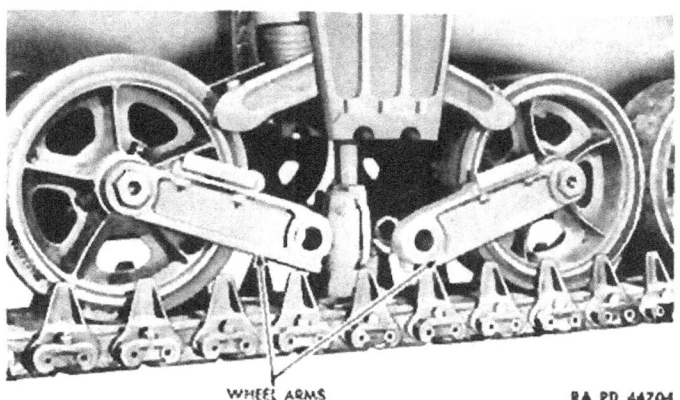

Figure 123. Wheels and arms ready for removal preparatory to removing volute springs.

(4) **Remove Wheels and Arms.** Lift wheels and arms out from beneath bogie assembly, and away from tank.

(5) **Lower Spring Seat.**

(a) Lower jack with the spring seat plate and springs.

(b) If jack has not sufficient travel to decompress springs, push two bars through the arm gudgeon pin holes and lower spring seat plate onto these. Then remove blocking, or lower the jack screw. Lift plate off bars. With jack, remove bars and continue to lower jack and decompress springs.

(6) **Remove Jack and Springs.**

       Wood block   Hammer

(a) Remove jack from beneath plate.

(b) If spring seat plate and springs do not pull out on removal of jack, knock them out with a hammer or a wood block.

Note: When replacing springs, always replace both springs; never use one old and one new spring. Three types of bottom spring seats are being used. Should seat C95130 fail, replace it with seat C95289 on all but limited service vehicles.

137. VOLUTE SPRING INSTALLATION (figs. 122, 123 and 124).

a. **Equipment.**

       Hammer, soft   Wrench, open-end, 1¼-inch

       Jack, hydraulic, 12-ton   Wrench, open-end, 1⅛-inch

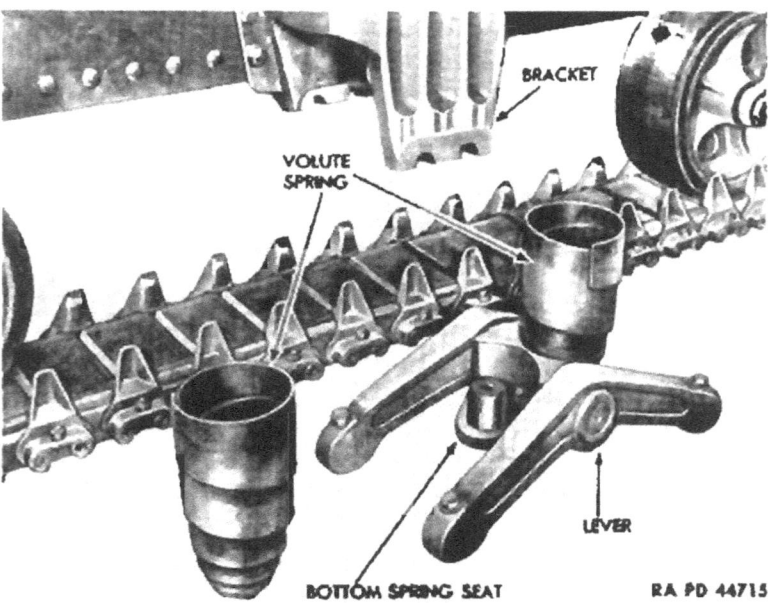

Figure 124. Volute springs, lever, and bottom seat removed and partially disassembled.

b. Procedure.

(1) Install Springs.

Jack, hydraulic, 12-ton

Install springs and spring seat plate. With a 12-ton hydraulic jack, compress springs by raising jack until thrust of springs has been taken up.

(2) Install Arm Gudgeons.

Hammer, soft

Install bogie arms and spacer plates; install arm gudgeons (soft hammer). Line up gudgeons so that grooves are in line for installation of locking bolts.

(3) Installation of Gudgeon Bolts.

Hammer, soft   Wrench, open-end, 1⅛-inch
Wrench, open-end, 1 1/16-inch

Install gudgeon locking bolts (soft hammer). Install nuts on gudgeon locking bolts (1 1/16-inch and 1⅛-inch open-end wrenches).

(4) Removal of Jack. Lower and remove jack, allowing levers to resume normal position on wheel arms.

138. TRACKS (fig. 126). Each track on the vehicle has 79 individual rubber or steel link assemblies. Link assemblies are reversible and can be turned over to lengthen the life of the track. Two parallel pins, projecting from either end of each link assembly, form anchorage points for the connectors, which link the blocks together on both sides to form a continuous track. The connectors are held to the pins by wedges, which fit into milled slots on the pins. These slots face outward and are vertical to the flat side of the link assembly. The wedges are tapered on each end, so that when they are pulled up between the two pins by the bolt, which is an integral part of the wedge and which projects up through the connector, they cause a

Figure 125. Example of loose track

Figure 126. Example of proper track tension.

15° angle between adjacent blocks. This 15° angle tends to make the track curve up sharply behind the rear bogie, around the idler, and on around the sprocket. Steel perpendicular end plates, cast integral with the connectors, serve as guides to keep the track in alignment with the bogie wheels, idlers, track support rollers, and

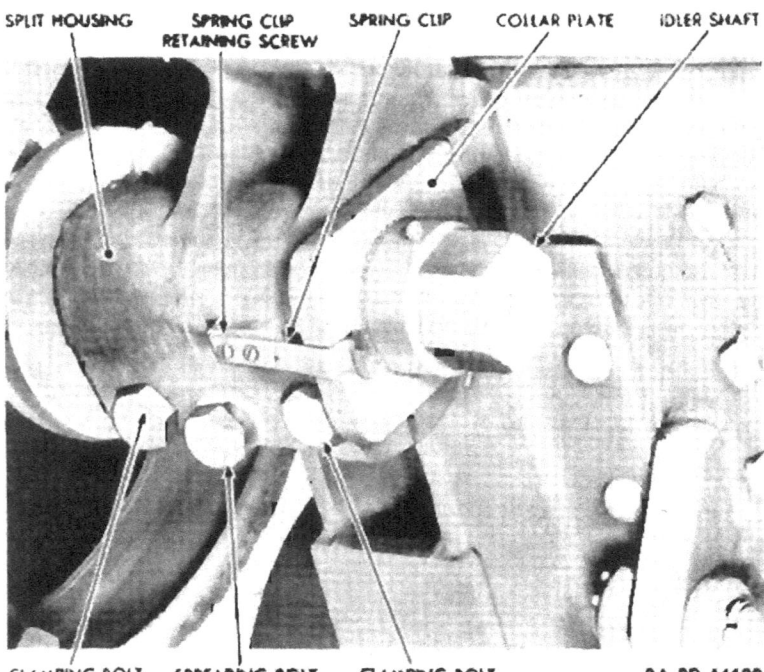

Figure 127. Clamping and spreader bolts of rear idler.

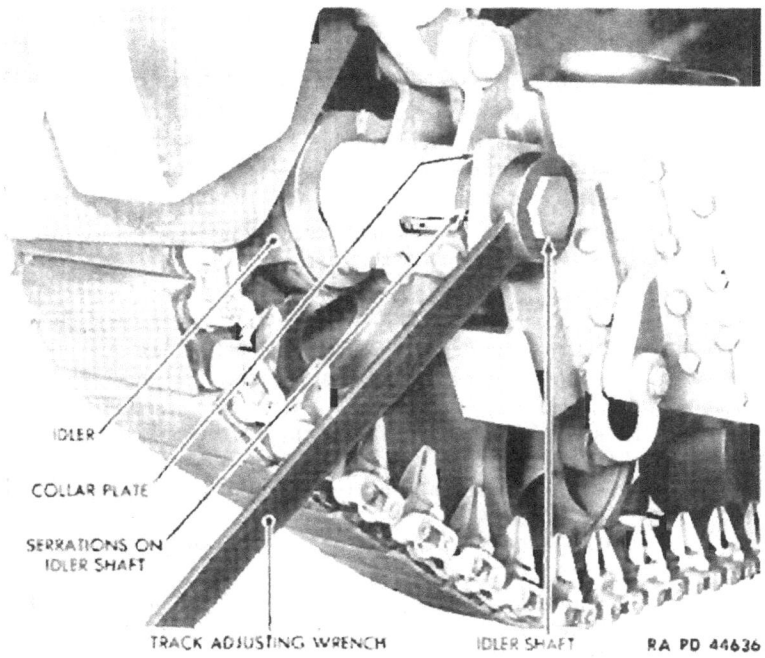

Figure 128. Track adjusting wrench in position.

drive sprockets. The sprocket drives the track through the sprocket teeth, which engage the track between adjacent connectors. Since the condition and effectiveness of the track can definitely limit or increase the performance ability of the vehicle, it is essential that it be inspected, adjusted, and maintained in the best possible condition.

139. LOOSE TRACK ADJUSTMENT.
a. Equipment.

Hammer Wrench, idler adjusting
Straightedge Wrench, socket, 1½-inch

b. Check the track daily for excessive sag. If pronounced sag is present, correct tension should be restored to eliminate the possibility of the track being damaged or thrown off because of looseness. Figure 125 shows a track very definitely in need of tightening, while figure 126 shows the same track adjusted to eliminate the sag.

c. Procedure.

(1) Loosen two outside clamping bolts which clamp the split housing to the idler spindle (1½-inch socket wrench) (fig. 127).

(2) Turn spreader bolt counterclockwise to open up housing (1½-inch socket wrench) (fig. 127).

Note: Because this bolt has a left-hand thread, it is turned counterclockwise.

(3) Raise clip at end of housing, and tap the collar plate off serrations on spindle to position illustrated (fig. 128). When tapping collar plate off serrations in the spindle, the track adjusting wrench should be on the adjusting nut, well blocked to hold the idler gear in position.

Note: In tightening track, more than one man is required. Also use hydraulic jack to help turn track adjusting wrench.

(4) Tighten track by using idler adjusting wrench on hex at end of spindle, raising the handle of the wrench until track sag does not exceed ½ to ¾ inch. Sag can be measured by placing a straightedge across the track at the two front support rollers and adjusting the tension until a sag of ½ to ¾ inch is measured at a point midway between the second and third track support rollers, counting from the front of the vehicle.

(5) Drive the collar plate back on the serrations and under the clip (fig. 127).

(6) Turn the spreader bolt clockwise (1½-inch socket wrench) (fig. 127).

(7) Pull both outside clamping bolts up tight, then tighten down spreader bolt sufficiently to hold it from jarring loose (1½-inch socket wrench) (fig. 127).

(8) Place housing collar plate clip back in position and snap it over collar.

140. TRACK REMOVAL (figs. 129 and 130).

a. Equipment.

Bar, track, 27-inch
Cable, tow
Crowbar
Hammer, sledge
Hammer, soft
Mover, prime
Wrench, socket, ⅞-inch

b. Procedure.

(1) **Remove Wedge Nuts.**

Hammer, soft   Wrench, socket, ⅞-inch

(a) Release track tension at the idler (par. 139).

(b) Loosen and remove wedge nuts from corresponding outside and inside connectors that are midway between front sprocket and front bogie wheel (⅞-inch socket wrench). Tap wedges out of connectors, being careful not to injure the threads.

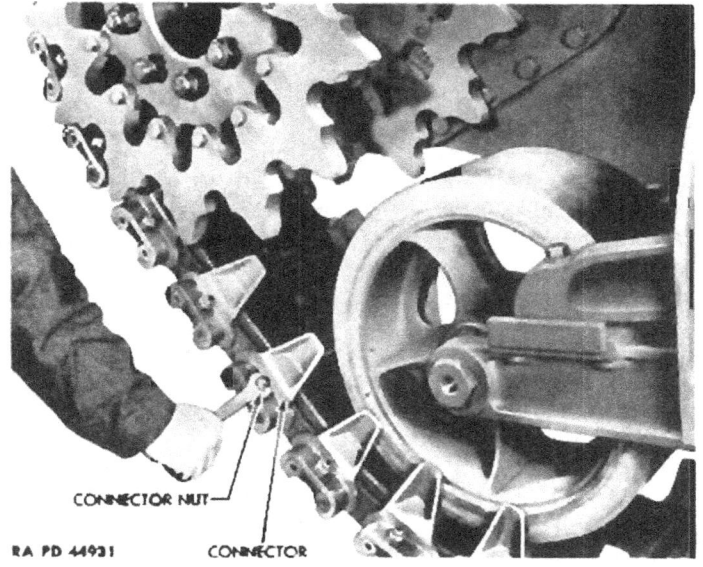

Figure 129. Wedge nut removal.

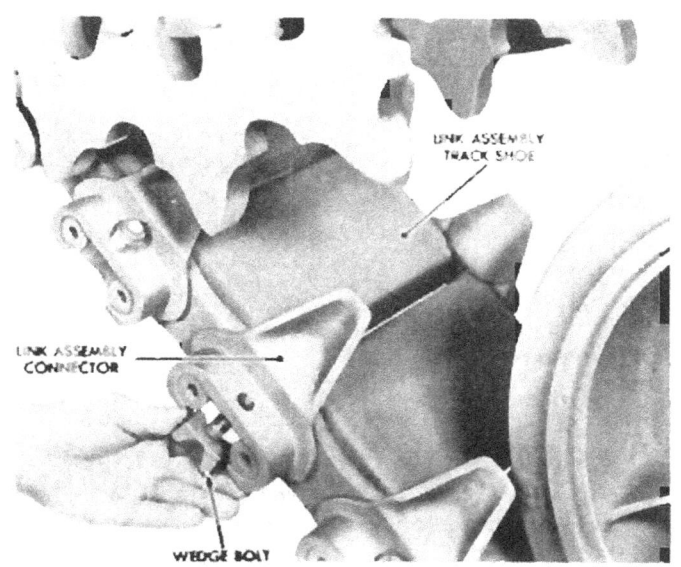

Figure 130. Disconnecting track shoe connector.

(2) **Remove Connectors.**

    Bar, track, 27-inch   Hammer, sledge

(a) Working from opposite side of track from connectors, and with point of track bar against flange of connector, drive the inside connector almost off track pins (sledge hammer).

(b) With inside connector still on, to prevent binding of outside connector on pins, drive outside connector off (sledge hammer).

(c) Drive inside connector off the rest of the way (sledge hammer).

(3) **Remove Track.**

    Cable, tow   Mover, prime
    Crowbar

(a) Move upper portion of track to the rear over the drive sprocket and support rollers, by means of a crowbar.

(b) Pull vehicle off tracks (prime mover).

141. TRACK INSTALLATION

a. Equipment.

    Cable, tow   Hammer, sledge
    Crowbar   Mover, prime
    Fixture, track connecting   Wrench, socket, ¾-inch

b. Procedure.

(1) **Place track in position.**

    Cable, tow   Mover, prime

(a) Hook a prime mover to the vehicle with a tow cable. Lay out the track in front of the vehicle in a straight line with the sprocket and bogie wheels.

(b) Pull vehicle forward on track until forward end of track is between driving sprockets and first bogie wheel.

(c) Connect a cable to the rear end of one track. Run the cable over the idler, across the top of the support rollers, and above the driving sprocket. Hook the cable to the prime mover, then pull the track up and into position for connection.

(2) **Connect Track.**

    Crowbar   Hammer, sledge
    Fixture, track connecting   Wrench, socket, ¾-inch

(a) Work ends of track as close together as possible (crowbar).

(b) Place track connecting fixture in position, then jack the ends of the track together (track connecting fixture) (fig. 131).

(c) Drive track shoe connectors on track shoe pins (sledge hammer).

(d) Install wedges and wedge nuts (¾-inch socket wrench) (figs. 129 and 130). Two to three threads must extend beyond the end of

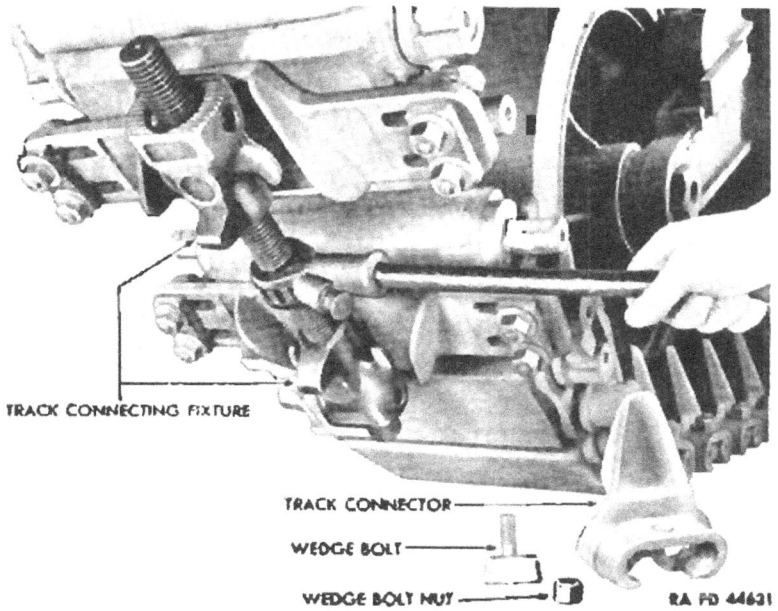

Figure 131. Connecting track.

the nuts to assure that safety fixture of nut has a full grip. After assembly of wedge, there must be at least (⅙-inch clearance between top of wedge and the track shoe connector.

(3) **Adjust Track Tension.** Adjust the track tension by means of the idler (par. 139).

142. DEAD RUBBER TRACK LINK ASSEMBLY REPLACEMENT.

a. Equipment.

Bar	Hammer, sledge
Chains, track	Hammer, soft
Fixture, track connecting	Wrench, box, ¾-inch

b. A dead track link assembly is one in which the flexible bond between the pin and the forged frame of the link assembly has failed, leaving the pin free to turn more than it should, and tending to make the block travel straight instead of curving up sharply behind the rear bogie, around the idler, and on around the sprocket. Dead blocks are easily recognizable (fig. 133) and must be replaced immediately, as they may result in a broken track.

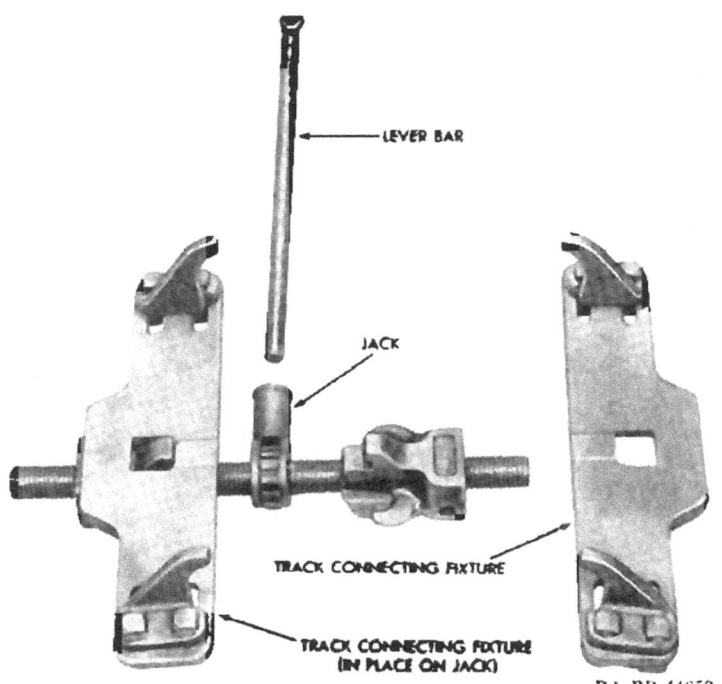

Figure 132. Detail of track connecting fixture.

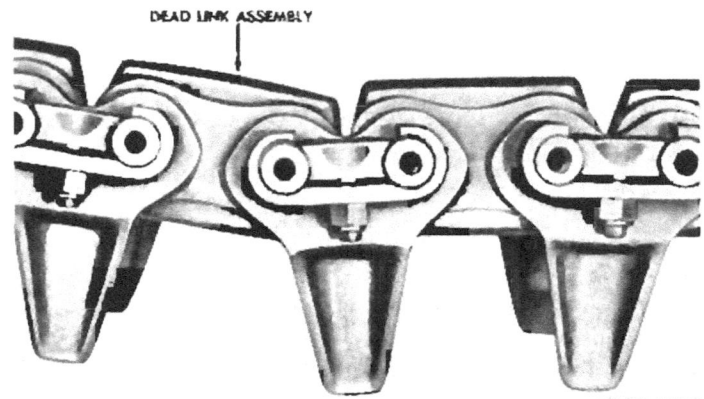

Figure 133. Dead track link assembly dropping out of line on top of track.

c. Procedure.

(1) Move vehicle so that link assembly to be replaced is midway between driving sprocket and front bogie wheel. Set parking brake.

(2) Release track tension (par. 139).

(3) Remove connector nuts (figs. 129 and 130) on two outside and two inside connectors adjacent to link assembly to be replaced (⅞-inch box wrench). Tap out wedges, being careful not to damage threads (soft hammer).

(4) Break the track (par. 140) and remove link assembly by driving off four connectors (sledge hammer).

(5) Install new link assembly to one end of track by tapping connectors into place (hammer). Install wedges in connectors, and install wedge nuts in place (⅞-inch box wrench).

Note: Inspect wedges and wedge nuts, and replace if worn or damaged. Whenever possible, replace defective rubber track blocks with used blocks in good condition, showing approximately the same amount of wear as the track. If no worn blocks are available, cut the tread side of the new blocks to the approximate height of the adjoining rubber blocks with a knife or a hacksaw. This must be done to prevent vibration and early failure of the track.

(6) Connect track (par. 141b(2)).

143. TURNING LINK ASSEMBLIES.

a. Where rubber tracks are to be turned, or several link assemblies must be replaced, the entire track must be removed from the vehicle. Remove track (par. 140).

b. Turn link assemblies by removing all connectors and turning over the link assemblies. To equalize wear on connectors, turn them end for end, and install them on the opposite side of track.

c. Connect track (par. 141).

144. NEW TRACK INSTALLATION.

a. Equipment.

Bar

b. Procedure. Following is the procedure for installing a new track when the old track is still on vehicle:

(1) Break the old track (par. 140) under sprocket, and roll track back off vehicle, using a bar in sprocket hub nuts.

(2) Lay out new track in front of old one and connect the two.

(3) Tow vehicle onto new track until about 16 inches of new track protrudes beyond front bogie wheel.

(4) Disconnect old track, roll new one up over the idlers, track support rollers, and around the sprocket, then connect track (par. 141).

145. REVERSING TRACK.

a. To equalize the driving wear on the track connectors and thus lengthen the operating life of the track, the entire track can be reversed, or turned end for end. This will shift the driving wear to the opposite ends of the connectors. Under good ground conditions, the shift can best be accomplished by breaking tracks at the front (par. 140**b**(1) and (2)), just below the sprocket, working the tracks up and off the sprockets, and pulling vehicle completely off both tracks. The following field method can be used, however, when it is an advantage because of mud, soft ground, or other conditions, to keep vehicle on part of track at all times, and to move vehicle under its own power.

b. Break the right track (par. 140**b**(1) and (2)) at the rear, just below the idler, and, using the left track for traction, move vehicle ahead until the end of the right track comes off sprocket. Break the right track at the center, turn the free section around end for end and drive connectors on enough to hold the track together.

c. Move vehicle back until the bogie wheels are on the reversed section of the track. Reverse other half of track and connect track (par. 141**b**(2)), this time driving connectors completely on and pulling down the wedges.

d. Move vehicle forward until the front bogie wheel is on the fourth track block from the end. Attach towing cable by means of a chain to the other end of track, bringing the free end of the cable up over the idler and support rollers and around the hub of the sprocket. Insert a short bar through the sprocket and the towing eye of the cable; then, using the sprocket hub as a windlass, pull the track forward to the sprocket with engine power of vehicle.

Note: Both steering levers should be left free for this operation.

e. Remove cable from sprocket drum, work track over sprocket, and connect track (par. 141).

f. Repeat operations on left track.

146. TRACK IDLERS (fig. 134).

a. Description. Two large steel idler wheels are mounted on the end opposite the driving sprockets to support the tracks. They are provided with an eccentric adjustment for the purpose of adjusting the tension of the tracks.

b. Adjustment. See paragraph 139.

c. Lubrication. A lubrication fitting adaptable to a grease gun is installed in the hub of the idler. It is also equipped with a relief fitting. See Lubrication Guide for lubrication instructions (fig. 16).

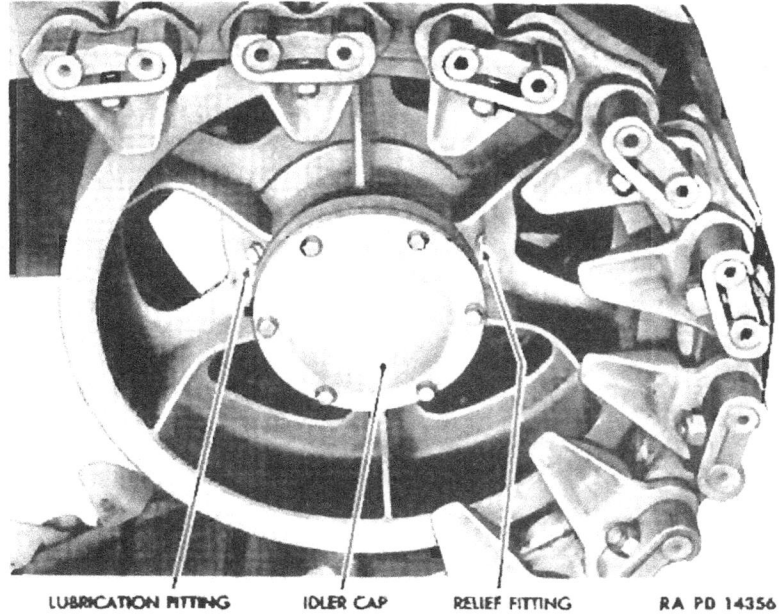

LUBRICATION FITTING   IDLER CAP   RELIEF FITTING   RA PD 14356

Figure 134. Idler.

d. Idler Wheels Removal (with Track Removed).

Pliers   Wrench, socket, ⅝-inch
Wrench, open-end, 2¾-inch

(1) Remove idler cap by removing six cap screws (⅝-inch socket wrench).
(2) Take out split pin securing wheel nut (pliers) and remove nut (2¾-inch open-end wrench).
(3) Remove wheel.

e. Idler Wheel Installation.

Pliers   Wrench, socket, ⅝-inch
Wrench, open-end, 2¾-inch

(1) Before installing wheel, clean bearings, oil retainers, and spacer.
(2) Install wheel nut and split pin (2¾-inch open-end wrench, pliers).
(3) Install idler cap and six cap screws (⅝-inch socket wrench).
(4) Pack inner and outer bearings with proper grade and quantity of grease (fig. 16 and notes, see sec. IV.).

147. GROUSERS (fig. 135).

a. Description. There are 13 grousers, or lugs, provided for each track for use when traversing loose dirt, deep sand, or snow and ice.

One grouser is provided for every sixth track block and each grouser must be installed separately.

b. Installation (fig. 135).

Wrench, socket, 15/16-inch.

(1) Place grouser over track so that it is across opening between two track blocks, and slide two pins on inside grouser end plate into holes in track block pins.

(2) Install pins on the outside grouser end plate in the corresponding holes in outer track block pins. Fasten the plate to grouser with special bolt and lock washer.

(3) Tighten bolt (15/16-inch socket wrench).

Note: To install grousers on the lower part of track, move vehicle so that section of track is on top.

c. Removal (fig. 135).

Wrench, socket, 15/16-inch

(1) Remove the bolt and lock washer securing outside grouser plate to grouser.

(2) Remove outside grouser plate and pins.

(3) Lift up grouser and push toward the inside of track to remove pins of inside grouser plate and grouser.

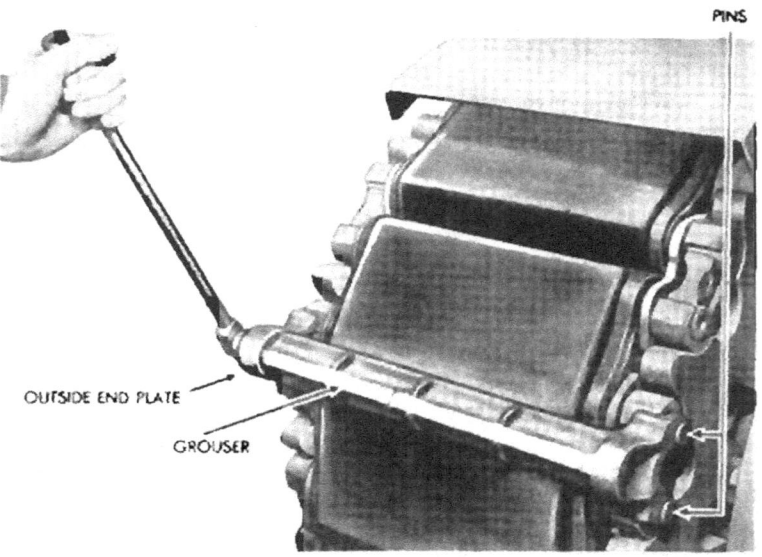

RA PD 14355

Figure 135. Installing grouser.

SECTION XII
ELECTRICAL SYSTEM

148. BATTERY TEST, REMOVAL, AND INSTALLATION.

a. General. Due to the large number of electrically operated accessories, a 24-volt electrical system is installed in the 155-mm Gun Motor Carriage M12 and Cargo Carrier M30.

b. Description.—Two 12-volt storage batteries are connected in series to maintain the voltage of the system at 24 volts. The batteries are installed in the battery compartment, located in the right sponson directly between the front sponson box and right fuel tank (fig. 4.)

c. Maintenance.

(1) **Care.** Check and tighten battery terminals and terminal posts frequently, clean and coat with seasonal grade general purpose grease. Check the battery fluid level once a week and after every long run. Maintain the level to ⅜ inch above the plate assemblies by adding distilled water. Take a specific gravity reading every 25 hours and exchange a battery having a specific gravity of 1.225 or less at 80° F. for a fully charged one.

(2) **Capacity and Temperature Data.** At temperatures below freezing, the load on the battery becomes greater and the relative capacity of the battery is reduced. For this reason, when low temperatures prevail, it will be necessary to maintain the specific gravity of the battery electrolyte at 1.250 or higher, and to replace the battery when its gravity reading is below that point. The following data show the capacity of the batteries and the relative freezing point of the electrolyte:

Capacity	Actual specific gravity	Freezing temperature
Battery charged	1.285	−96° F.
Battery ¼ discharged	1.255	−60° F.
Battery ½ discharged	1.220	−31° F.
Battery ¾ discharged	1.185	−8° F.
Battery normally discharged	1.150	+5° F.
	1.100	+18° F.

(3) To determine the actual specific gravity of the electrolyte, it is necessary to check the temperature of the solution with a thermometer. If the temperature is normal (80° F.) the specific gravity reading will be correct. However, if the temperature is above or below 80° F.,

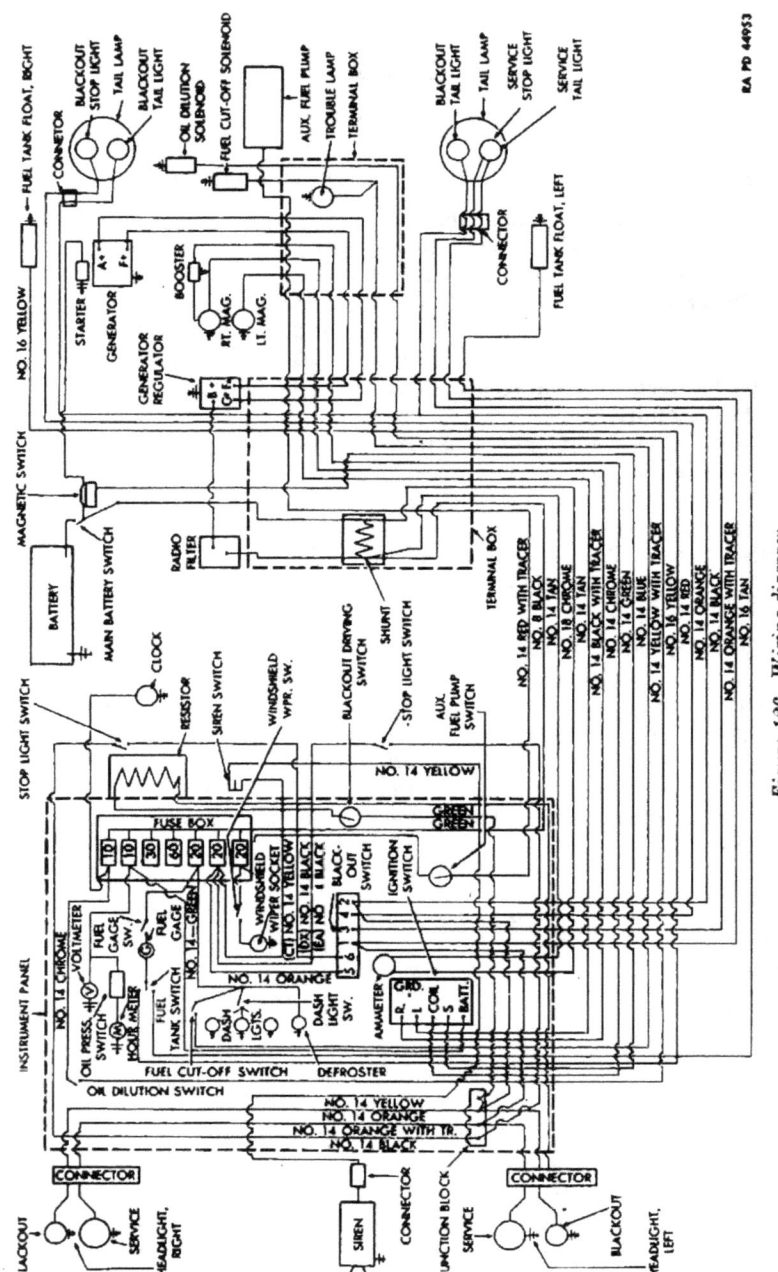

Figure 136. Wiring diagram.

it will be necessary to make an allowance to determine the actual specific gravity; due to the fact that the liquid expands when warm, and the same volume weighs less than when it is at normal temperature. The reverse is also true, and when the temperature is below normal or 80° F., the liquid has contracted and the same volume weighs more than it does when normal. To correct the specific gravity reading of the hydrometer to make it correspond to the temperature of the electrolyte, add to or subtract 0.004 from the hydrometer reading for each variation of 10° in temperature from the normal of 80° F. For example, when the specific gravity, as shown by the hydrometer reading is 1.290 and the temperature of the electrolyte is 60° F., it will be necessary to subtract 8 points or 0.008 from 1.290 which gives 1.282 as the actual specific gravity. If the hydrometer reading shows 1.270, at a temperature of 110° F., it will be necessary to add 12 points or 0.012 to the reading which gives 1.282 as the actual specific gravity. Refer to figure 137.

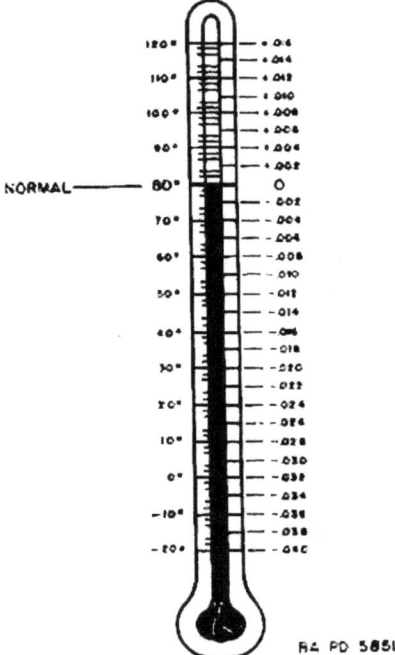

Figure 137. Hydrometer correction chart.

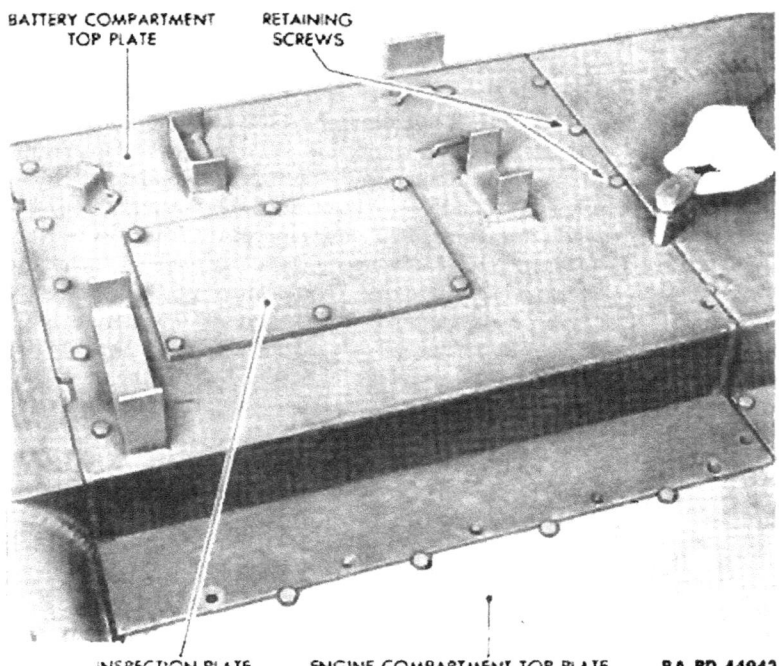

Figure 138. Battery compartment top plate removal.

d. Battery Removal (figs. 138 and 139).

Wrench, open-end, ⅜-inch Wrench, socket, ⅝₆-inch
Wrenches, open-end, ¾-inch (2) Wrench, socket, ¾-inch

(1) Open battery switch (fig. 14).

(2) Remove cap screws which secure front, inner, and rear sides of battery compartment top plate (⅝₆-inch socket wrench).

(3) Remove cap screws which secure outer side of battery compartment top plate (¾-inch socket wrench).

(4) Lift off battery compartment top plate.

(5) Remove the two battery bracket hold-down bolts (two ¾-inch open-end wrenches).

(6) Disconnect battery cables at the battery (⅜-inch open-end wrench). (Check legibility of markings.)

(7) Slide battery back toward rear of vehicle to clear solenoid starter switch, and remove.

e. **Battery Installation** (figs. 138 and 139).

Petrolatum
Wrench, open-end, ⅜-inch
Wrenches, open-end, ¾-inch (2).
Wrench, socket, ⁹⁄₁₆-inch
Wrench, socket, ¾-inch

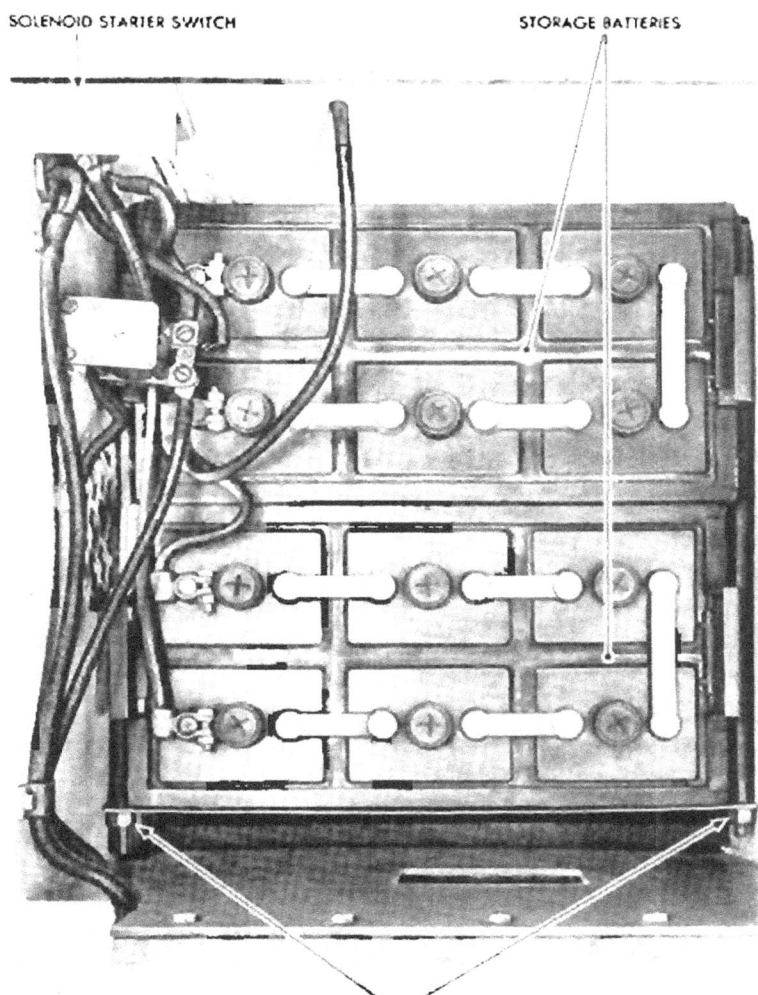

Figure 139. Position of storage batteries.

(1) Place batteries in position in battery compartment. Terminals face toward engine compartment.

(2) Connect battery cables to the proper battery terminals (⅝-inch open-end wrench). Make sure posts and terminals are clean before installation. After cable clamps have been installed, coat with seasonal grade general purpose grease. Terminals must be tight.

(3) Install battery bracket hold-down bolts (two ⅝-inch open-end wrenches).

(4) Place battery compartment top plate in position. Install cap screws which secure plate at front, inner, and rear sides (⅝₆-inch socket wrench). Install cap screws which secure outer side of top plate (⅝-inch socket wrench).

149. BATTERY SWITCH.

a. Description. The battery switch knob is located inside the driving compartment, on the front engine bulkhead behind the commander (fig. 14). It connects through the bulkhead to the battery switch, which is mounted inside the battery compartment above the batteries (fig. 140). Function of the battery switch is to cut off the

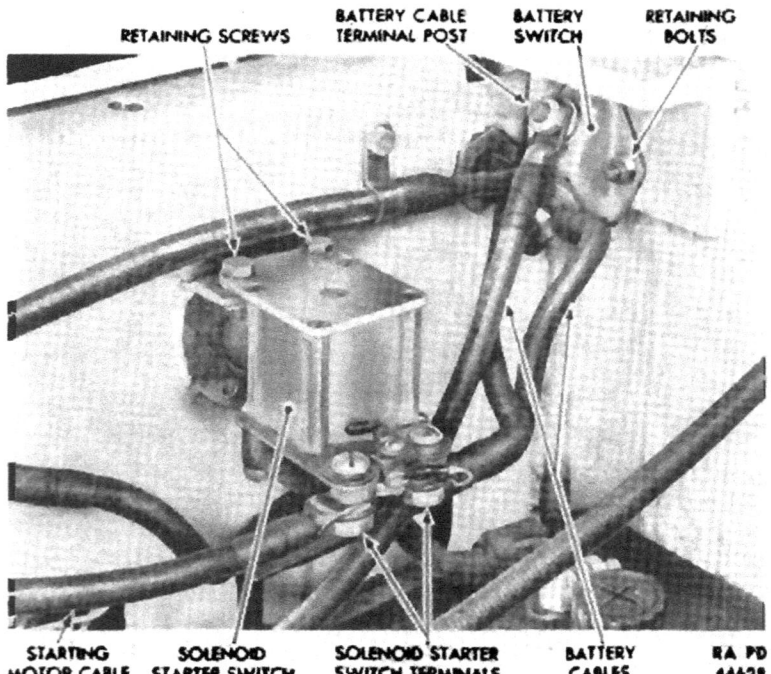

Figure 140. Installed battery switch and solenoid starter switch.

battery current at its source. To open the circuit, turn and pull out the battery switch.

b. Equipment.
Screw driver
Wrenches, open-end, 9/16-inch (2).
Wrench, open-end, 3/8-inch

c. Procedure.
If the battery switch becomes damaged in any way, or no longer makes a good contact, it must be replaced.

(1) Remove screw which holds battery switch knob, and remove knob (screw driver).

(2) Remove battery compartment top plate (par. 148).

(3) Remove terminal nuts and leads, and label for correct installation (3/8-inch open-end wrench).

(4) Remove battery switch retaining bolts and lift off switch (two 9/16-inch open-end wrenches). (Hold bolts from inside driving compartment.)

150. BATTERY SWITCH INSTALLATION (fig. 140).

a. Equipment.
Screw driver
Wrenches, open-end, 9/16-inch (2).
Wrench, open-end, 3/8-inch

b. Procedure.
(1) Place battery switch in position and install retaining bolts (two 9/16-inch open-end wrenches). (Hold bolts from inside driving compartment.)

(2) Install terminal leads and nuts. Make sure terminals are correctly installed (3/8-inch open-end wrench).

(3) Install battery compartment top plate (par. 148).

(4) Install battery switch knob and retaining screw (screw driver).

151. SOLENOID STARTER SWITCH (fig. 140).

a. Equipment.
Wrench, open-end, 9/16-inch
Wrench, open-end, 7/16-inch

b. Procedure.
(1) **Description.** A solenoid starter switch is mounted on the wall inside the battery compartment. It eliminates the necessity of running heavy cables up to the instrument panel and, therefore, helps eliminate voltage drop. This switch is actuated by a toggle type starter switch located on the instrument panel.

(2) **Starter Switch Removal** (fig. 140).

Wrench, open-end, 5/16-inch   Wrench, open-end, 7/16-inch

(a) Open battery switch.

(b) Disconnect leads (5/16-inch open-end wrench). Tag each lead as it is removed, to assure correct reinstallation.

(c) Remove two screws that hold solenoid to bracket (7/16-inch open-end wrench).

(d) Remove solenoid.

(3) **Starter Switch Installation** (fig. 141).

Wrench, open-end, 5/16-inch   Wrench, open-end, 7/16-inch

(a) Install solenoid in place and install two screws that hold it to bracket (7/16-inch open-end wrench).

(b) Connect leads (5/16-inch open-end wrench).

152. GENERATOR REGULATOR (figs. 141 and 147).

a. Description. To prevent overcharging of batteries, the generator is equipped with a generator regulator. It is of the three-unit detachable type, and is mounted on the engine bulkhead above the fire extinguishers, behind the driver. It includes a voltage regulator, current limitator, and reverse current relay.

(1) The voltage regulating unit maintains the output of the generator at a constant predetermined voltage of 28.4 volts. The current output of the generator is automatically varied in accordance with the state of charge of the battery, and the amount of current being used throughout the vehicle. Thus, the proper charge is delivered to the battery at all times without danger of overcharging.

(2) The current limitator unit limits the maximum current output to a value slightly in excess of the rated capacity of the generator.

(3) The reverse current relay or cut-out, or main switch, prevents the battery from discharging through the generator when the generator is at rest, or when it is not developing its normal voltage.

b. Inspection and Adjustments. When properly installed and operated, the generator control units should not require adjustment. If the voltage as indicated by the voltmeter is consistently above or below normal, replace the generator regulator top (fig. 141). To replace it, release the quick releasing clasp and take the box top off the box base and install new top (fig. 141). In pulling the box top from the box base, the operation is similar to removing a tube from a radio set. *Caution:* Before attempting to replace the generator regulator, make sure the battery switch is open.

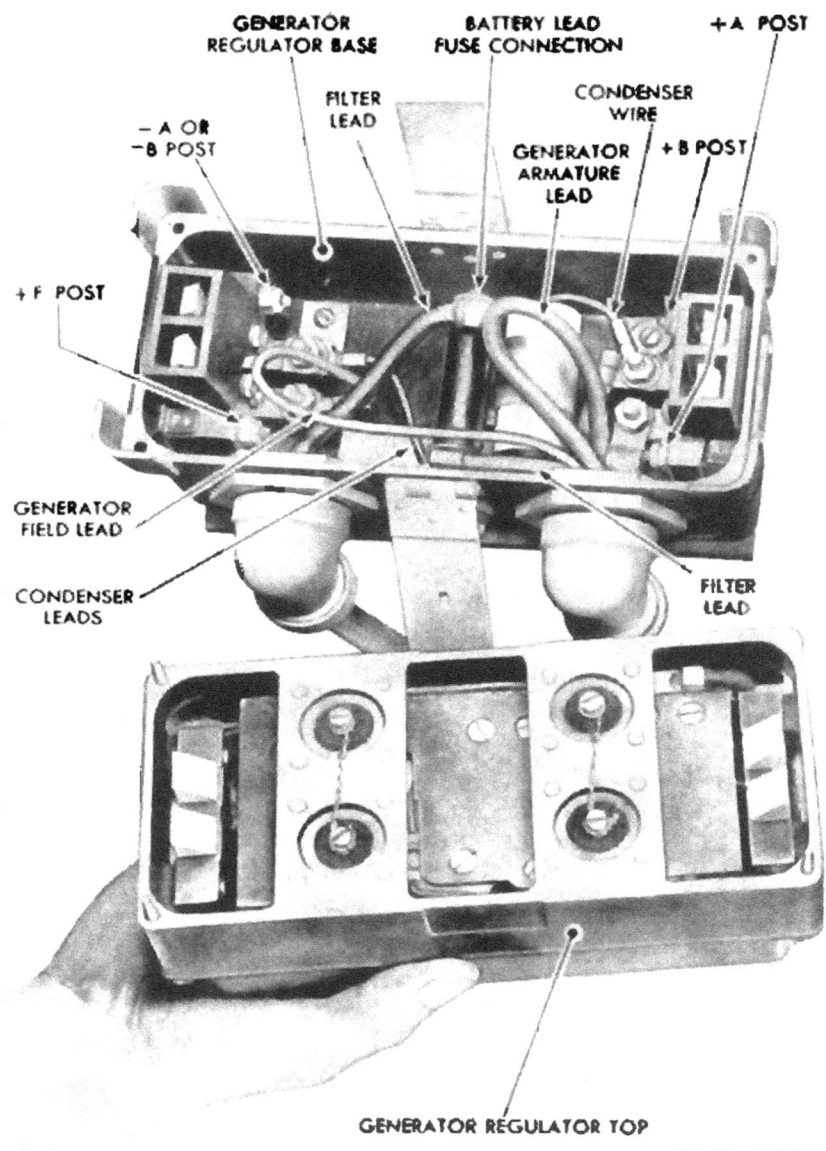

Figure 141. Generator regulator.

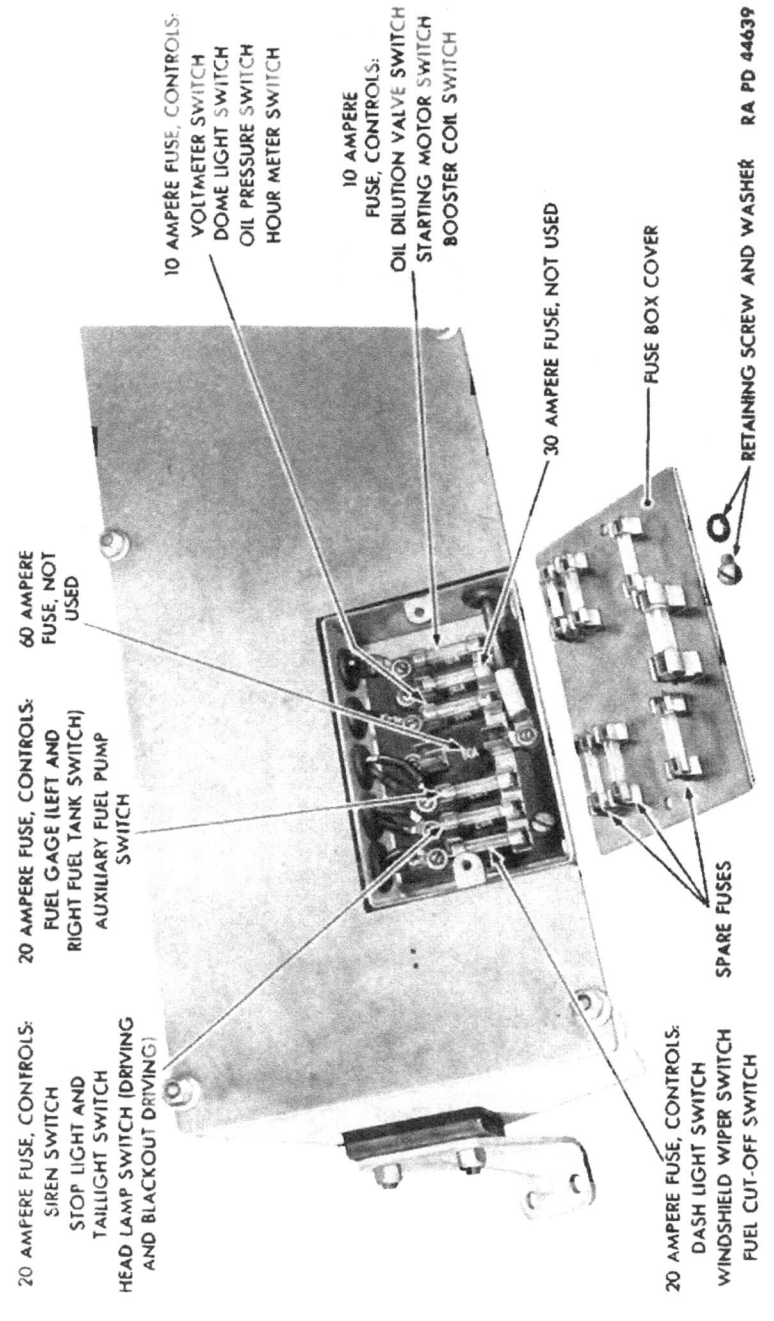

Figure 142. Fuse box, showing circuits controlled by fuses.

153. FUSE BOX (fig. 142). A fuse box, designed to prevent overloading of the circuit, is located on the under side of the left end of the instrument panel. In the event of an overload (or short circuit), the fuse burns through, breaking the circuit. When this happens, remove the two screws which secure the bottom plate to the fuse box and install a new fuse. If a fuse or fuses continue to burn through under normal load, look for a short circuit in the line or lines. The fuse box, together with the circuits controlled, is illustrated (fig. 142).

154. INSTRUMENT PANEL (figs. 12 and 13).

   a. The instrument panel is located across the front of the driving compartment (fig. 12) and contains various electrical and nonelectrical instruments, meters and switches (fig. 13).

   b. Voltmeter. A voltmeter, located on the instrument panel, indicates the voltage in the circuit. It is connected through a 10-ampere fuse.

   c. Ammeter. The ammeter is located on the instrument panel and indicates the amount of current in amperes flowing to or discharging from the battery. The amount of current will vary depending on engine speed.

   d. Hour meter.
(1) The hour meter indicates the total number of hours the engine has been in operation. The meter is installed in connection with a pressure switch. When the required pressure is reached, the meter will start operating and will continue to operate whether the engine idles or is under full load, until the engine is stopped.
(2) The pressure switch is installed in the oil pressure system by means of a T-connection near the oil pressure gauge, behind the instrument panel. It can be removed by disconnecting the pressure line.

   e. Ignition Switch. The ignition, or magneto switch, located on the instrument panel has four positions as indicated on the switch. With switch turned to extreme left, both magnetos are OFF; with switch to right of center, right magneto is ON; to left of center, left magneto is ON; to extreme right, both magnetos are ON. Directly beneath the magneto switch are the booster and starter switches. These are mounted parallel to each other and swing outward in the OFF position. Direction of movements for starting the engine is indicated.

   f. Blackout Switch. The blackout switch is of the push-pull type. It is located on the upper left-hand corner of the instrument panel. To operate the blackout switch, the spring button on the side of the switch must be depressed. The switch has three positions in addition to OFF. The first operates the blackout lights; the second operates only the service lights, and the third operates the service

255

stop lights only. When the blackout switch is in the first or BLACKOUT position, a switch directly beneath the blackout switch may be pulled out to operate the blackout driving light on the left side of the vehicle (when so equipped).

g. Auxiliary Fuel Pump, Windshield Wiper, Oil Dilution and Fuel Cut-off Switches. These are all toggle switches located on the instrument panel, and are so indicated. (Refer to par. 81a concerning auxiliary fuel pump and par. 11e concerning oil dilution.)

h. Clock. Mounted in the center of the panel to the left of the primer pump is an 8-day clock. A second-hand is also provided. The stem winder is at the base of the clock.

i. Dash Light Switch. The dash light switch is of the push-pull type, located on the lower right section of the instrument panel.

j. Tachometer. The speed of revolution of the engine crankshaft is indicated by the tachometer. It is driven by a flexible, encased shaft connecting to the tachometer drive in the rear face of the accessory case.

k. Speedometer. The speedometer is driven by a flexible shaft connecting from the spiral gear on the output shaft of the transmission. Two odometers, one showing trip and the other showing total mileage, are incorporated in the head of the speedometer.

l. Oil Pressure Gauge. This gauge records the pressure in the engine oil manifold through a ¼-inch copper tube.

m. Oil Temperature Gauge. This gauge indicates the engine oil temperature by means of a temperature-sensitive device located on a fitting in the oil pump.

n. Fuel Primer Pump. The fuel primer pump is located on the instrument panel and is discussed in paragraph 77.

o. Resistance Box. Four screws mount the ventilated blackout driving switch resistance box on the top right under side of the instrument panel box. The unit has a resistance of 7.3 ohms.

155. INSTRUMENT MAINTENANCE (fig. 13). All instruments and switches that become inoperative should be exchanged for serviceable ones.

156. INSTRUMENT REMOVAL (fig. 13).

a. Equipment.

 Screw driver   Wrench, open-end, ⁷⁄₁₆-inch.
 Wrench, open-end, ½-inch.

b. Procedure.

(1) Open battery switch (fig. 14).

(2) Remove six nuts and washers which secure bottom cover of instrument panel. Lift off cover (⁷⁄₁₆-inch open-end wrench).

(3) Remove screws on face of panel which hold instrument in place, and remove nuts and lock washers (screw driver and ⅜-inch open-end wrench).

(4) Remove wires from connections and mark for replacement.

Note: In removal of tachometer, disconnect drive shaft, and in removal of oil pressure gauge disconnect oil lines.

(5) Remove instrument from bottom of instrument panel.

(6) The oil temperature gauge line cannot be removed from the gauge but must be disconnected from the oil pump, pulled ahead into the driving compartment, and removed from panel with gauge.

(7) The magneto switch, mounted on the instrument panel, is held by a clip across the back of the switch, which is secured to the back of the panel by two ⅜-inch screws. If these screws are removed (⅜-inch spin-tight wrench), the switch can be pulled through the panel toward the driving compartment and wires can be removed with a screw driver.

(8) The siren switch is secured to the floor by two screws which can be removed with a ⁷⁄₁₆-inch socket wrench.

157. INSTRUMENT INSTALLATION (fig. 14).

a. Equipment.
Screw driver
Wrench, open-end, ⅜-inch.
Wrench, open-end, ⁷⁄₁₆-inch.

b. Procedure.

(1) Place new instrument in position in panel.

(2) Install screws, lock washers, and nuts which hold instrument in place (screw driver and ⅜-inch open-end wrench).

(3) Connect leads, oil lines, or drive shaft, as the case may be.

(4) Place bottom cover in position and install six screws and washers which secure cover to instrument panel (⁷⁄₁₆-inch open-end wrench).

(5) To install the magneto switch, pull wires through hole in instrument panel toward driving compartment, and attach to switch (screw driver). Slide switch into panel. Replace clip across back of switch and install two screws (⅜-inch spin-tight wrench).

(6) To install siren switch, place switch in position on floor and install two screws (⁷⁄₁₆-inch socket wrench).

158. TERMINAL BOX (figs. 30 and 31).

Two terminal boxes are provided. The front terminal box is mounted on the engine bulkhead, above the automatic fire extinguishers, in the recess behind the driver. It is the terminus for circuits which lead to the instrument panel and equipment mounted on the front of the vehicle. The rear terminal box is mounted inside the engine compartment on the upper left of the rear engine bulkhead. It is the terminus for circuits which lead to the engine and to equipment mounted on the rear of the vehicle.

159. SOLENOIDS.

a. The fuel cut-off valve (fig. 21) and the oil dilution valve (fig. 20) are operated by means of solenoids, controlled by two toggle switches on the instrument panel (fig. 13).

b. When properly installed, these solenoids should not require attention. If the solenoid is inoperative, replace it with one in good condition.

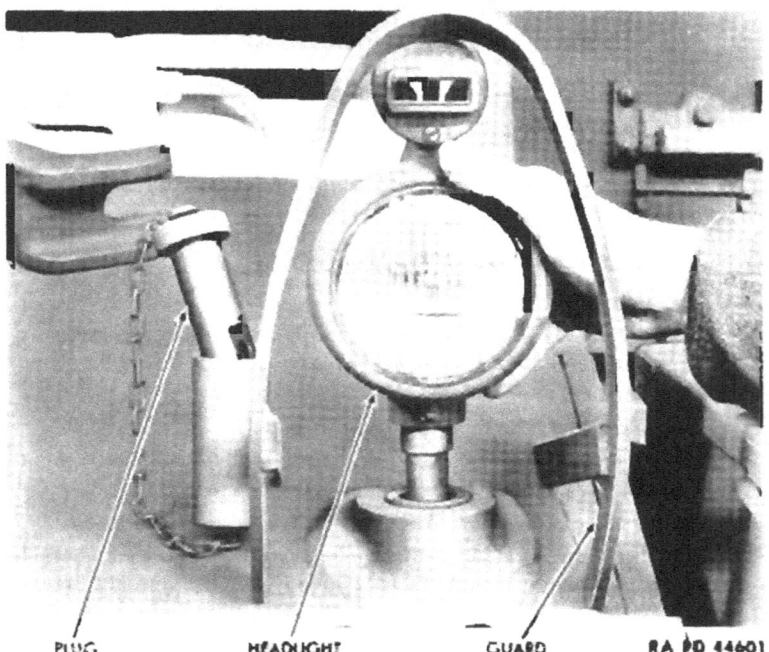

Figure 143. Headlight removal.

160. LIGHTS.

a. Headlights (fig. 143). Two sealed beam headlights, each 50-candlepower, are provided. They are located on the front end of the vehicle (fig. 1), one beneath each indirect vision door. They are focused by means of a focusing screw back of the lights. Lights are removed by pulling out a catch from inside the driving compartment beneath the light, and at the same time lifting up on the light. A protective plug, loosely mounted in a bracket next to the light, is inserted in the hole from which the light was removed.

b. Blackout Lights (fig. 143). A blackout light is located above each headlight. Three-candlepower lamps are used in them.

c. Taillights (figs. 144 and 145). The taillights are combination lights. The rear left is a combination stop, service, and blackout light. The lower lens is the service and service stop light, and the upper lens is the blackout light. In the right rear light, the lower

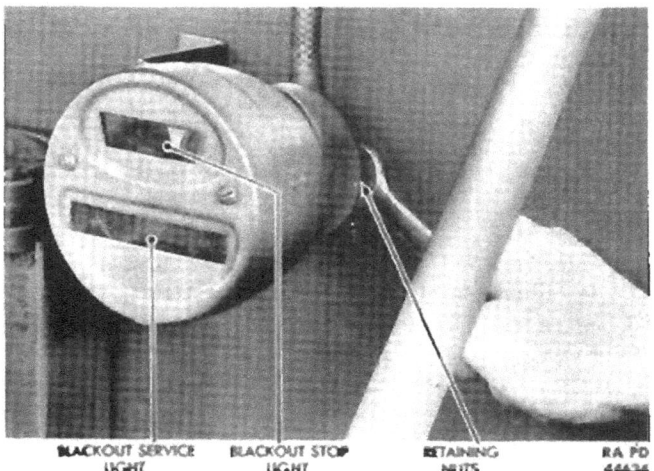

Figure 144. Disconnecting blackout stop and service light.

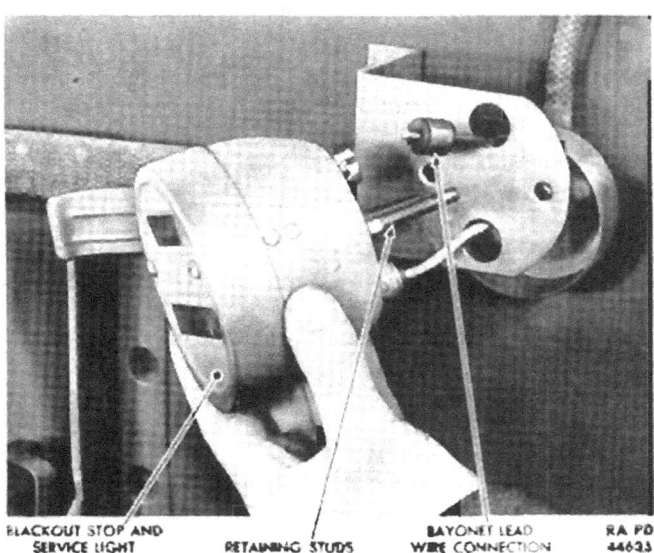

Figure 145. Blackout stop and service light removal.

lens is the blackout light, and the upper lens is the blackout stop light. Both stop lights are controlled by stop light switches connected with the steering hand levers. No stop signaling will be seen until both levers are pulled, indicating a slowing or full stop. Taillights are removed by removing the two retaining nuts at the back of the light (⅜-inch open-end wrench) and pulling out the bayonet lead wire connections.

d. Dash Lights. Two dash lights are provided. Each is provided with a 3-candlepower lamp.

e. Fuel Gauge Lamp. The fuel gauge is illuminated by a 3-candlepower lamp.

f. Lamps. All lamps are of bayonet-type base and are replaceable except in headlights and taillights, in which they are a part of the sealed beam units.

(1) Sealed beam lamp units are replaced by removing the retaining ring (screw driver) and gasket replacing the faulty unit with a new one, and installing gasket and retaining ring.

(2) To replace blackout lamps, remove door (screw driver), lens, and gasket. Press lamp in and remove with a counterclockwise one-quarter turn. Turn in new lamp and install gasket, lens, and door.

(3) All other lamps are removed by a one-quarter turn in the socket.

161. SIREN (fig. 146).

  a. Equipment.
    Pliers   Wrench, open-end, ⁷⁄₁₆-inch

  b. Procedure.

(1) **Description.** The siren is located on the front plate of the vehicle between the two indirect vision doors, and is operated by a foot button at the driver's left foot. If the siren is defective, replace it with a serviceable unit.

(2) **Siren Removal.**
    Pliers   Wrench, open-end, ⁷⁄₁₆-inch

  (a) Open battery switch (fig. 14).

  (b) Disconnect conduit knurled nut (pliers), and pull out electric plug connection.

    (c) Remove two retaining nuts (⁷⁄₁₆-inch open-end wrench).

    (d) Remove siren.

(3) **Siren Installation.**
    Pliers   Wrench, open-end, ⁷⁄₁₆-inch

  (a) Install new siren in place.

  (b) Install retaining nuts (⁷⁄₁₆-inch open-end wrench).

  (c) Plug in electric plug connection, and tighten conduit knurled nut (pliers).

  (d) Close battery switch (fig. 14).

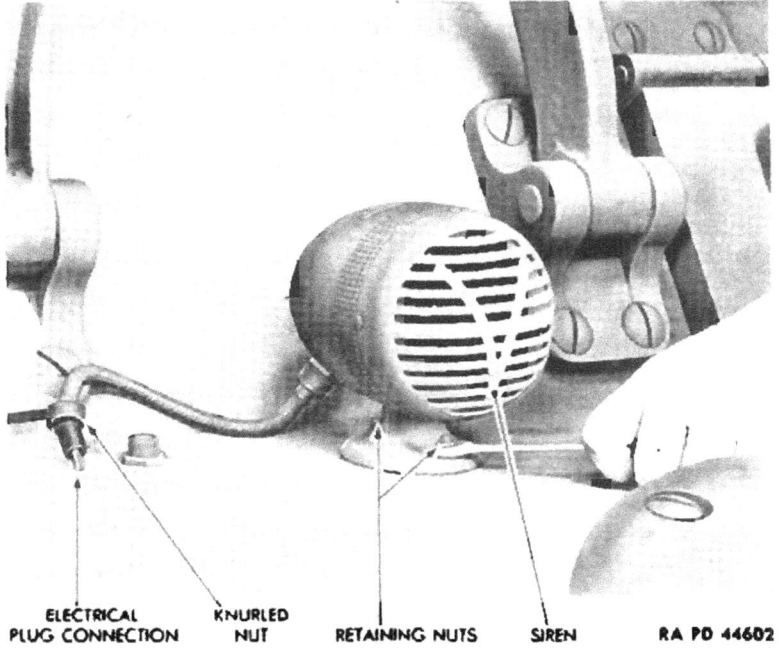

Figure 146. Siren removal

162. TROUBLE SHOOTING.

a. General. Trouble shooting procedure in the electrical system is largely concerned with checking the wiring to make sure that it is serviceable. All grounded wires or wires that have open or short circuits must be replaced with wires of the same size and color.

b. Test Light. To test wiring requires a test light such as a 24-volt instrument panel lamp, a socket, two pieces of insulated wire (each about 18 inches in length), a clamp, a metal prod approximately 1 inch long, and tape. Install the lamp in the socket and connect one of the wires to the lead on the socket. Attach the metal clamp to the other end of the wire. Place tape over connection. Solder the end of the second wire to the side of the lamp socket. To the other end of this wire connect the metal prod and place tape over the connection.

Note: If a prod and a clamp are not available, scrape approximately an inch of insulation from each of the wires on the ends to which the clamp and the prod should be attached. If no soldering equipment is available, scrape off several inches of insulation from the end of the wire which should be soldered to the side of the lamp socket; wrap the end around the outside of the lamp socket and place tape over the connection.

c. General Trouble Shooting. If the unit does not work, when the switch for any electrical unit is turned on, make the following check:

(1) Check battery switch to make sure it is closed.

(2) Depress starter, and watch ammeter and voltmeter to make sure current is coming from the battery. If ammeter does not show a discharge, the battery may need to be charged or replaced.

(3) Check for blown-out fuse, ground, loose connection, and open or short circuit.

(4) Test any electrical unit not operating properly for cause of failure; a defective light, for example, would be removed and tested. Replace any defective units.

(5) Check wiring and replace defective wires.

(6) Check with test light at input terminal of unit to determine if current is flowing to unit. If circuit is operating, replace unit. If no current is at unit, check lead from current source to unit, to locate open ground or short (d and e below).

d. Jumper. Trouble shooting on the electrical system requires a jumper, which can be made by hooking up metal clamps on either end of a piece of insulated wire about 18 inches long. If clamps are not available, scrape off 1 inch of insulation from each end of the wire.

e. Testing of Wire. To test a wire for an open circuit or to determine whether it is grounded on the vehicle, use the following procedure:

(1) Disconnect both ends of wire.

(2) Hook up one end of test light to a positive lead and connect other end to one free end of the wire to be tested. If light burns, the wire is grounded on vehicle and should be replaced. If light does not burn, the wire is not grounded. To determine if a wire has an open circuit, proceed to next step.

(3) Ground the free end of wire being tested on the vehicle. If light burns, the wire is all right. If light does not burn, the wire has an open circuit and should be replaced.

f. Replacing Defective Wire.

(1) With current off, disconnect defective wire at both ends.

(2) Attach end of old wire to end of new wire, soldering them securely.

(3) Pull new wire in place by slowly drawing old wire out.

(4) Disconnect the two wires.

(5) Hook up new wire at both ends.

Note: In replacing wire, be sure to install new wire of the same size and color as old wire.

g. Generator. If ammeter registers discharge when engine is being run at 1,000 revolutions per minute, and the lights and other electrical accessories are on, the generator, generator regulator, generator filter, or generator circuit is faulty. To determine the cause of the trouble, proceed according to the following:

(1) Run engine at 1,000 revolutions per minute.

(2) Determine if circuit from generator regulator to battery is good. This is done as follows:

  (a) Loosen the quick-releasing clamp on the generator regulator and remove the top box.

  (b) Connect test light from +B (positive) to −B (negative) or −A (negative) in the control box base (fig. 141). If light burns, the fuse and the circuit from the control box to the battery are good, and it will be necessary to proceed to (3) below. If the light does not burn, proceed to (c) below.

  (c) Connect the test light from the battery lead fuse connection to −B (negative) (fig. 141). If the light burns, the circuit to the battery is good. Remove and check fuse. If the fuse is defective, remove condenser lead from the +B (positive) post, and connect one lead of the test light to the battery lead fuse connection, and the other test light lead to the condenser lead removed from the +B (positive) post. If the light burns, install a new condenser and fuse. If the light does not burn, install a new fuse and connect the condenser in the circuit. Install a serviceable top box assembly.

(3) Determine if generator is supplying current to control box. This is done as follows:

  (a) Connect the test light from +A (positive) to ground (fig. 141). With a jumper, make a momentary connection between the +F (positive) and +A (positive) posts. If the light burns, install a serviceable top box assembly. If the light does not burn, however, the generator circuit is not supplying current to the control box, and it will be necessary to proceed to (b) below.

  (b) Check wiring in circuit from generator regulator to generator, using a test light. If wiring is defective, replace. If it is good, continue on with (4) below.

(4) Determine if the generator is supplying current to its own terminals. This is done as follows:

  (a) Disconnect leads at generator terminals.

  (b) Connect test light from +A (positive) to ground. Make momentary connection between +A (positive) and +F (positive) with a jumper. If light burns, the generator is good. Replace the top box of the generator regulator with a serviceable unit. If the light does not burn, replace the defective generator with a serviceable unit.

h. Generator. If voltmeter needle goes above 28.4 volts, the control box has failed, or the voltmeter is defective. Replace the defective unit or units with serviceable units.

I. Starter Solenoid. If engine will not turn over, and the voltmeter reads 28 or over, trouble in the starter solenoid or in the starter solenoid circuit is indicated. Check circuit for defective wires. Check ground connection on solenoid for loose ground nuts. Check bridge between solenoid terminals, using a test light. If tests reveal no trouble, the solenoid starter switch is defective and should be replaced (par. 151).

J. Booster. If engine will not start, it may be due to failure of the booster. To check, use following procedure:

(1) One man should turn on the booster switch.

(2) Another man, at the engine compartment, should listen for the buzz of the points on the booster. If there is no buzz, proceed to (3) below.

(3) Check wiring (**d and e** above). If defective wire is found, it must be replaced. If no defective wire is found, the booster has failed. Replace the defective booster coil with a serviceable unit (par. 53c(1) and (2)).

SECTION XIII
FIRE EXTINGUISHERS

163. FIXED FIRE EXTINGUISHER SYSTEM (figs. 147, 148, and 149).

a. Description.

(1) Two 10-pound fire extinguishers are mounted in special brackets in the recessed compartment directly behind the driver's seat. They are used solely for extinguishing fires in and around the engine and are connected directly to tubing and discharge nozzles located in the front and rear of the engine compartment. In use, they force carbon dioxide gas around the engine, from top and bottom, and fill the engine compartment, thus smothering the fire. In case of fire, shut off the engine.

(2) The cylinders are connected to the supply tubing by means of a double-check tee, which tee prevents the loss of gas into crew compartment should one cylinder be operated while the other cylinder is removed for weighing or recharging.

(3) In addition to the built-in or fixed system for the engine compartment just described, two 4-pound portable extinguishers are provided for small fires in or about the vehicle. Full operating instructions are found on extinguisher name plate.

b. Operation. The fire extinguisher system is entirely manual in operation. It is, therefore, imperative that there be as little delay as

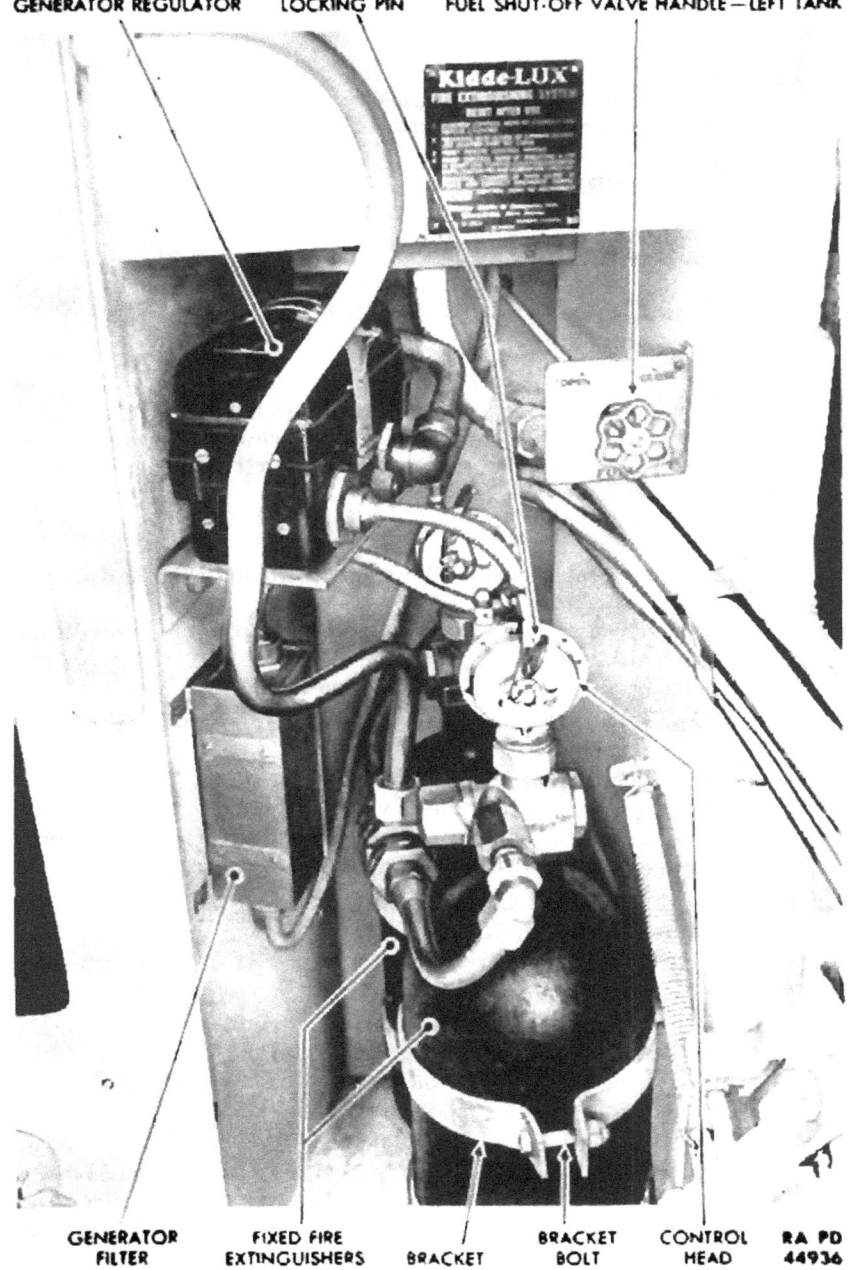

Figure 147. Installation of fixed fire extinguishers.

possible in discharging the gas, as its effectiveness is greatly increased by catching the fire in the beginning.

164. FIXED FIRE EXTINGUISHER CYLINDER REMOVAL
(figs. 148 and 149).

a. Equipment.
Pliers   Wrench, open-end, $\frac{9}{16}$-inch
Screw driver   Wrench, open-end, $1\frac{1}{4}$-inch

b. Procedure.

(1) Removing Control Head (fig. 146).
Pliers   Wrench, open-end, $\frac{9}{16}$-inch
Screw driver   Wrench, open-end $1\frac{1}{4}$-inch

(a) Remove lock wire and screws which secure control head cover (screw driver and pliers). Lift off cover.

(b) Disconnect cable tube at control head ($\frac{9}{16}$-inch open-end wrench).

(c) Disconnect cable from block within control head (screw driver). Pull cable out of control head.

(d) Remove control head by turning swivel nut off valve housing ($1\frac{1}{4}$-inch open-end wrench), and lift off head. *Caution:* Never handle cylinder until control head has been removed.

(2) Disconnecting Connector Tube to Double Check Tee (fig. 148).

Wrench, open-end, $1\frac{1}{4}$-inch.

Disconnect the tube that connects valve outlet to double check tee, removing it at the valve outlet ($1\frac{1}{4}$-inch open-end wrench). *Caution:* Never remove cylinder with connector tube attached.

(3) Loosening Bracket Clamps (fig. 145).

Wrench, open-end, $\frac{9}{16}$-inch

Loosen bolts that hold bracket clamps to cylinder ($\frac{9}{16}$-inch open-end wrench).

(4) Lifting Cylinder Out of Bracket. Lift cylinder from bracket. *Caution:* Any cylinder containing gas under high pressure is as dangerous as a loaded shell. Never drop or strike cylinders, or handle roughly. Do not expose cylinder to unnecessary heat.

165. FIXED FIRE EXTINGUISHER CYLINDER INSTALLATION.

a. Equipment.
Pliers   Wrench, open-end, $\frac{9}{16}$-inch
Screw driver   Wrench, open-end, $1\frac{1}{4}$-inch

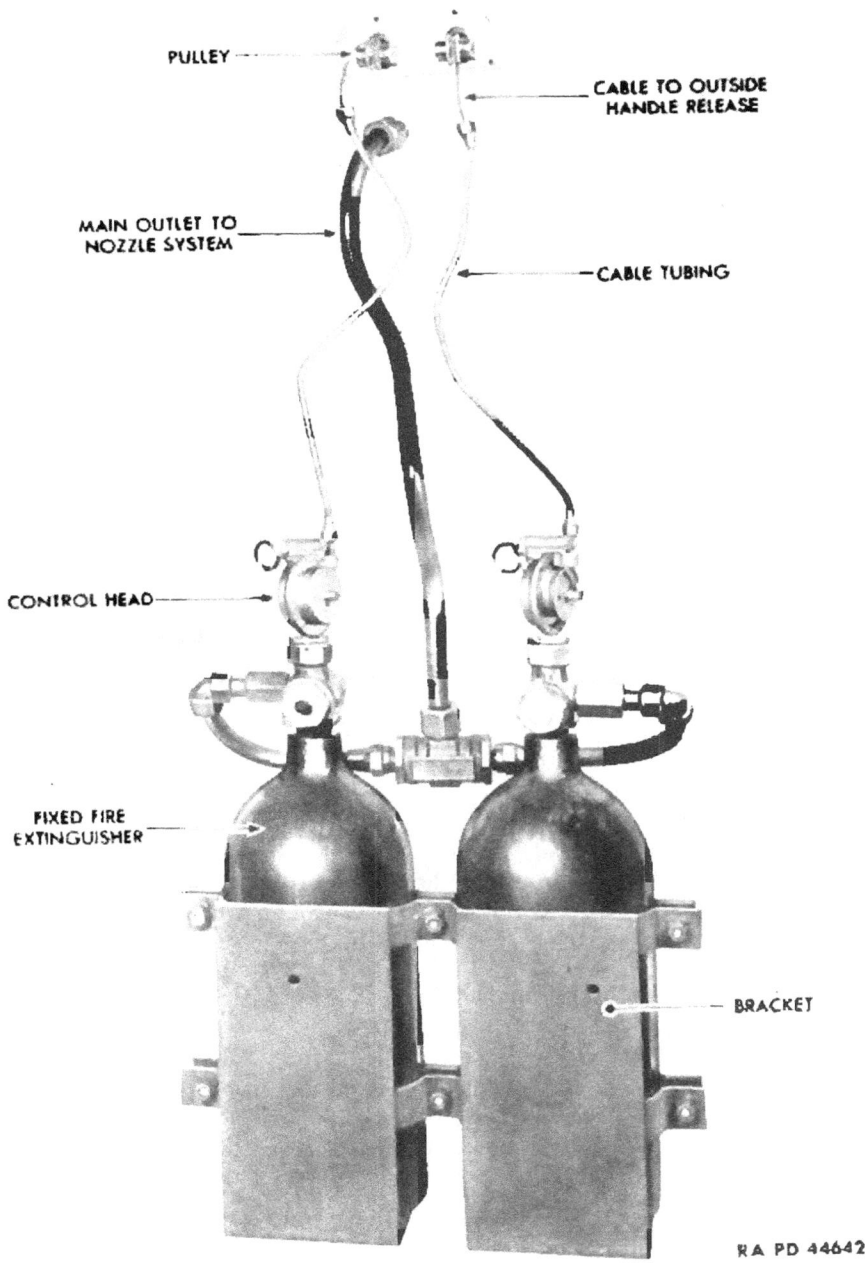

Figure 148. Fixed fire extinguishers.

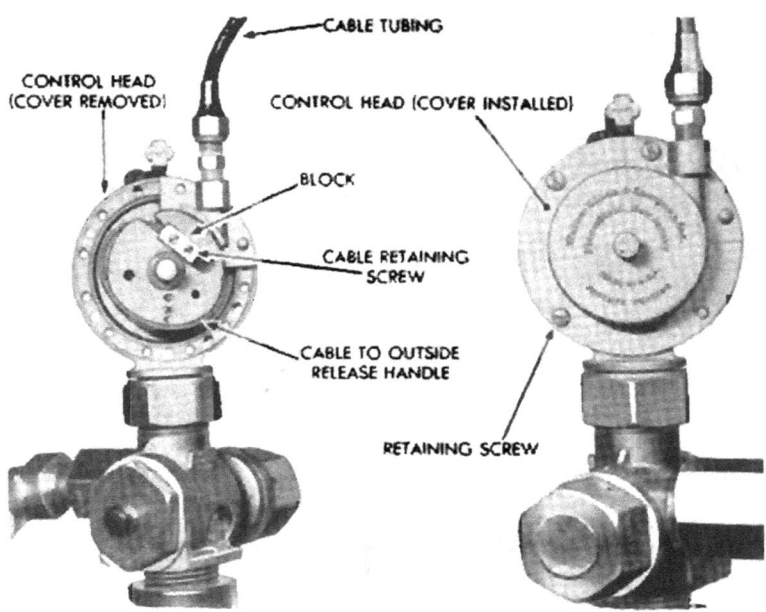

Figure 149. Fixed fire extinguisher control head.

b. Procedure.

(1) Installing Cylinder.
    Wrench, open-end, ⁹⁄₁₆-inch   Wrench, open-end, 1¼-inch

(a) Set cylinder (with control head removed) in bracket. Do not tighten clamping bolts at this point.

(b) Install connector tube on valve outlet (1¼-inch open-end wrench).

(c) Tighten bracket clamping bolts (⁹⁄₁₆-inch open-end wrench).

(2) Installing Control Head.
    Pliers   Wrench, open-end, ⁹⁄₁₆-inch
    Screw driver   Wrench, open-end, 1¼-inch

(a) Set control head in position on valve housing, and tighten down swivel nut (1¼-inch open-end wrench).

(b) Insert outside fire extinguisher handle cable into control head, around circular center section (fig. 149). Place cable in attaching block, and tighten cable retaining screws (screw driver).

(c) Install control head cover and retaining screws (screw driver). Install lock wire in screws (pliers).

(d) Install and tighten cable tube retaining nut (⁹⁄₁₆-inch open-end wrench).

166. PORTABLE FIRE EXTINGUISHERS.

a. Two portable 4-pound fire extinguishers are provided in the vehicle, one on either side of the crew compartment.

b. Portable extinguishers are operated by pulling the trigger while directing the discharge cone toward the fire. The position of the trigger determines the rate of discharge. The extinguisher should be carried in the left hand and the hose or cone in the right. Direct the discharge at the base of the flame, with the cone as close to the flame as the operator can safely hold it. Increase the rate of discharge from the extinguisher as the fire is put out.

167. FIRE EXTINGUISHER CARE.

a. Any cylinder containing gas under high pressure is as dangerous as a loaded shell. The extinguisher cylinders should never be dropped, struck, handled roughly, or exposed to unnecessary heat.

b. After use, the extinguisher should be immediately exchanged for one that is fully charged. Every 4 months, weigh each extinguisher. If the extinguisher weighs less than $3\frac{1}{2}$ pounds (portable) or $9\frac{1}{2}$ pounds (fixed), it should be exchanged for a fully charged one.

c. Care should be taken to see that extinguishers are always securely fastened inside the vehicle, and that other equipment does not interfere with the accessibility of controls or ease of operation of the fixed fire extinguisher system.

d. Periodic Inspections. The fire extinguisher system requires no more than ordinary care to insure its proper operation. As the system is for emergency use, it must be kept in operating condition at all times; therefore, frequent inspection should be made to insure that apparatus is intact. Check red cap on safety outlet of valve. If not intact, cylinder has been prematurely discharged due to high temperature and must be recharged immediately. The following inspections will be performed:

(1) **50-Hour Inspection.** Inspect entire system for any mechanical damage. Make certain that shielded nozzles are free of all foreign matter.

(2) The system may be put into operation from the outside of the vehicle, or from within the driving compartment. To operate the system, either pull out the outside handles on the engine compartment top plate (fig. 24), or pull the pull-out safety pin from the control head of one of the cylinders, breaking the wire seal. When operating the system from within the driving compartment, turn control counterclockwise to release gas (fig. 147). The further control is turned, the faster gas will be released. Do not open second cylinder except in emergency, and then only after first cylinder has been discharged. Purpose of second cylinder is to provide protection in case of second fire, after first cylinder has been discharged.

e. **Principle.**

(1) The fire extinguisher system uses carbon dioxide as the extinguishing agent. Carbon dioxide (not to be confused with carbon monoxide) is not poisonous but is suffocating.

(2) "Fast" fires, such as those involving gasoline or oil, are quickly extinguished by flooding the area with carbon dioxide gas. This reduces the oxygen content and creates an inert atmosphere which smothers the fire. "Slow" or "deep-seated" fires, such as fires in baled cotton and similar substances, are extinguished by prolonged action of a high concentration of carbon dioxide. In addition to its smothering action, carbon dioxide further aids in extinguishing fire by its cooling effect.

(3) Since a person cannot breathe, but will suffocate in an atmosphere of carbon dioxide, caution must be taken before entering any space filled with this gas. Thoroughly ventilate the space into which the gas has been discharged to make certain that all portions contain only fresh air. Should it be necessary for a person to enter a space before it is thoroughly ventilated, he may do so for a short period by holding his breath.

(4) Should a person be overcome by carbon dioxide, it is essential that he be rescued from the space containing the gas within 5 minutes. To revive a person so overcome, give him plenty of fresh air and apply artificial respiration, as in the case of drowning.

CHAPTER 3
ARMAMENT

SECTION I
GENERAL

168. SCOPE.

a. This part of the manual contains a brief description of the functioning and operation of the armament mounted on the 155-mm Gun Motor Carriage M12 and Cargo Carrier M30.

b. It also lists the authorized ammunition for the armament carried in these vehicles.

c. For detailed information on the care, preservation, malfunction, maintenance, assembly, and disassembly of the 155-mm guns M1918 and M1, and 155-mm gun mount M4, see TM 9-345 and TM 9-350. For information on the .50 caliber Browning machine gun, HB, M2, see FM 23-65.

169. CHARACTERISTICS.

a. The armament on the 155-mm GUN MOTOR CARRIAGE M12 is employed against emplaced battery and other ground objectives. The armament on the Cargo Carrier M30 is employed against aircraft. The 155-mm gun mount is located in the 155-mm gun motor carriage so that the center of gravity will be as low as possible. The 155-mm gun may be elevated to 30° and depressed to 5°. The maximum horizontal range of the 155-mm gun is approximately 20,000 yards. The .50 caliber Browning machine gun, HB, M2, flexible is mounted on a mount that may be moved as desired along a raised circle track at the rear of the Cargo Carrier M30.

b. The sighting equipment used with the 155-mm gun is the telescope M53 and telescope mount M40. The caliber .50 machine gun M2 is equipped with the conventional sights. The maximum range of the caliber .50 machine gun is approximately 7,200 yards.

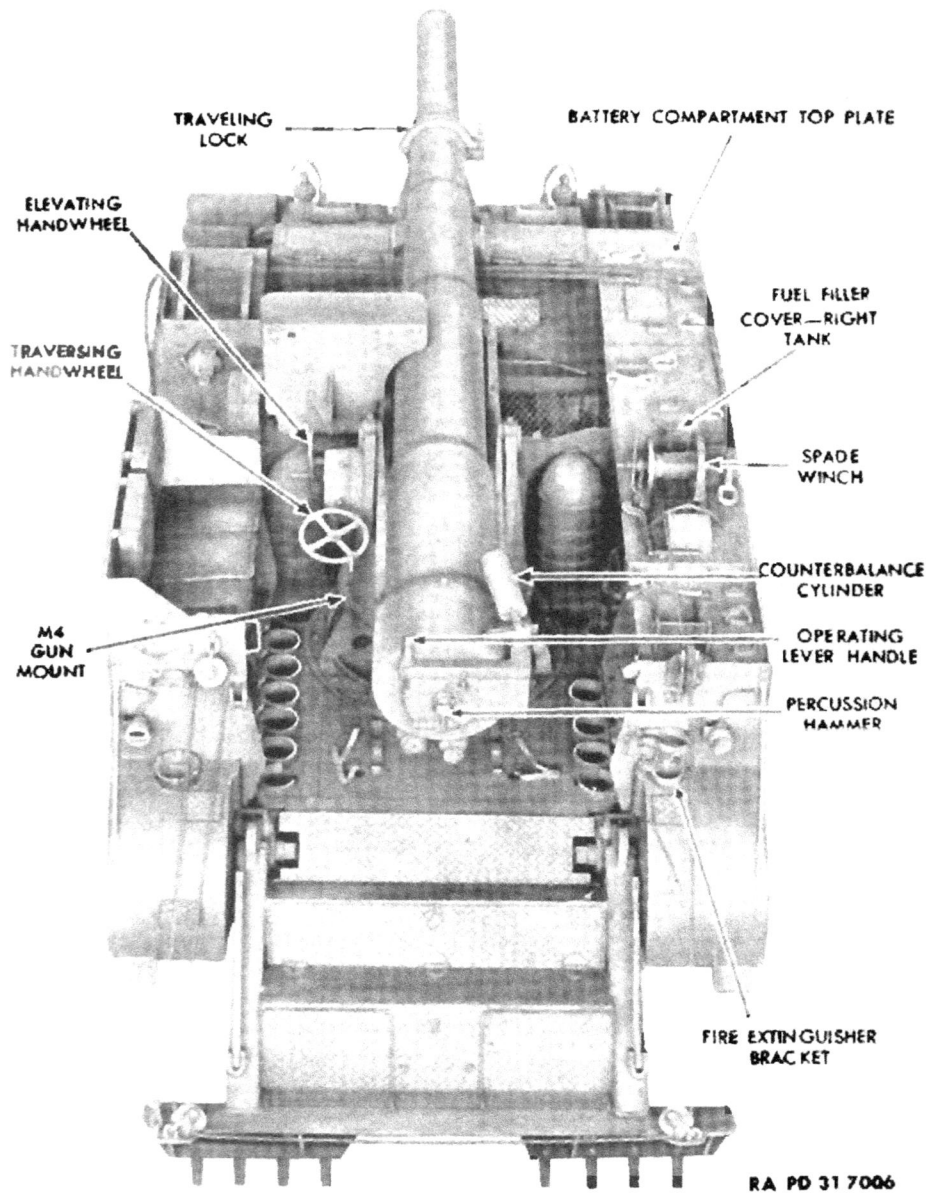

Figure 150. 155-mm gun motor carriage—top rear.

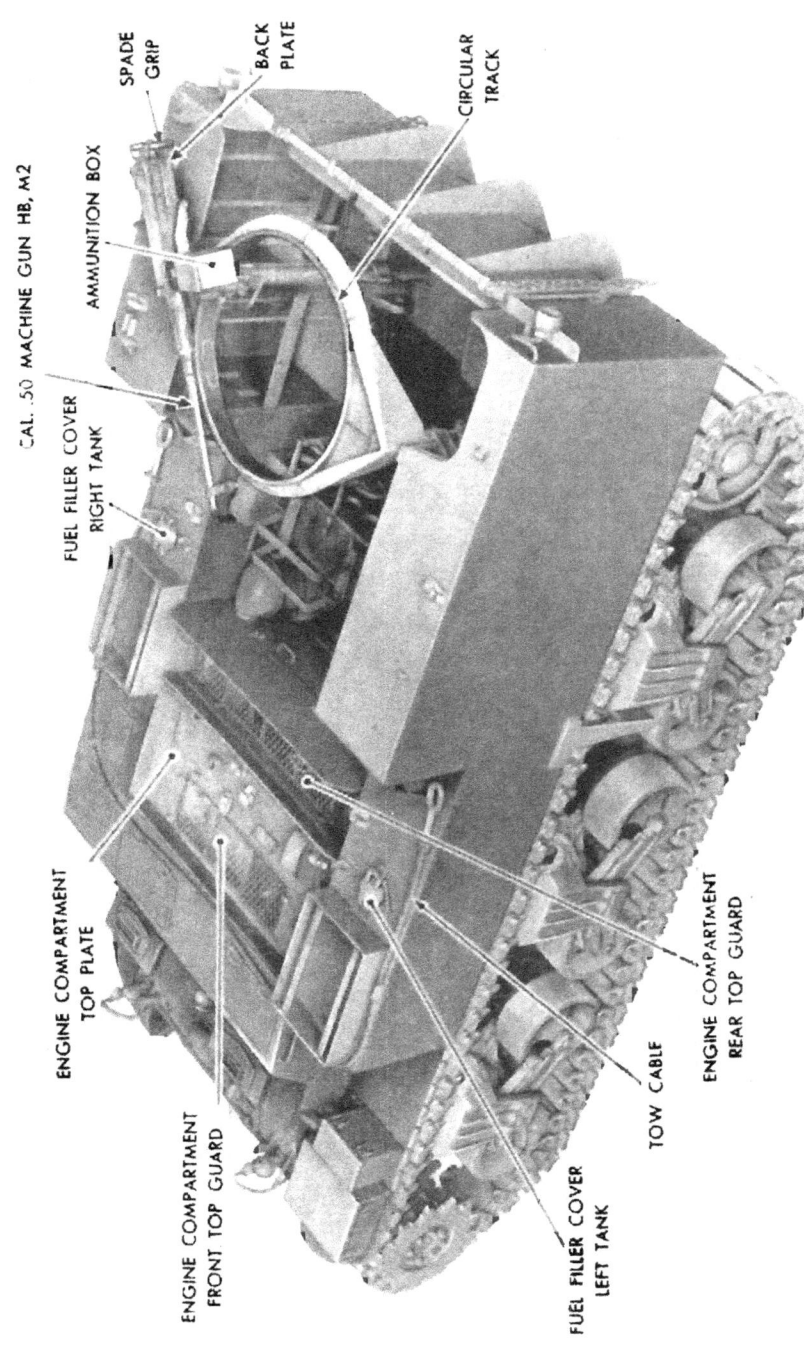

Figure 151. Cargo carrier—top.

170. DATA.

a. Weights, Dimensions and Ballistics, 155-mm Gun.

Weight of a 155-mm gun, complete. pounds.. 8,715.
Caliber, 155-mm gun............inches. 6.102.
Length (muzzle to rear face of breech ring)
                                    inches... 232.87.
Chamber:
    Diameter.....................do   6.693.
    Length, breech closed to base of projectile................do... 37.087.
    Capacity.............cubic inches.. 1,329.
Rifling:
    Number of grooves..................... 48.
    Twist, right-hand, uniform, 1 turn in 29.89 calibers (inclination 6°).
    Travel of projectile in bore....inches . 185.
Weight of projectile............pounds . 95.
Weight of full powder charge........do.... 25¼.
Maximum powder pressure per square inch
                                    pounds . 31,500.
Maximum range with supercharge....yards.. 20,000.
Rate of fire (with supercharge) .... 4 rounds per minute, not to exceed 40 rounds.
Weight of recoil mechanism with elevating sector and piston rod nuts.......pounds.. 3,114.

b. Maneuvers, 155-mm Gun.

Range movement in elevation............ 5° to 30° (533½ mils).
Traverse to right or left from midposition.... 14° (247.8 mils).

c. .50 Caliber Browning Machine Gun HB, M2.

Weight of gun without barrel.......pounds . 54.
Weight of barrel (45-inch) approximately
                                    pounds.. 30.
Over-all length of barrel..........inches.. 40.
Rate of fire (cyclic)....................... 400 to 500.
Muzzle velocity........feet per second.. 2,500 to 3,000.
Sight (graduated to yards)................. 2,600.

SECTION II
DESCRIPTION AND FUNCTION

171. 155-MM GUN. The following step-by-step description covers the complete firing cycle of the gun and starts with the breechblock closed and the gun chamber empty. The operating lever (fig. 3) performs the function of first rotating the breechblock in the breech ring until the threads are disengaged and then swinging the mechanism as a whole around the hinge pin until it is locked in an open position.

a. The gun is loaded by inserting the projectile into the bore of the gun and ramming it home by means of a hand rammer. The powder charge is inserted in the gun chamber. The breechblock is then closed by means of the operating lever which is locked by the breechblock carrier lever catch. A primer is inserted into the slot of the primer holder and the firing mechanism is screwed into the firing

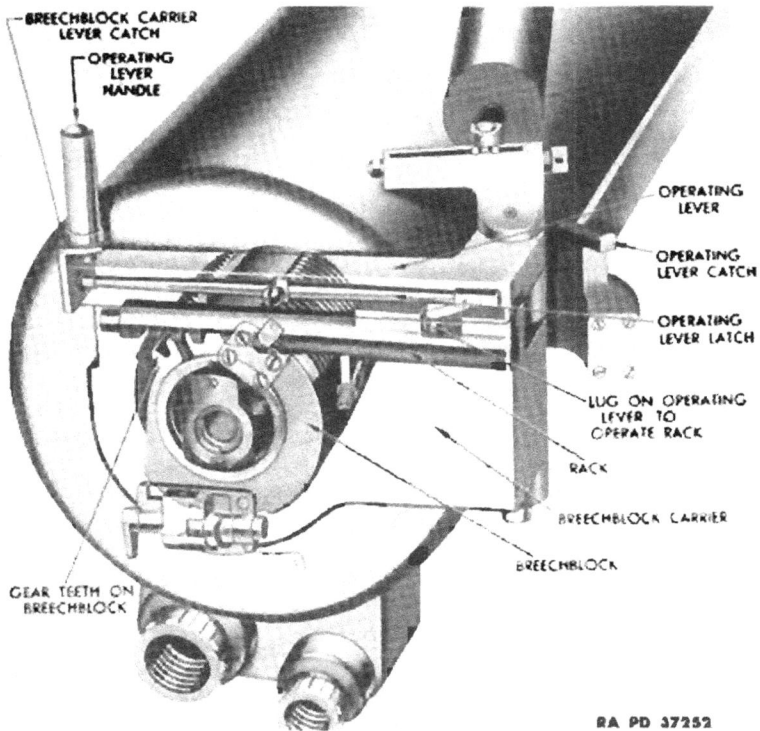

Figure 152. Phantom view of breechblock carrier, operating lever and operating lever handle.

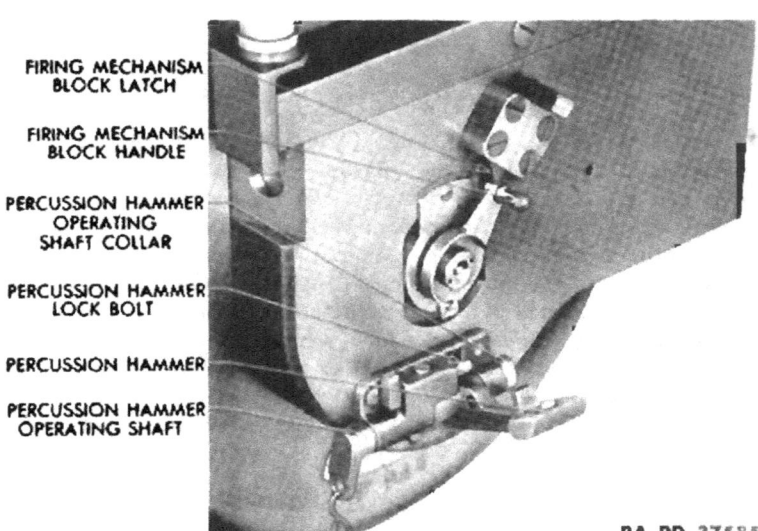

Figure 153. Firing mechanism block latch and percussion mechanism.

mechanism housing until it is locked by the firing mechanism block latch.

b. The gun is now ready to fire. This is accomplished by a pull on the lanyard which is attached to the percussion hammer operating shaft. The rotation of the percussion hammer operating shaft, to which the percussion hammer is attached, causes the hammer to strike the primer in the firing mechanism and ignite the propelling charge.

c. As soon as the gun fires, the reaction of the expanding gases causes the gun to recoil in its mount. The movement of recoil compresses the nitrogen gas within the recoil mechanism and builds up sufficient pressure to return the recoiling mass back into firing position.

d. The firing mechanism must be removed from the firing mechanism housing in order that the breechblock may be opened for loading or cleaning purposes. This is accomplished by lifting up on the handle of the firing mechanism block latch and unscrewing the firing mechanism counterclockwise from its housing.

172. 155-MM GUN MOUNT M4.

a. The 155-mm Gun Mount M4 is designed to support a cradle which is suspended by its trunnions resting in trunnion bearings of the top carriage. The gun is carried, and slides in recoil and counterrecoil in guiding slots formed in the upper portion of the cradle. The lower portion of the cradle houses the recoil and counterrecoil systems.

On the left side and to the front of the cradle is a replenisher cylinder which automatically fills the recoil cylinder with oil.

b. The recoil system is of hydro-pneumatic variable recoil type. The purpose of the recoil system is to control the backward thrust of the gun created when fired, and to check the movement of the recoiling mass. With this type of recoil system, the length of recoil is automatically shortened as the angle of elevation of the gun is increased. The recoil system is distinctly separated from the counterrecoil system. The recoil mechanism is connected through its piston rods to the lower lug of the breech ring. The piston rods move in recoil with the gun.

c. The counterrecoil system is for the purpose of returning the gun to battery position after firing. The counterrecoil system is housed in the counterrecoil cylinder and the recuperator cylinder of the cradle.

d. The recuperator cylinder houses the floating piston, which separates the compressed nitrogen in the forward end of the cylinder from the oil at the rear of the floating piston. The counterrecoil cylinder has direct communication with the recuperator cylinder.

e. In recoil, the oil between the counterrecoil piston and the counterrecoil cylinder head is forced through the communicating orifices into the recuperator cylinder, where it forces the floating piston forward, and builds up sufficient pressure in the nitrogen to return the gun to battery from its recoil position.

f. There is a small amount of oil separating the floating piston from the regulator valve when the gun is at rest. This oil is known as the oil reserve. Should the oil be reduced through leakage, the floating piston would bear against the regulator valve, and damage to the mechanism would occur if the gun was fired. An oil index in the recoil cylinder rear head indicates, by its position, whether or not such reserve oil is present.

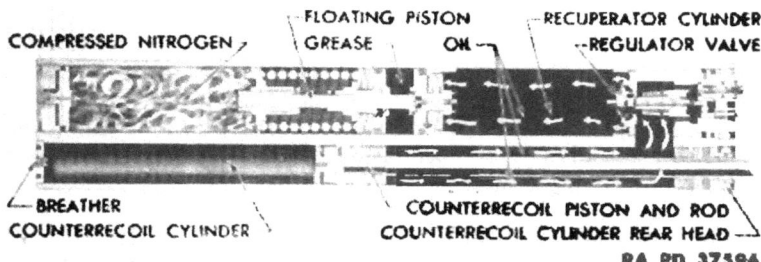

Figure 154. Sectional view of counterrecoil and recuperator cylinders, showing movement of oil when gun is fired.

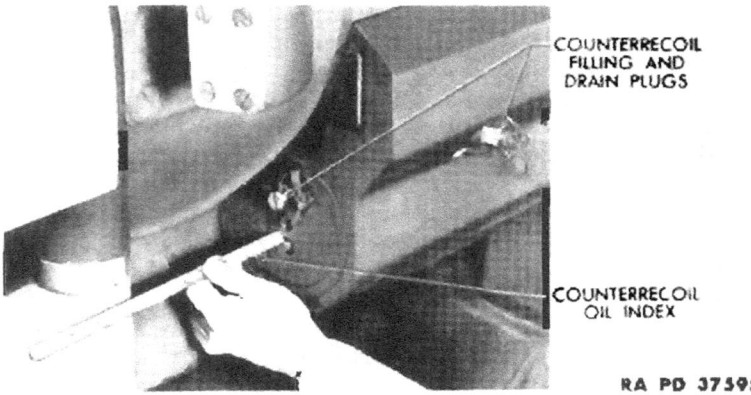

Figure 155. Measuring oil index to determine amount of counterrecoil reserve oil.

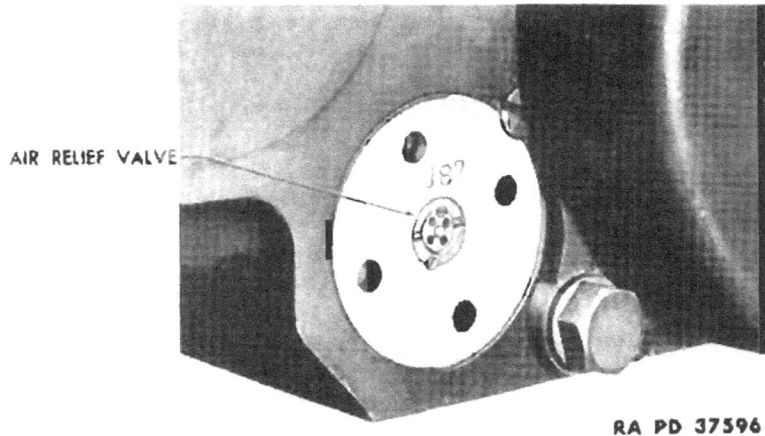

Figure 156. Air relief valve in counterrecoil cylinder front head.

g. If there is a full reserve, the oil index will project 5-mm (0.19). If there is no reserve, the oil index will disappear into the cylinder head, and the reserve oil must be reestablished before firing.

h. An air relief valve is provided in the counterrecoil cylinder front head to allow air trapped in the cylinder to escape when the gun returns to the firing position.

173. CALIBER .50 BROWNING MACHINE GUN, HB, M2. The .50 caliber Browning machine gun, HB, M2 is a recoil-operated, belt-fed, air-cooled machine gun. In recoil operation, the rearward force of the expanding powder gas (kick) furnishes the operating energy.

The moving parts, while locked together at the moment of explosion, are left free within the receiver to be forced to the rear by the recoil. This movement is controlled by means of various springs, cams, and levers, and utilized to perform the necessary mechanical operations of unlocking the breech, extraction and ejection of the empty case, feeding in of the new round, as well as cocking, locking, and firing the mechanism. A retracting slide is provided for initial loading and as a means of operating the gun by hand. The retracting slide handle remains stationary in its forward position while the gun is firing, thus eliminating all moving parts outside of the receiver.

SECTION III
OPERATIONS

174. PLACING 155-MM GUN IN FIRING POSITION.

a. Remove gun covers and store them out of the way in the proper place assigned for them; likewise, store the other equipment not required for the operation of the gun.

b. Remove the traveling lock by loosening the wing nuts which secure upper half of the traveling lock. Rotate the upper half up and off the gun, then push traveling lock down on the hull.

c. Place spade in firing position. Free the top support arms of spade assembly by removing the locking pins. Release winch ratchet arm and brake to lower spade to ground. Back up vehicle to force spade prongs into ground. When in its proper position, the spade should be flush with the ground and the steps, in a horizontal position. Place track spades in position. Remove front track spades from their assigned stowage position on the right front fender. Open spades and place one under the front of each track.

d. Remove Gun Jack. Remove locking pin from gun jack at rear of gun. Loosen jack with bar then remove jack.

e. Unload Tools and Equipment. Unload tools and accessory equipment and place them on tarpaulin. In bad weather, use half of the tarpaulin to protect them if another tarpaulin is not available for the purpose.

f. Examine Bore of Gun. Open the breech by means of the operating lever, examine the bore of the gun to see that it is clean and free from foreign matter, that the breechblock operates freely, and that the counterbalance mechanism is properly adjusted to overcome the effect of gravity in opening and closing of the breech mechanism.

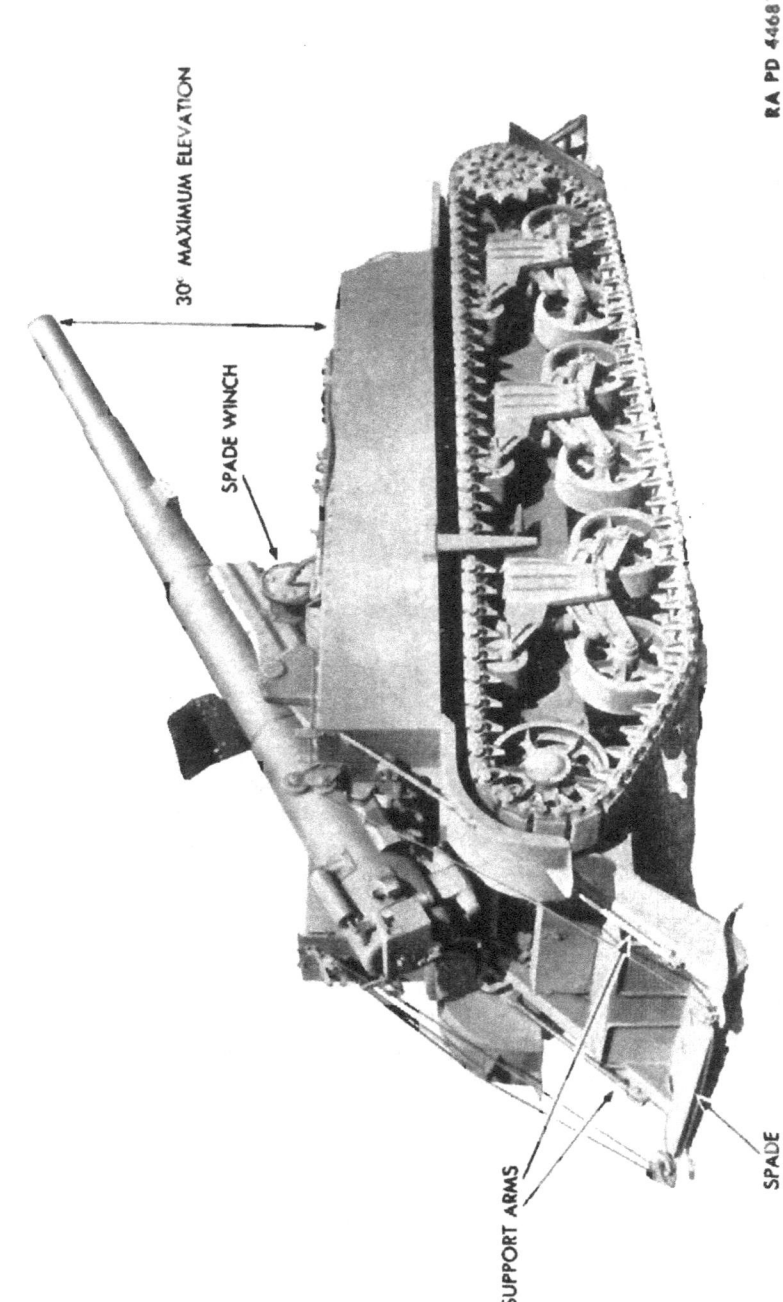

Figure 157. 155-mm gun motor carriage in firing position—gun at maximum elevation.

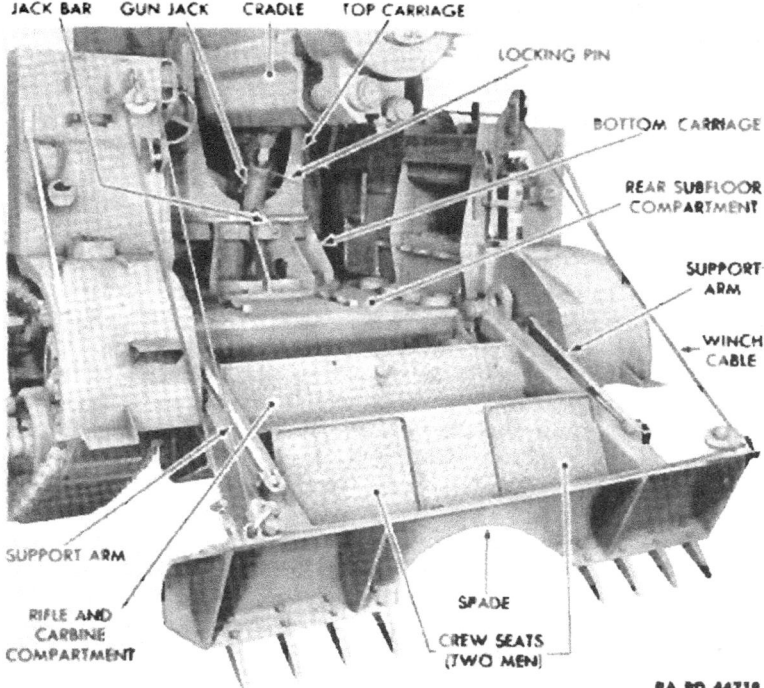

Figure 158. Installation of gun mount and spade assembly — rear view of vehicle.

175. CHECK AND SERVICE RECOIL MECHANISM.

a. Check Liquid in Recoil System.

(1) The position of the replenisher piston indicates the amount of liquid in the replenisher cylinder. To measure the position of the replenisher piston, insert a centimeter or inch scale into the opening in the center of the rear face of the replenisher, and push it until it comes in contact with the replenisher piston. Read on the scale the graduation which is flush with the rear face of the replenisher.

Note: When it is known that rapid fire is to take place, the replenisher should be drained until the reading of the replenishing piston is 200-mm (7.87 inches).

(2) When the replenisher piston is at a point 100-mm (3.93 inches) or less from the rear face of the replenisher, oil should be removed from the recoil cylinder before firing is continued.

Note: When it is necessary, in an emergency, to continue firing without interruption, firing may be permitted until the reading is down to 60-mm (2.36 inches).

(3) When the replenisher piston has moved to a point of 200-mm (7.87 inches) or more, oil should be added.

b. Filling Recoil Cylinder. Before filling the recoil cylinder with oil, test the operation of the replenisher piston by inserting a scale through the opening of the replenisher piston guide and against the replenisher piston, then releasing oil from the recoil cylinder by means of the filling and drain valve release screwed into the recoil cylinder drain hole. If movement of the replenisher piston takes place, proceed with the filling of the recoil cylinder as follows:

(1) Unscrew the filling and drain plug of the recoil cylinder filling hole (on left side of the replenisher), and screw the union of the pump coil into the filling hole loosely. The union should be set up without the use of a wrench, except for the final tightening. Extreme care must be taken to prevent any injury to the threads of the filling hole, as any damage may put the entire cradle out of commission. As the filling and drain valve release is already in place, work the pump and force oil through until it flows from the filling and drain valve release free from air bubbles; then remove the filling and drain valve release and continue to work the pump until the rear end of the replenisher piston is 150-mm (5.9 inches) from the rear face of the replenisher. Remove the filling pipe union and replace the filling and drain plugs.

(2) The oil screw filler may also be used in filling the recoil cylinder with oil. The oil screw filler requires careful handling by the operator to avoid its being broken off in the filling hole, due to improper turning of the handle. Only an experienced man should be allowed to use the oil screw filler, and it should only be used when no pump is available.

(3) In filling the recoil cylinder with the oil screw filler, remove the filling plug from the replenisher. Before screwing the oil screw filler into the filling hole, it must be filled with oil as follows: Unscrew the the screw assembly of the filler as far as the threads permit. Unscrew the screw nut from the body and remove the nut and screw. Hold the filler vertically, close the opening at the nozzle with a finger, and pour the filler three-fourths full of perfectly clean recoil oil. Replace the filler screw and screw on the screw nut, invert the filler, and give the screw a turn or more to remove all air contained in the filler.

(4) Screw the filler into the recoil cylinder filling hole of the replenisher with great care, to avoid damage to the threads. While still loose, give a few turns to the screw to force out any air which may be in the filling hole; then tighten against the gasket. Turn the screw carefully with both hands on the handle, so that there will be no tendency to push the filler to one side. Screw the piston in as far as it will go. Continue the above operation until the replenisher piston is 150-mm (5.9 inches) from the rear face of the replenisher. Unscrew the oil filler and replace the filling plug.

176. CHECK RESERVE OIL IN COUNTERRECOIL SYSTEM.

   a. The position of the oil index, which is directly below the recuperator filling valve, indicates the quantity of oil in the counterrecoil or recuperator cylinder. To measure the position of the oil index, hold one end of a centimeter or inch scale firmly against the cylinder head of the counterrecoil cylinder, with the edge of the scale parallel and close to the oil index. Read on the scale the graduation opposite the end of the oil index.

   b. The normal position of the oil index is 5-mm (0.20 inch) out from the rear face of the cradle.

177. TO FILL COUNTERRECOIL CYLINDER WITH OIL PUMP M3.

   a. The position of the oil index, which is directly below the filling and drain plug, governs all filling and draining of the system. The normal position of the oil index is 5-mm out from the rear face of the cradle.

   b. When the counterrecoil of the gun, or the position of the oil index, indicates that there is too small a quantity of oil in the recuperator, it will be necessary to drain off the reserve oil before refilling. This is accomplished by inserting the filling and drain valve release in the recuperator rear cylinder head. The reserve oil will spurt out in a stream and suddenly drop to a trickle. The amount of reserve oil which will escape will be approximately 1 quart.

   c. At this point, the flow of oil should be stopped by unscrewing the filling and drain valve release. It will be noted that the oil index has moved out of sight before all of the reserve oil has been released. If the oil index has not moved, tap it gently with a small piece of wood, as it may be frozen.

   d. To fill the counterrecoil system, remove the plug from the filling hole located on the right side of the cradle. Clean the union and screw loosely into the filling hole. Work the pump a few strokes to clear the pipe and connection of air, and screw the union firmly against the gasket of the filling hole.

   e. Screw in the filling and drain valve release, and give the pump a few more strokes. Observe whether air bubbles appear in the escaping oil. If the oil is free from air bubbles, remove the filling and drain valve release; if not free, continue to pump oil through until bubbles disappear. If air bubbles persist, this will indicate gas escaping by the floating piston. Report to the ordnance maintenance company.

   f. With the filling and drain valve release removed, give the pump 225 strokes, which will fill the cylinder with the required amount of oil. It will be noted that 67 full strokes of the pump will cause the oil index to move out to its maximum projection of from 5-mm to 6-mm beyond the rear face of the cradle, but 158 more strokes on the lever are required to introduce the necessary reserve oil. Detach the pump and replace both plugs.

178. TO FILL RECUPERATOR WITH OIL SCREW FILLER. Follow the same procedure in the release of the reserve oil as outlined for filling the system by means of the pump, see paragraph 177. Fill the oil screw filler with recoil oil, as outlined in paragraph 175. Screw the oil screw filler into the filling hole located on the right side of the cradle. Force five screw fillers full of oil into the recuperator cylinder after the oil index starts to move. Remove the oil screw filler, and replace the filling plug.

179. TO MEASURE LENGTH OF RECOIL. Coat the edge of the cradle with hard grease and adjust the recoil pointer until the point just touches the greased surface of the cradle. After the gun has fired and returned to battery, measure the distance between the point of the recoil pointer and the far end of its trace on the greased surface.

180. LENGTH OF RECOIL. Under normal conditions, the length of recoil should be as follows:

Charge		Quadrant elevation in mils				
		100	200	300	400	500
Super	Inches	66±2	66±2	58±2	45±1	45±1
Normal	Inches	58±2	58±2	52±2	43±1	43±1

If the measured length of recoil does not fall within the tolerance given in this table, notify the ordnance personnel.

181. TO TRAVERSE GUN. The traversing handwheel is located on the left side of the top carriage at the rear of the sighting gear casing. Traversing is accomplished by turning the traversing handwheel either to the right or the left. Right and left traverse of 14° from zero is provided and is controlled by stops.

182. TO ELEVATE GUN. The elevating handwheel is located on the left side of the top carriage at the left side of the sighting gear casing. Elevating is accomplished by turning the handwheel in a clockwise direction; depression, by turning the handwheel in a counterclockwise direction. Elevation of 30°, and depression of 5° are provided for, and controlled by stops.

183. TO LOAD GUN.

a. Preliminary. Release the percussion hammer by unlocking the percussion hammer lock bolt. Remove the firing mechanism and open the breech. Lock the percussion hammer lock bolt with the percussion hammer in the released position. The percussion hammer lock bolt will not be unlocked until after the breech has been closed and locked and the piece is ready to be fired.

b. Inspect and Clean Gun. Swab the powder chamber and breech recess. If a charge has been fired, wipe off the powder residue from the obturator spindle, gas check pad, gas check seat, and the threaded sector of the breech recess and breechblock with a cloth slightly dampened with light lubricating oil. Clean the primer vent with the vent cleaning bit. Inspect the bore for burning fragments of powder bags or other objects and for bore injuries.

c. Prepare the Projectile. Verify the type, weight, and lot number, and examine carefully for defects. Inspect the rotating band with special care and remove any burs with a file. Clean the entire surface with a piece of waste or with a sponge and water. Sand or dirt on the projectile might cause premature detonation when the piece is fired.

d. Fuze the Projectile.

(1) Unscrew the eyebolt lifting plug from the fuze socket. Insert the designated fuze, being careful that it is fitted with its felt or rubber washer. Screw it in place by hand. Give the fuze its final seating with the fuze wrench. No great force should be used. Set the fuze.

(2) If there is any difficulty in screwing the fuze in place, the fuze should be removed and another inserted. If the same trouble is encountered with the second fuze, the shell should be rejected.

e. Bring up Projectile. Bring up the projectile on the loading tray. Place a prepared projectile on the loading tray. Grasp the handles of the tray and raise it with the front slightly above the rear. Get the proper grip as a shell may be dropped easily if the tray is not carried in the proper position.

Note: The projectile will not be brought to the rear of the recoil pit until after the gun has returned to battery.

f. Load Projectile.

(1) Insert the lip on the bottom of the loading tray in the recess at the bottom of the breech recess. Lower the rear of the loading tray to bind the tray in place. Support the rear of the tray while the projectile is being rammed.

(2) **Ram the Projectile.** Place the rammer head squarely against the base of the projectile, push it slowly until it has cleared the threads

of the breech recess, then ram it home with a powerful stroke. Uniformity of ramming is essential to accuracy of fire.

g. Prepare the Powder Charge. Open the powder containers. Remove the powder charges from the containers. Remove the protector cap from the igniter pad at the base of the charge. When normal charge is indicated, cut the tying straps and remove the increment section.

h. Load Powder Charge.

(1) Bring the prepared powder charge up to the breech immediately after the projectile has been rammed. An exposed powder charge will not be near the gun at any other time. Place the charge in the chamber, igniter end to the rear and lashed end to the front. Push it in until the base of the charge is flush with the rear end of the chamber. Close the breech.

(2) To insure transmission of the flash from the primer to the charge, the obturator spindle must come in contact with the base of the charge when the breech is closed. In the breech, the obturator spindle head will push the charge forward to its final position.

184. FIRING THE GUN.

a. Make sure vehicle battery switch is turned off. Hold the firing mechanism in one hand, slide the primer into U-shaped groove of the primer holder. Screw the firing mechanism into firing mechanism housing until it is latched by the firing mechanism block latch.

b. Firing is accomplished by a quick pull or snap of the lanyard attached to the percussion hammer shaft, causing the percussion hammer to strike the firing pin. While traveling, or for reason of safety, lock the percussion hammer in a neutral position with the percussion hammer lock bolt pin.

Note: The breechblock must be closed before assembling the firing mechanism, and the firing mechanism must be removed before the breechblock can be opened.

c. Misfires.

(1) **General.** A misfire occurs when the piece fails to fire. Misfire is caused by—

(a) Failure of the primer to fire.

(b) Failure of the propellant charge to ignite when the primer fires. When a misfire occurs, all personnel must remain clear of the path of recoil. The piece must be kept pointed at the target or at a safe place in the field of fire.

(2) **Primer Failures.**

(a) If the primer is not heard to discharge, make at least three attempts to fire it, pulling the lanyard with considerable snap. If a special device is available which permits the lifting of the firing mechanism block latch and removal of the firing mechanism by a person entirely clear of the path of recoil, remove the primer and examine it after 2 minutes have elapsed since the last attempt to fire. After 2 minutes, insert a new primer and resume firing. If such a special device is not available, the primer must not be removed, nor the breechblock opened, until 10 minutes have elapsed since the last attempt to fire.

(b) When removing the firing mechanism, the operator will note whether or not the firing mechanism was fully screwed in place. (The primer will not be hit properly unless the firing mechanism is screwed as far beyond the latch as possible.) If the primer is found to have fired, proceed as in (3) below. If the primer has not been fired, examine the percussion head and—

1. If the head has been properly indented, handle the primer carefully and dispose of it quickly, because of the possibility of a primer hangfire. Insert a new primer and make another attempt to fire.
2. If the head has not been properly struck, and if the firing mechanism was properly seated, the firing mechanism should be inspected for the following faults: dirty or gummy parts, broken firing pin or firing pin spring, loosened firing pin housing or primer holder. After correction of the fault, insert a new primer and make another attempt to fire.

(3) **Propelling Charge Failure.** If the primer is heard to discharge, but the propelling charge has failed to explode, no attempt will be made to remove the primer or to open the breech, until 10 minutes have elapsed after the firing of the primer. After 10 minutes, remove the primer, run a cleaning bit through the vent, insert another primer and repeat the attempt to fire the piece. Failure of the propelling charge to ignite indicates an abnormal condition of the charge, such as a missing igniter end of charge against the projectile, wet ignition charge, or ignition charge folded over and not accessible to the flash of the primer.

185 TO UNLOAD.

a. **Service Rounds.** No unloading rammer is provided with this matériel for use in unloading service rounds of ammunition. When it is desired to unload the piece, the projectile may be fired out of the gun, after it has been determined that the field of fire is clear.

b. **Dummy Projectile.** To unload the dummy projectile, lower the gun to a convenient elevation (about 150 mils) and with the loading tray in place, remove the projectile with the dummy projectile extractor. Place the hook of the extractor in the recess in the base of the dummy projectile, and engage the hook on the shoulder; then jerk the projectile to release the band stuck in the forcing cone. Push the projectile forward and repeat if necessary. Use the extractor to guide the projectile on to the loading tray.

   c. **To Remove a Fuze from a Shell.** If, for any reason, a projectile which has been fuzed is not to be fired, the fuze will be removed. The operation of inserting the fuze is reversed. *Caution:* If the adapter starts to unscrew with the fuze, the unscrewing must be stopped at once and the shell disposed of as directed by the executive.

186. TO PLACE GUN IN TRAVELING POSITION.

   a. **Fix Gun Position.** Bring the gun to a horizontal position in the center of the traverse.

   b. **Clean and Oil Gun.** Clean, thoroughly dry, and cover with a thin coat of light oil, the bore, powder chamber, breech recess, breechblock, and firing mechanism; oil the top and bottom carriages. Lock the percussion hammer in traveling position.

   c. **Stow Tools, Equipment, and Ammunition.** Return sighting equipment, firing tools, and accessory equipment to their proper stowage position in chests, and in brackets, and clips on the vehicle. Store the ammunition, and close, and store the powder containers.

   d. **Place Gun Jack in Position** (fig. 10). Place gun jack in position. Turn jack bar to tighten jack, then turn down locking ring and insert locking pin.

   e. **Stow Track Spades.** Remove track spades from under the front tracks, fold legs, and stow in assigned position on right front fender.

   f. **Install Gun Covers.** Install the breech, muzzle, and piston rod covers

   g. **Install Traveling Lock.** Rotate traveling lock up to a vertical position. Clamp upper half of lock around gun. Tighten wing nut.

   h. **Raise Spade.** Move the vehicle forward to free the spade from the earth. Crank up spade winch, and lift spade to traveling position. Insert spade support arms through brackets and insert locking pin.

187. CALIBER .50 BROWNING MACHINE GUN.—To prepare the gun for action, proceed as follows:

   a. **To Half Load.** Lock bolt latch release down, push double loop end of belt into feedway until the first round is held by belt holding pawl, and pull bolt completely to the rear and release it.

   b. **To Load.** Load is executed the same as half load except that the bolt is pulled to the rear and released twice.

   c. **To Unload.**

(1) Lift cover, remove belt, retract bolt, and look or feel in the feedway, T-slot, and chamber to make certain that gun is unloaded.

(2) Release bolt and lower cover.

(3) Press trigger or sear mechanism to relieve tension on firing pin spring.

d. To clear gun.
(1) Execute that part of unload described in **c.** (1) above.
(2) Place a wooden clearing block between face of bolt and rear end of barrel and let bolt go forward.
(3) Report, "Clear."

e. To Fire.
(1) **Semiautomatic action.** By unlocking bolt latch release and alternately pressing trigger and bolt latch release, semiautomatic fire can be obtained. This method is used primarily in targeting a barrel.

(2) **Automatic action.** Lock bolt latch release down, and press and hold trigger down. (To fire single-shot, automatic action, slide the thumb off quickly after pressing trigger.)

SECTION IV
SIGHTING EQUIPMENT

188. GENERAL. The following equipment is required for sighting and aligning the 155-mm Gun Motor Carriage M12:

a. The Telescope Mount M40 with telescope M53 places the gun in azimuth and elevation in direct fire against fast-moving targets. An instrument light illuminates the telescope reticle.

b. The Quadrant Sight M1918 or M1918A1 with Panoramic Telescope M6 places the gun in azimuth and elevation in indirect fire, primarily against distant targets. The 14-inch extension, when used, raises the panoramic telescope line of sight above the shield. The Instrument Light M9 illuminates the instrument.

c. Both of these tests of sights are mounted on the left trunnion, with the Telescope M53 outside the quadrant sight. See figures 159 and 160.

d. Two Aiming Posts M1, which are placed in a direct line to provide an aiming point to detect any lateral displacement of the piece during firing.

e. Two Aiming Post Lights M14, are used with the aiming posts for night firing.

f. The gunner's Quadrant M1 (or the gunner's Quadrant M1918) is for measuring the elevation of the piece, and laying the piece to a given elevation.

g. The bore sight is used to align the sights with the bore of the gun.

h. The testing target is used during the bore sighting operation.

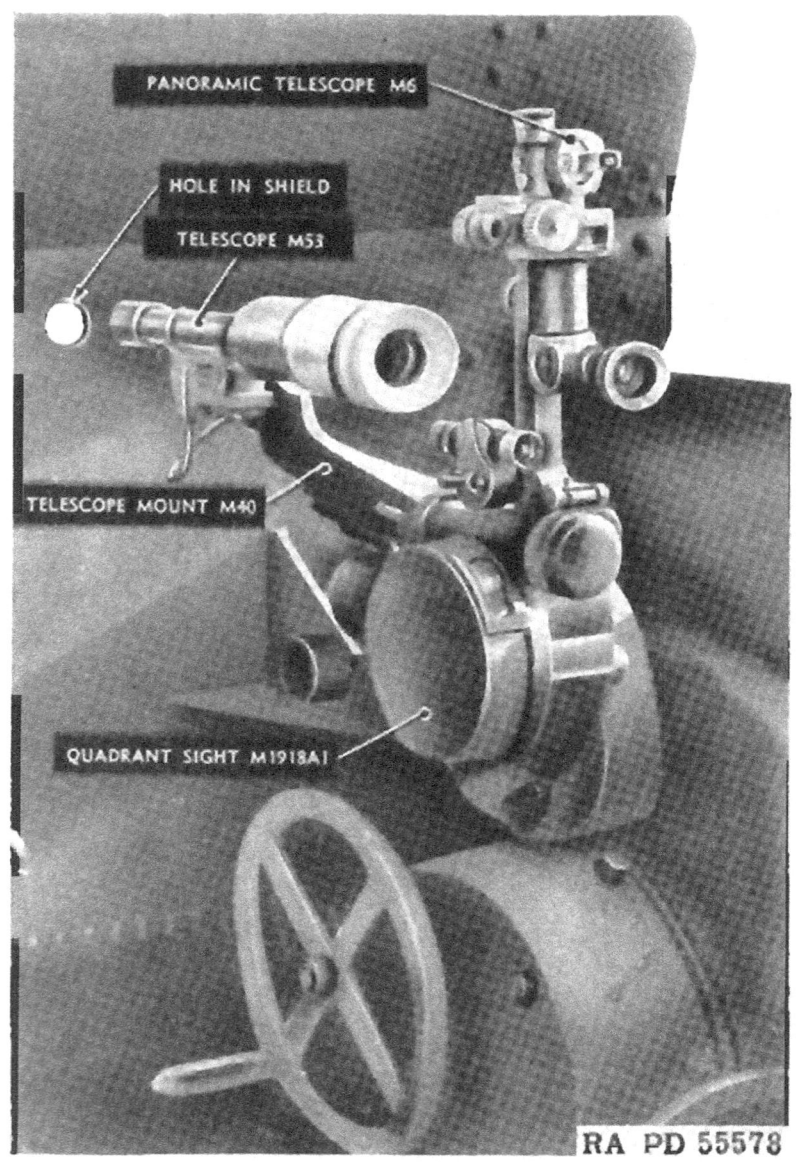

Figure 159. Arrangement of direct and indirect sighting equipment—left rear.

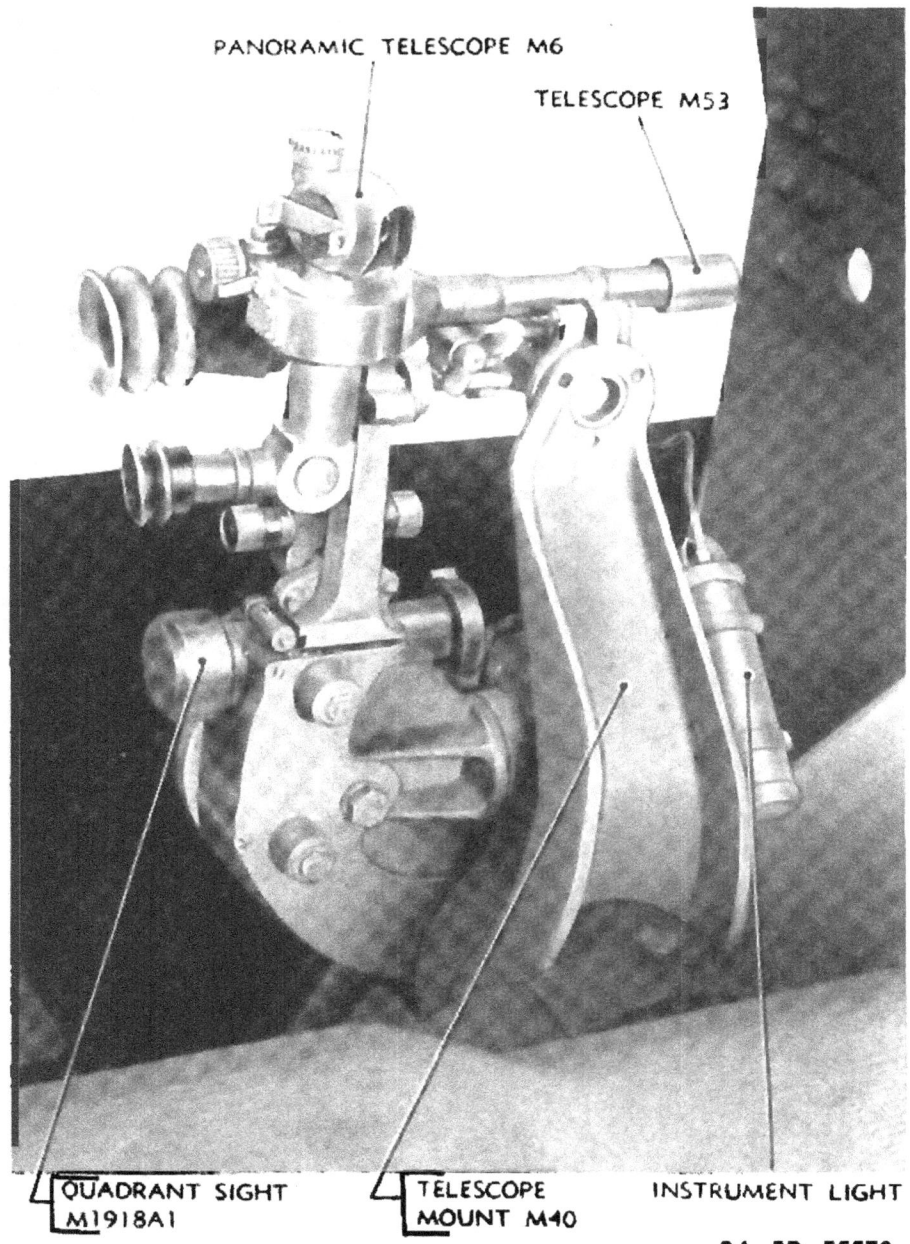

Figure 160. Arrangement of direct and indirect sighting equipment—right rear.

SECTION V
AMMUNITION

189. GENERAL.

a. General. Ammunition issued for the 155-mm Gun Motor Carriage M12 and Cargo Carrier M30 consists of the following:
(1) Ammunition for 155-mm Gun, M1918MI.
(2) Cartridges for .50 caliber Browning machine gun, HB.
(3) Cartridges for the basic weapons carried on the vehicles: (Personal equipment carried on, but not supplied with the vehicle.)
    (a) Caliber .45 Thompson submachine gun M1928A1.
    (b) Caliber .30 U. S. rifle M1903 w/grenade launcher M1.
    (c) Caliber .45 automatic pistol M1911A1.
(4) Hand and rifle grenades.

b. The ammunition for 155-mm Gun, M1918M1 is of the separate loading type. The loading of a complete round requires three operations—loading the projectile, loading the propelling charge, and inserting the primer. The components of the complete round, fuze projectile, propelling charge, and primer, are shipped separately The fuze is assembled to the projectile just prior to firing.

190. AUTHORIZED AMMUNITION.

The ammunition authorized for use with the weapons mounted or carried on the 155-mm Gun Motor Carriage M12, and Cargo Carrier M30, and other ammunition carried on these vehicles, are listed in table I. (See also figs. 161, 162, and 163.) It will be noted that the nomenclature (standard nomenclature) completely identifies the ammunition as to type and model.

TABLE I.—*Authorized rounds*

AMMUNITION FOR 155-MM GUN M1918MI

Nomenclature of projectile	Fuze		Propelling charge type
	Type and model	Action	
Service ammunition [a]			
Projectile, AP, 100-pound, M112, w/fuze, BD, M60, 155-mm guns, M1917–17A1–18M1, and M1A1.	BD M60	Delay	Base and increment.[1]
Shell, gas, persistent, HS, M104, unfuzed, 155-mm guns, M1917–17A1–18M1, M1, and M1A1 (adapted for fuze, PD, M51, w/booster, M21, or M51A1, w/booster, M21A1).	PD M51A1, w/booster M21A1.	SQ and delay	Base and increment.[1]

See footnotes at end of table.

Nomenclature of projectile	Fuze		Propelling charge type
	Type and model	Action	
Service ammunition—Con.			
Shell, gas, persistent, HS, Mk. VIIA1, unfuzed, 155-mm guns, M1917–17A1–18M1 (adapted for fuze, PD, M51, w/booster, M21 or M51A1, w/booster, M21A1).	PD M51A1, w/booster, M21A1	SQ and delay	Base and increment.[1]
Shell, HE, M101, unfuzed, 155-mm guns, M1917–17A1–18M1, M1 and M1A1 (adapted for fuze, PD, M51, w/booster, M21, or M51A1, w/booster, M21A1; or fuze, time, mechanical, M56, w/booster, M21A1).	PD M15A1, w/booster, M21A1 or or Mechanical time, M67, w/booster, M21A1.	SQ and delay, time.	Base and increment.[1]
Shell, HE, Mk. III, unfuzed, 155-mm guns, M1917–17A1–18M1 (adapted for fuze, PD, Mk. III, Mk. IV, M35 Star, M46, or M47).	PD M46 or PD M47.	SQ, delay	Base and increment.[1]
Shell, HE, Mk. IIIA1, unfuzed, 155-mm guns, M-1917–17A1–18M1 (adapted for fuze, PD, M51, w/booster, M21, or M51A1, w/booster, M21A1) or fuze, time, mechanical, M67; w/booster, M21A1).	PD M51A1, w/booster, M21A1 or Mechanical time, M67 w/booster, M21A1.	SQ and delay, time.	Base and increment.[1]
Shell, smoke, FS, M104, unfuzed, 155-mm guns, M1917–17A1–18M1–M1, and M1A1 (adapted for fuze, PD, M51, w/booster, M21, or M51A1, w/booster, M21A1).	PD M51A1, w/booster, M21A1.	SQ and delay	Base and increment.[1]
Shell, smoke, FS, Mk. VIIA1, unfuzed, 155-mm guns, M1917–17A1–18M1 (adapted for fuze, PD, M51, w/booster, M21, or M51A1, w/booster, M21A1).	PD M51A1 w/booster, M21A1.	SQ and delay	Base and increment.[1]
Shell, smoke, phosphorus, WP, M104, 155-mm guns, M-1917–17A1–18M1, M1, and M1A1 (adapted for fuze, PD, M51, w/booster, M21, or M51A1, w/booster, M21A1).	PD M51A1, w/booster, M21A1.	SQ and delay	Base and increment.[1]
Shell, smoke, phosphorous, WP, Mk. VIIA1, 155-mm guns, M1917–17A1–18M1 (adapted for fuze, PD, M51, w/booster, M21 or M51A1, w/booster, M21A1).	PD M51A1 w/booster, M21A1.	SQ and delay	Base and increment.[1]

See footnotes at end of table.

Nomenclature of projectile	Fuze		Propelling charge type
	Type and model	Action	
Service ammunition—Con.			
Shrapnel, Mk. I, fuzed, 155-mm gun, or howitzer, M1917-17A1-18M1.	M1907M.	Time and percussion.	Base and increment.[1]
Practice ammunition [3][4]			
Shell, empty, for sand loading, 95-pound, Mk. III, unfuzed, 155-mm guns, M1917-17A1-18M1 (adapted for inert PDF Mk. IV or M47).	PD M47, inert, none.		Base and increment.[1]
Dummy ammunition [4][5]			
Projectile, dummy, 95-pound, Mk. I, 155-mm guns, M1917-17A1-18M1.	M1907M, inert, none.		Base and increment.[2]
Subcaliber ammunition			
Shell, fixed, practice, Mk. II, w/fuze, practice, M38, 37-mm gun, M1916.	M38, impact base practice.		
Shell, fixed, sand loaded, Mk. I, 100 percent service charge, 37-mm gun, M1916.			

AP—Armor-piercing. BD—Base detonating. SQ—Superquick.
HE—High explosive. PD—Point detonating. LE—Low explosive.

[1] The service or practice charges are designated Charge, propelling, 155-mm guns, M1917-17A1-18M1, and Charge, propelling, NH powder, 155-mm guns, M1917-17A1-18M1.
[2] The dummy charge is designated Charge, propelling dummy (21-pound base with 5½ pounds increment), Mk. I, 155-mm guns, M1917-17A1-18M1.
[3] The service or practice primer is designated Primer, percussion, 21-grain, Mk. IIA1.
[4] A fired service primer is used with dummy ammunition for drill purposes.
[5] In addition to the components shown, one primer is required for each full round of separate loading ammunition.

AMMUNITION .50 CALIBER BROWNING MACHINE GUN, HEAVY BARREL M2

Service ammunition

Cartridge, armor-piercing, caliber .50, M2.
Cartridge, ball, caliber .50, M2.
Cartridge, tracer, caliber .50, M1.

Blank ammunition

Cartridge, blank, caliber .50, M1.

Dummy ammunition

Cartridge, dummy, caliber .50, M1.
Cartridge, dummy, caliber .50, M2.

AMMUNITION FOR .45 CALIBER THOMPSON SUBMACHINE GUN M1928A1, AND .45 CALIBER AUTOMATIC PISTOL M1911 and M1911A1

Service ammunition

Cartridge, ball, caliber .45, M1911.

Dummy ammunition

Cartridge, dummy, caliber .45, M1921.

AMMUNITION FOR .30 CALIBER U. S. RIFLE, M1903 (WITH GRENADE LAUNCHER, M1)

Service ammunition

Cartridge, armor-piercing, caliber .30, M2.
Cartridge, ball, caliber .30, M1.
Cartridge, ball, caliber .30, M2.
Cartridge, rifle grenade, caliber .30, M3*.
Cartridge, tracer, caliber .30, M1.

Blank ammunition

Cartridge, blank, caliber .30, M1909.

Dummy ammunition

Cartridge, dummy, caliber .30, M1906 (corrugated).

AMMUNITION FOR .30 CALIBER CARBINE, M1

Service ammunition

Cartridge, carbine, caliber. .30, M1.

Dummy ammunitions

Cartridge, carbine, dummy, caliber .30, M1.

Grenades

Service Grenades and Fuzes

Fuze, detonating, hand grenade, M6A2.
Grenade, hand, fragmentation, Mk. II, with hand grenade igniting fuze, M10A2.
Grenade, hand, incendiary, M14, with hand grenade igniting fuze, M200A1**.
Grenade, hand, offensive, Mk. IIIA1, unfuzed (adapted for hand grenade detonating fuze, M6A2).
Grenade, hand, smoke, HC. M8, with hand grenade igniting fuze, M200A1**.
Grenade, rifle, HE, M9.
Grenade, AT, M9A1.

Practice and training grenades

Grenade, hand, training, Mk. IA1.
Grenade, AT, practice, M11.
Grenade, AT, practice, M11A1.

---

*Special blank cartridge for use only in .30 caliber rifle in projecting antitank grenades.
**Procurement from Chemical Warfare Service.

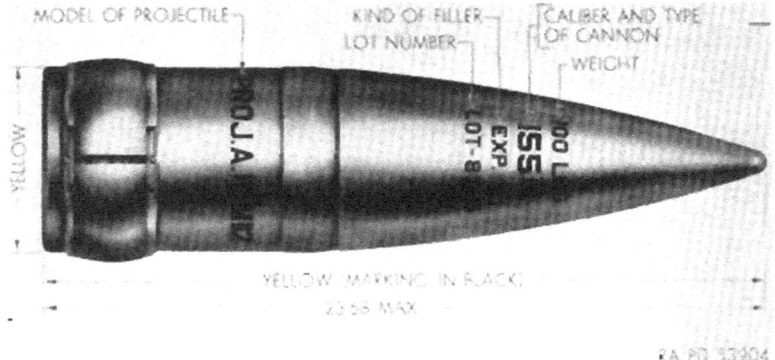

Figure 161. Projectile, A. P., 100-pound, M112, w/fuze, B. D., M60, 155-mm guns, M1917–17A1–18M1, M1, and M1A1.

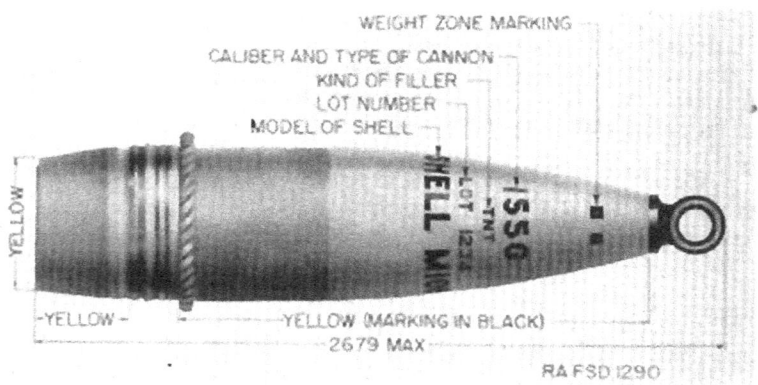

Figure 162. Shell, HE, M101, unfuzed, 155-mm guns, M1917–17A1–18M1, M1, and M1A1 (adapted for fuze, P. D., M51, w/booster, M21, or M51A1, w/booster, M21A1, or fuze, time, mechanical, M67, w/booster, M21A1).

191. STOWAGE OF AMMUNITION. Ammunition is stowed or carried on 155-mm Gun Motor Carriage, M12, or on personnel assigned thereto, as indicated in table II and on Cargo Carrier M30, or on personnel assigned thereto, as indicated in table III. Items of ammunition are stowed on the vehicles in their packing containers. These tables have been prepared to serve as a guide only, and cover a typical load without reference to any possible future provision for mounting brackets or stowage boxes.

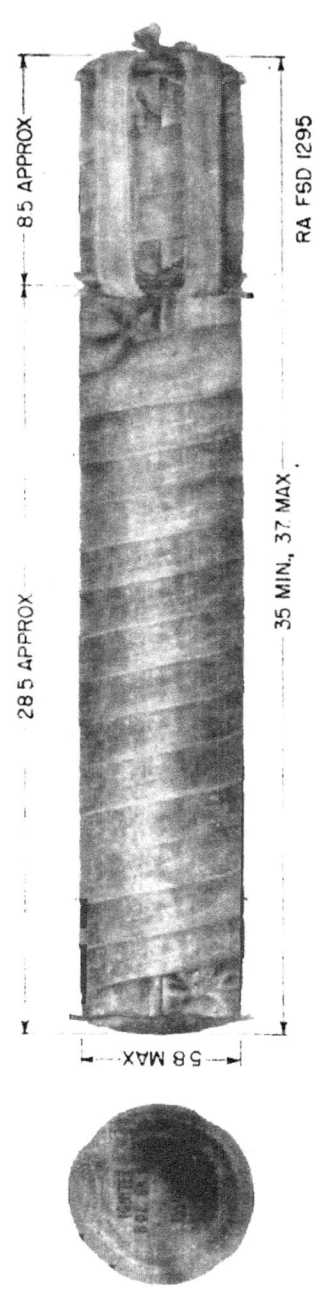

Figure 163. Charge, propelling, NH powder, 155-mm guns, M1917–17A1–18M1.

TABLE II.—*Ammunition stowage chart for 155-mm gun motor carriage M12*

Weapons	Ammunition	Ammunition stowage positions
1-Gun, 155-mm, M1918MI.	10 projectiles, 155-mm	At rear of carriage, 6 on left side, and 4 on right side of gun.
	10 propelling charges, 155-mm gun.	At rear of carriage, center.
	25 fuzes, artillery	In 1 box.
	50 primers, artillery	In 1 can.
1-Gun, submachine, caliber .45, Thompson, M1928A1.	360 rounds, caliber .45	In 30-round drums.
4-Carbine, caliber .30, M1.	300 rounds, caliber .30 carbine.	In belts of crew members.
1-Rifle, U.S., caliber .30, M1903 w/launcher, grenade, M1.	60 rounds, caliber .30 rifle	In 5-round clips in bandoleer of one crew member.
	10 grenade, AT, M9A1	In box, with 11 rifle grenade cartridges.

TABLE III.—*Ammunition stowage chart for cargo carrier M30*

Weapons	Ammunition	Ammunition stowage positions
	40 projectiles, 155-mm	In center of carriage in racks.
	40 propelling charges, 155-mm gun.	To the rear of projectiles.
	50 fuzes, artillery	In 2 boxes.
	50 primers, artillery	In 1 can.
1-Gun, machine, caliber .50, Browning, HB (flexible).	1000 rounds, caliber .50	In link belts.
1-Gun, submachine, caliber .45, Thompson, M1928A1.	360 rounds, caliber .45	In 30-round drums.
1-Pistol, automatic, caliber .45, M1911A1.	21 rounds, caliber .45	In belt of one crew member.
4-Carbine, caliber .30, M1.	360 rounds, caliber .30 carbine.	In belts of crew members.
1-Rifle, U. S., caliber .30, M1903, w/ launcher, grenade, M1.	60 rounds, caliber .30 rifle	In 5-round clips in bandoleer of one crew member.
	Rifle grenades: 10 grenade, AT, M9A1. Hand grenades: 4 fragmentation, Mk. II. 4 smoke, HC, M8 2 offensive, Mk. IIIA1, w/fuze. 2 incendiary, M14	In box with 11 rifle grenade cartridges.

APPENDIX
REFERENCES

1. STANDARD NOMENCLATURE LISTS.

Carrier, cargo, T14 (medium tank M3 chassis), and Carriage, motor, 155-mm gun, M12 (T-6) (medium tank M3 chassis)—Parts and equipment	SNL G-158.

Ammunition.

Ammunition instruction material for antiaircraft, harbor defense, heavy field, and railway artillery, including complete round data	SNL P-8.
Ammunition instruction material for grenades, pyrotechnics, and aircraft bombs	SNL S-6.
Ammunition, revolver, automatic pistol, and submachine gun	SNL T-2.
Ammunition, rifle, carbine, and automatic gun	SNL T-1.
Charges, propelling, separate loading, 6-inch to 240-mm, inclusive, for harbor defense, heavy field, and railway artillery	SNL P-2.
Firing tables and trajectory charts	SNL F-69.
Fuzes, primers, blank ammunition, and miscellaneous items for antiaircraft, harbor defense, heavy field, and railway artillery	SNL P-7.
Grenades, hand and rifle, and fuzing components	SNL S-4.
Packing materials used by field service for small-arms service ammunition	SNL T-5.
Projectiles, separate loading, 6-inch to 240-mm, inclusive, for harbor defense, heavy field, and railway artillery, including complete round data	SNL P-1.
Service fuzes and primers for pack, light, and medium field artillery	SNL R-3.

Armament.

Carbine, caliber .30, M1 and M1A1—Parts and equipment	SNL B-28.
Gun, 155-mm, M1918M1; and mount, gun, 155-mm, M4	SNL D-36.

Gun, machine, caliber .50, Browning, M2, heavy barrel, fixed and flexible, and ground mounts— Parts and equipment	SNL A-39.
Launcher, grenade, M1 and M2	SNL B-39.
Pistol, automatic, caliber .45, M1911 and M1911A1— Parts and equipment	SNL B-6.
Rifle U. S., caliber .30, M1903, M1903A1 and M1903A3— Parts, equipment and appendages	SNL B-3.

Miscellaneous.

Tools, maintenance, for repair of automatic guns, automatic gun antiaircraft matériel, automatic and semiautomatic cannon and mortars—Individual items and parts	SNL A-35.

Sighting Equipment.

Circle, aiming, M1	SNL F-160.
Light, aiming post, M14	SNL F-220.
Post, aiming, M1	SNL F-35.
Quadrant, gunner's, M1	SNL F-140.
Quadrant, gunner's M1918	SNL F-13.
Sights, bore (small arms and field artillery)	SNL F-10.
Sight, quadrant, M1918 and M1918A1	SNL F-24.
Tables, firing, graphical	SNL F-237.
Telescope, BC, M1915 and M1915A1	SNL F-9.
Telescope, M53	SNL F-235.
Telescope, panoramic, M6	SNL F-22.

Cleaning and preserving.

Cleaning, preserving, and lubricating materials; recoil fluids, special oils, and miscellaneous related items	SNL K-1.
Soldering, brazing and welding material, gases and related items	SNL K-2.
Current Standard Nomenclature Lists are as tabulated here. An up-to-date list of SNL's is maintained as the "Ordnance Publications for Supply Index"	OPSI.

2. EXPLANATORY PUBLICATIONS.

Armament.
Army Regulations.

Qualifications in arms and ammunition training allowances	AR 775-10.
Range regulations for firing ammunition for training and target practice	AR 750-10.

Field Manuals.

Automatic pistol, caliber .45, M1911 and M1911A1	FM 23-35.

Browning machine gun, caliber .50, HB, M2 (mounted in combat vehicles)	FM 23-65
Camouflage	FM 5-20.
Decontamination of armored force vehicles	FM 17-59.
Defense against chemical attack	FM 21-40.
Grenades	FM 23-30.
U. S. carbine, caliber .30, M1	FM 23-7.
U. S. rifle, caliber .30, M1903	FM 23-10.

Technical Manuals.

155-mm gun matériel, M1917, M1918 and modifications	TM 9-345.
Ammunition, general	TM 9-1900.
Auxiliary fire-control instruments (field glasses, eyeglasses, telescopes and watches)	TM 9-575.
Ordnance maintenance—155-mm Guns M1917, M1917A1, and M1918M1; Carriages M1917, M1917A1, M1918, M1918A1, M2 and M3; Limbers M1917, M1917A1, M1918, M1918A1, and M3	TM 9-1345
Small-arms ammunition	TM 9-1990
Targets, target materials, and rifle range construction	TM 9-855

Automotive Matériel.

Automotive power transmission units	TM 10-585.
Chassis, body, and trailer units	TM 10-560.
Fire prevention and safety precautions	TM 10-360.
Military motor transportation	TM 10-505.
Motor transport inspections	TM 10-545.
Motor transport technical service bulletin	No. Z-11.
Ordnance maintenance—Accessories for Wright R975-EC2 engines for medium tanks M3 and M4 and related gun motor carriages (now published as TM 9-1750D)	TM 9-1750D.
Ordnance maintenance—Power train unit, three-piece differential case, for medium tanks M3, M4, and modifications (now published as TM 9-1750)	TM 9-1750.
Ordnance maintenance, Wright whirlwind engine model R975EC-2 (now published as TM 9-1751)	TM 9-1730C.
Sheet metal work body, fender and radiator repairs	TM 10-450
Storage of motor vehicle equipment	AR 850-18.
The motor vehicle	TM 10-510.

Care and Preservation.

Chemical decontamination materials and equipment	TM 3-220.

Cleaning, preserving, lubricating, and welding materials and similar items issued by the Ordnance Department ... TM 9-850.
Communication.
Radio fundamentals ... TM 11-455.
Radio set SCR-610 ... TM 11-615.
The radio operator ... TM 11-454.
Maintenance and Inspection.
Automotive lubrication ... TM 10-540.
Cleaning, preserving, lubricating, and welding materials and similar items issued by the Ordnance Department ... TM 9-850.
Detailed lubrication instructions for ordnance matériel ... OFSB 6 series.
Echelon system of maintenance ... TM 10-525.
Hand, measuring and power tools ... TM 10-590.
Maintenance and repair ... TM 10-520.
Tune-up and adjustment ... TM 10-530.
Ordnance maintenance procedure, material inspection and repair ... TM 9-1100.
Miscellaneous.
Automotive electricity ... TM 10-580.
List of training publications ... FM 21-6.
Military motor vehicles ... AR 850-15.

3. FIRING TABLES.

Firing tables and trajectory charts ... SNL F-69.
Projectile, A. P., M112, w/fuze, B. D., M60, 155-mm guns ... FT 155-W-1.
Shell, empty, for sand loading, 95-pound, Mk. III. FT 155-B-5.
Shell, fixed, practice, Mk. II, w/fuze, base, practice, M38, 37-mm gun; M1916 ... FT 37-0-1 (Abridged).
Shell, gas, persistent, HS, M104, unfuzed, 155-mm guns, M1917-17A1-18MI, M1, and M1A1 ... FT 155-U-1.
Shell, gas, persistent, HS, Mk. VIIA1, unfuzed, 155-mm guns, M1917-17A1-18MI ... FT 155-U-1.
Shell, HE, M101, unfuzed, 155-mm guns, M1917-17A1-18MI, M1 and M1A1 ... FT 155-U-1.
Shell, HE, Mk. III, unfuzed, 155-mm guns, M1917-17A1-18MI ... FT 155-B-5.
Shell, HE, Mk. IIIA1, unfuzed, 155-mm guns, M1917-17A1-18MI ... FT 155-U-1.
Shell, smoke, FS, M104, unfuzed, 155-mm guns, M1917-17A1-18MI, M1, and M1A1 ... FT 155-U-1.

Shell, smoke, FS, Mk. VIIA1, unfuzed, 155-mm
    guns, M1917–17A1–18MI_____ _____ FT 155-U-1.
Shell, smoke, phosphorus, WP, M104, 155-mm
    guns, M1917–17A1–18MI, M1, and M1A1_____ FT 155-U-1.
Shell, smoke, phosphorus, WP, Mk. VIIA1, 155-
    mm guns, M1917–17A1–18MI_____ ___ FT 155-U-1.
Shrapnel, Mk. I, fuzed, 155-mm gun, or howitzer,
    M1917–17A1–18MI____ ___ _____ _ ____ FT 155-C-2

INDEX

	Paragraph	Page
Accessories, engine	36–70	80
Air cleaners:		
Changing oil in	62	141
Maintenance	63	143
Types	61	141
Ammunition:		
Authorized	190	292
155-mm gun	189	292
Stowage	191	296
Armament:		
Characteristics	169	271
Data	170	274
Description and function	171–173	275
Operations	174–187	279
Arrangement, content	2	1
Auxiliary fuel pump	81, 82, 83	170, 172, 172
Battery:		
Switch, installation	149, 150	250, 251
Test, removal, and installation	148	245
Bogie arm and lever	133, 134	229, 229
Bogie wheel:		
Grease seals and bearing installation	132	229
Grease seals and bearing removal	131	228
Installation	135	229
Removal	130	228
Bogies	129	226
Booster coil	53	129
Brake:		
Band:		
Installation	120	214
Removal	119	213
Parking, adjustment	121	215
Steering, adjustment	118	210
Carburetor:		
Installation	59	140
Removal	58	139
Carburetor air scoop	60	141
Characteristics	3, 169	1, 271
Chassis, inspecting in cold weather	28	55
Chemicals, matériel affected	32	57
Cleaners, air	61	141

INDEX

	Paragraph	Page
Clutch:		
Care and adjustment	106	191
Controls	106	191
Description	103	190
Operation	104	190
Release bearings	110	204
Trouble shooting	105	191
Clutch and flywheel:		
Inspection	108	202
Installation	109	202
Removal	107	197
Coil, booster	53	129
Content and arrangement	2	1
Controls	6	8
Cooling system:		
Description	101	189
Inspection and maintenance	102	190
Counterrecoil:		
Cylinder	177	283
System	176	283
Cowl, engine	67	150
Data	170	274
Data, tabulated	4	2
Description and function	171–173	275
Drive sprockets	125	220
Electrical system	148–162	245
Electrical system, inspecting in cold weather	27	54
Engine:		
Cowl	67	150
Gasoline, grades of	90	177
Inspecting in cold weather	28	55
Inspections during warm-up	8	15
Installation	70	153
Oil tank removal	99	185
Removal	10	89
Starting and warm-up	7	11
Starting in cold weather	26	53
Stopping	11	17
Engine and accessories	36–70	80
Equipment and tools stowage on 155-mm gun	21–22	34
Exhaust system:		
Description	41	107
Installation	43	109
Maintenance	41	107
Removal	42	108
Final drive	124	218
Fire extinguishers:		
Care	167	269
Cylinder installation, fixed	165	266
Fixed system	163	264
Portable	166	269

INDEX

	Paragraph	Page
Fuel bypass regulator valve:		
Installation	88	176
Removal	87	176
Testing	89	177
Fuel cut-off solenoid	84	173
Fuel lines and valves	86	175
Fuel pump	56	135
Fuel system:		
Description	71	158
Fuel tanks	74	158
Inspection	72	158
Trouble shooting	73	158
Installation	75	160
Removal	76	164
Fuse box	153	255
Gasoline fuel filter	85	174
Generator	55	134
Generator regulator	152	252
Governor removal and installation	64	144
Guns:		
Browning machine, caliber .50, HB, M2	173, 187	278, 288
Elevate	182	284
Firing	184	286
Load	183	285
Traveling position	186	288
Traverse	181	284
Unload	185	287
155-mm:		
Description	171	275
Placing in firing position	174	279
155-mm gun mount M1	172	276
Harness, ignition	52	126
Ignition harness	52	126
Inspection	72	158
Inspection and preventive maintenance:		
Purpose	13	20
Service:		
After operation and weekly	17	25
At halt	16	24
Before operation	14	21
During operation	15	23
Inspection, periodic	37, 38	86, 86
Instructions, operating	1–32	1
Instrument:		
Installation	157	257
Maintenance	155	256
Panel	154	255
Removal	156	256
Lights	160	258

INDEX

	Paragraph	Page
Lubrication:		
Guide	19	28
Propeller shaft	113	207
Reports and records	20	33
Lubrication system:		
Description	91	177
Inspections	92	177
Oil filter removal, cleaning and installation	94	179
Trouble shooting	93	177
Magnetos	48	112
Maintenance:		
Magnets:		
Bosch	50	121
Scintilla	49	112
Maintenance, preventive	13–17	20
Matériel affected by chemicals	32	57
Oil, changing in air cleaner	62	141
Oil cooler	96	182
Installation	98	186
Removal	97	186
Oil filter:		
Automatic	95	182
Removal, cleaning and installation	94	179
Oil pump:		
Description	44	109
Fingers trainer removal, cleaning, and installation	47	110
Installation	46	110
Removal	45	110
Oil tank:		
Installation	100	185
Removal	99	185
Operating:		
Instructions	1–32	1
Precautions	12	18
Vehicle	9	16
Operation:		
High temperatures	29	56
Slippery conditions	31	56
Under unusual conditions	25–31	52
Operations and controls	5–12	8
Organization:		
Preventive maintenance	34	64
Tools and equipment	35	73
Panel, instrument	154	255
Plugs, spark	51	124
Power train:		
Brake band installation	120	214
Brake band removal	119	213
Description	116	209
Drive sprockets	125	220
Final drive	124	218

	Paragraph	Page
Power train—Continued.		
Installation	127	223
Parking brake adjustment	121	215
Removal	126	221
Steering brake adjustment	118	210
Steering lever:		
Installation	123	218
Removal	122	218
Transmission and differential lubrication	117	209
Precautions:		
Cold weather	25	52
Operating	12	18
Preventive maintenance, inspection	13–17	20
Preventive maintenance services, second echelon	34	64
Primer line removal and installation	80	170
Primer pump:		
Description and operation	77	168
Installation	79	170
Maintenance	77	168
Removal	78	168
Propeller shaft:		
Description	111	205
Inspections	112	205
Installation	115	208
Lubrication	113	207
Removal	114	207
Pump:		
Fuel:		
Auxiliary	81, 82, 83	170, 172, 172
Oil	44, 56	109, 135
Push rods	68	151
Recoil:		
Length	180	284
Measure	179	284
Recoil mechanism	175	281
Recuperator	178	284
References	App.	299
Rocker:		
Arms	66	147
Box cover	65	147
Rods, push	68	151
Scope	1, 33, 168	1, 58, 271
Sighting equipment	188	289
Siren	161	260
Solenoid starter switch	151	251
Solenoids	159	258
Spark plugs	51	124
Starter	54	133
Steering lever:		
Installation	123	218
Removal	122	218

	Paragraph	Page
Storage on cargo carrier, tools and equipment	23, 24	43
Stowage, tools and equipment on 155-mm gun	21, 22	34
Suspension and tracks	129–147	226
Tanks, fuel	74	158
Terminal box	158	257
Tools and equipment:		
Allocation of	35	73
Stowage on 155-mm gun motor carriage	21–22	34
Tracks:		
Dead rubber, link assembly replacement	142	239
Grousers	147	243
Idlers	146	242
Installation	141	238
Loose, adjustment	139	235
New, installation	144	241
Removal	140	236
Reversing	145	242
Turning link assemblies	143	241
Train, power	116–127	209
Trouble shooting	39, 162	86, 261
Valve clearance adjustment	69	151
Vehicle:		
Equipment	22, 24	37, 46
Maintenance instructions	33–167	58
Operating	9	16
Stopping	10	17
Tools	21, 23	34, 43
Volute spring:		
Installation	137	232
Removal	136	230

IN HIGH DEFINITION
NOW AVAILABLE!

COMPLETE LINE OF WWII AIRCRAFT FLIGHT MANUALS

WWW.PERISCOPEFILM.COM

©2013 Periscope Film LLC
All Rights Reserved
ISBN#978-1-937684-39-6
www.PeriscopeFilm.com

www.ingramcontent.com/pod-product-compliance
Lightning Source LLC
Chambersburg PA
CBHW060414170426
43199CB00013B/2135